Climatic Variability

Climatic Variability

Impacts on Agriculture and Allied Sectors

M. Datta
ICAR Research Complex for NEH Region
Tripura Centre, Tripura

D. Daschaudhuri
ICAR Research Complex for NEH Region
Tripura Centre, Tripura

S.V. Ngachan
ICAR Research Complex for NEH Region
Umiam Centre, Meghalaya

NEW INDIA PUBLISHING AGENCY
New Delhi – 110 034

NEW INDIA PUBLISHING AGENCY
101, Vikas Surya Plaza, CU Block, LSC Market
Pitam Pura, New Delhi 110 034, India
Phone: + 91 (11)27 34 17 17 Fax: + 91(11) 27 34 16 16
Email: info@nipabooks.com
Web: www.nipabooks.com

Feedback at feedbacks@nipabooks.com

Citation: Datta M (mdatta2@rediffmail.com, mdatta54@gmail.com), Daschaudhuri D (dhiman.neh@gmail.com) and Ngachan SV (svngachan@rediffmail.com) 2014, Climatic Variability: Impacts on Agriculture and Allied Sectors.

ISBN: 978-93-81450-94-9

Composed, Designed and Printed in India

डा. एस. अय्यप्पन
सचिव एवं महानिदेशक
Dr. S. AYYAPPAN
SECRETARY & DIRECTOR GENERAL

भारत सरकार
कृषि अनुसंधान और शिक्षा विभाग एवं
भारतीय कृषि अनुसंधान परिषद
कृषि मंत्रालय, कृषि भवन, नई दिल्ली 110 001
GOVERNMENT OF INDIA
DEPARTMENT OF AGRICULTURAL RESEARCH & EDUCATION
AND
INDIAN COUNCIL OF AGRICULTURAL RESEARCH
MINISTRY OF AGRICULTURE, KRISHI BHAVAN, NEW DELHI 110 001
Tel.: 23382629; 23386711 Fax: 91-11-23384773
E-mail: dg.icar@nic.in

Foreword

Agricultural systems worldwide over the last 40-50 years have responded to the effects of the interacting driving forces of population increase, income growth, urbanization and globalization on food production, markets and consumption. To these forces can be added the twin elements of climate variability and climate change which direct effects on both food production and food security. Climate continually change and there is evidence for the effect of recent accelerated warming on many biological systems. Not least of these are the effects on the geographic distributions of pests and pathogens, with potentially serious implications for food security. However, cropping systems will also change in response to climate, with consequent impacts on their interactions with pests and pathogens,

The first was that climate change would likely increase the number of people at risk of hunger compared with reference scenarios with no climate change. Projected climate change impacts depended strongly on future social and economic development. Additionally, the magnitude to climate change impacts was projected to be smaller compared to the impact of social and economic development. In 2006, the global estimate for the number of people undernourished was 820 million under the SRES A1, B1, and B2 scenarios, projections for the year 2080 showed a reduction in the number of people undernourished of about 560-700 million people, with a global total of undernourished people of 100-240 million in 2080. By contrast, the SRES A2 scenario showed only a small decrease in the risk of hunger from the 2006 level. The smaller reduction under A2 was attributed to the higher projected future population level in this scenario. India is expected to be the most populous country in the world by 2025 and feeding the population is likely to be one of the serious challenges the country will face in the coming decades. As per Global Food Security Index (GFSI) 201, in which among 105 nations, India was ranked 66, which is a moderate position.

Changes in mean temperature and precipitation patterns influence pest and pathogens incidence and distribution. Climate change is therefore likely to modify the crop and livestock disease spectrum in some regions and turn pathogens or pests, considered unimportant today, into potential new threats in the future. Evidence on the measured climate change on crops, livestock and their associated pests and pathogens is starting to be documented; however studies are still scarce and scattered. It is essential to better understand such complex interactions and to achieve a more mechanistic inclusion of pest and diseases effects in crop and livestock models and early warning systems. Taking up this challenge is the crucial step forward towards more realistic predictions of crop production and to allow a more efficient assistance in the development of robust adaptation strategies and food security policies. Climate change can affect pathogen and pest dynamics in multiple ways. Crucial is the question whether the effect of climate change outnumbers other factors affecting the epidemiology of pests and pathogens, and if so, how pathosystem management should be modified.

I do congratulate the authors of ICAR Research Complex for making a compilation and editing the proceedings of the National Workshop into a book edited "Climatic Variability: Impacts on Agriculture and Allied Sectors". The edited book has got 5 major divisions such as Weather and Agromet Service, Pest and Diseases, Climate Influences over Agricultural and Horticultural crops, Climatic Influences over Animal husbandry and Fishes and ICT application in Agriculture. I do hope that various avenues of climatic variability and its impact on agricultural productivity are dealt in this volume and the information compiled herein can be made use in monitoring agricultural planning particularly in the North East India

Dated the 22nd January, 2013
New Delhi

(S. Ayyappan)

डॉ. लक्ष्मण सिंह राठोड़
मौसम विज्ञान के महानिदेशक

Dr. L. S. Rathore
Director General of Meteorology &
Permanent Representative of India with W.M.O.

भारत सरकार
पृथ्वी विज्ञान मंत्रालय
भारत मौसम विज्ञान विभाग
मौसम भवन, लोदी रोड़,
नई दिल्ली-110 003

Government of India
Ministry of Earth Sciences
India Meteorological Department
Mausam Bhawan, Lodi Road,
New Delhi-110 003

Foreword

The euphoria surrounding the 'Green revolution' was questioned in the wake of the energy crisis and growing awareness of long-term environmental consequences since the beginning of the 1980s yet another threat to agriculture has attracted much attention. Many climatologists predict significant global warming in the coming decades due to increasing atmospheric carbon dioxide and other trace gases.

Beyond what is known about greenhouse gases and the climate system however lie great uncertainties. How much warming will occur at what rate and according to what geographical and seasonal pattern? What secondary processes will the warming trend induce and what might be the physical and biological impacts of such processes? Will some areas benefit while other areas suffer and who might the winners and losers be? And if such damages are unavoidable what can be done to adapt to modify our systems so as to minimize or overcome them? These are important and complex questions and we have only begun to understand them and to develop methods for their analysis. The effects of changes in climate on the productivity of agriculture and allied sectors are likely to vary greatly from region to region across the globe.

The book entitled "Climatic Variability: Impacts on Agriculture and Allied Sectors" being an outcome of the National Workshop organized by ICAR Research Complex for NEH Region, Tripura Centre is a document edited by M. Datta, D. Daschaudhuri and S.V. Ngachan and covers various aspects *viz*. Weather and Agromet Service, Pest and Diseases, Climatic Influences over Agricultural and Horticultural Crops, Climatic Influences over Animal Husbandry / Fishes and ICT application in

Agriculture. I would congratulate the authors in organizing the National Workshop in the North east and bringing out this sort of valuable publication which would definitely pave a way for sustaining the productivity of crops and animals / fishes under variable climatic variability situations. Agromet Advisory Services being in operation country wide under the purview of Ministry of Earth Sciences, Govt. of India may be brought under use in the district level and in future in the block / Panchayat level to adequately focus the issues of climatic variability and suggest suitable measures for the benefit of the farming community.

L.S. Rathore
Director General of Meteorology &
Permanent Representative of India with W.M.O.

Preface

The potential impacts of climate change on agriculture are highly uncertain. The risks associated with climate change lie in the interaction of several systems with many variables that must be collectively considered. Agriculture (including crop husbandry, animal husbandry, forestry and fisheries) can be defined as one of the systems, and climate the other. If these systems are treated independently, this would lead to an approach which is too fragmentary. The issue is more global. It is now held as likely that human activities can affect climate, one of the components of the environment. Climate, in turn affects agriculture, the source of all food consumed by human beings and domestic animals. It must be further considered that not only climate may be changing, but that human societies and agriculture develop trends and constraints of their own which climate change impact studies must take into consideration. The robust conclusion that does emerge from impact studies is that climate change has the potential to change significantly the productivity of agriculture at most locations. Some currently highly productive areas may become much less productive. Some currently marginal areas may benefit substantially while others may become unproductive. Crop yield studies show regional variations of +20, 30 or more per cent in some areas and equal size losses in other areas. Most areas can expect change and will need to adapt, but the direction of change, particularly of precipitation, and required adaptations cannot now be predicted. Nor may it ever be possible to predict them with confidence. Current evidence suggests that pole ward regions where agriculture is limited by short growing seasons, are more likely to gain while subtropical and tropical regions may be more likely to suffer drought and losses in productivity. However, these broad conclusions hardly provide the basis for mapping out a long-term strategy for agricultural adaptation.

The book entitled "*Climatic Variability : Impacts on Agriculture and Allied Sectors*" is an outcome of the National Workshop organized by ICAR Research Complex for NEH Region, Tripura Centre in collaboration with National Initiatives on Climate Resilient Agriculture (NICRA), Directorate of Biotechnology and Department of Agriculture, Govt. of Tripura with financial support from Ministry of Earth Sciences, Govt of India, National Horticulture Board, New Delhi, Farm Information Division (Extension), Ministry of Agriculture & Cooperation, New Delhi. The

National workshop has a broad divisions of 5 different topics, namely, *Weather and Agromet Services*, *Pest and Diseases, Climatic Influences over Agricultural and Horticultural crops, Climatic Influences over Animal husbandry and Fisheries and ICT application in Agriculture*. There is a need to strengthen the Agromet Services presently in operation by India Meteorological Department and pest and disease surveillance along with prediction models and supported by ICT application can definitely give a respite to the farming community to combat the weather vagaries.

M. Datta
D. Daschaudhuri
S.V. Ngachan
ICAR Research Complex for NEH Region
Tripura Centre, Tripura

Contributors

A. Goswami
Division of Veterinary Parasitology
ICAR Research Complex for NEH Region
Umroi Road, Umian
Meghalaya

A.K. Singh
Programme Coordinator
Krishi Vigyan Kendra
Birchandramanu
South Tripura

Ajay Kumar
Dairy Cattle Nutrition Division
National Dairy Research Institute
Karnal – 132 001, Haryana

Amar Bahadur
Asstt Professor (Plant Pathology)
College of Agriculture, Lembucherra
Agartala – 799 210, Tripura

B.L. Meena
ICAR-RC-NEH Region Centre
Lembucherra – 799210, Tripura

B.C. Debbarme, Supt. (Sericulture)
DHHS, Takarjala
Tripura

B.S. Prakash
Division of Dairy Cattle Physiology
NDRI, Karnal – 132 001, Haryana

Banjul Bhattacharyya
Department of Agricultural Statistics
Bidhan Chandra Krishi Viswavidyalaya
Mohanpur – 741 235, West Bengal

Boopathi T.
ICAR Research Complex NEH Region
Mizoram Centre
Kolasib – 796 081, Mizoram

Britan Rahman
Krishi Vigyan Kendra-West Tripura
Chebri, Khowai – 799 207
Tripura (West)

C. Debnath
Scientist (Fishery)
ICAR Research Complex for NEH Region
Tripura Centre
Lembucherra – 799 210
Tripura

C.K. Kundu
Department of Agronomy
Bidhan Chandra Krishi Viswavidyalaya
Mohanpur, Nadia – 741 252, West Bengal

Chander Datta
Dairy Cattle Nutrition Division
National Dairy Research Institute
Karnal – 132 001, Haryana

D. Daschaudhuri
ICAR Research Complex for NEH Region
Tripura Centre, Lembucherra
Tripura – 799 210

D. Majumder
Department of Agricultural Statistics
Bidhan Chandra Krishi Viswavidyalaya
Mohanpur – 741 235
West Bengal

D. Sharmah
Krishi Vigyan Kendra
Birchandramanu, South Tripura

D. Saha
Sericulture Development Officer
DHHS, Agartala, Tripura

Debashis Bhowmik
Department of Agricultural Statistics
Bidhan Chandra Krishi Viswavidyalaya
Mohanpur, West Bengal – 741 235

Dipak Nath
Krishi Vigyan Kendra-West Tripura
Chebri, Khowai – 799 207
Tripura (West)

G. Bhagawati
ICAR Research Complex for NEH Region
Arunachal Pradesh Centre, Basar
West Siang District, Arunachal Pradesh

G. Sounda
Department of Agronomy
Bidhan Chandra Krishi Viswavidyalaya
Mohanpur – 741 235
West Bengal

Hemavati R.
ICAR Research Complex for NEH Region
Tripura Centre
Lembucherra – 799 210
Tripura

J. Parhi
College of Fisheries
CAU, Lembucherra
Tripura

Jai Prakash Sharma
Principal Scientist (Plant Pathology)
ICAR Reserach Complex for Eastern Region
Ranchi Centre, Uttar Jharkhand
Jharkhand

K.A. Pathak
ICAR-RC-NEH Region
Mizoram Centre, Kolasib – 796 081
Mizoram

K.C. Puzari
Principal Scientist, Mycology Research Section
Department of Plant Pathology
Assam Agricultural University (AAU)
Jorhat, Assam – 785 013

Kaushik Bhagawati
ICAR Research Complex for NEH Region
Arunachal Pradesh Centre, Basar
West Siang District
Arunachal Pradesh

L.K. Hazarika
Dean, Faculty of Agriculture
Assam Agricultural University (AAU)
Jorhat, Assam – 785 013

Lopamudra Sahoo
Scientist (Fishery)
ICAR Research Complex for NEH Region
Tripura Centre, Lembucherra – 799 210
Tripura

M. Das
Division of Veterinary Parasitology
ICAR Research Complex for NEH Region
Umroi Road, Umiam
Meghalaya

M. Datta
ICAR Research Complex for NEH Region
Tripura Centre, Lembucherra
Tripura – 799 210

M. Premjit Singh
Director of Extension Education
Central Agricultural University
Imphal – 795 004

M.R. Reddy
Project Directorate on Poultry
Hyderabad – 500 030
Andhra Pradesh

Mausumi Parya
Deparment of Agronomy
Bidhan Chandra Krishi Viswavidyalay
Mohanpur – 741 235
West Bengal

Md. Hedayetullah
Department of Agronomy
Bidhan Chandra Krishi Viswavidyalaya
Mohanpur, Nadia – 741 252, West Bengal

Monica Puniya
Dairy Cattle Nutrition Division
National Dairy Research Institute
Karnal – 132 001, Haryana

N.A. Janardhana Gowda
AICRP on Agrometeorology
University of Agricultural Sciences
G.K.V.K.
Bangalore, Karnataka

Nitin Singh
ICAR-RC-NEH Region
Mizoram Centre
Kolasib – 796 081, Mizoram

P. Das, Subject Matter Specialist
Krishi Vigyan Kendra
Golaghat, Assam

P.K. Chakraborty
Department of Agril., Meteorology & Physics
Faculty of Agriculture
Bidhan Chandra Krishi Viswavidyalay
Mohanpur – 412 35, West Bengal

Padmashri, H.S.
AICRP on Agrometeorology
University of Agricultural Sciences
G.K.V.K, Bangalore, Karnataka

Pavithra, B.V.
AICRP on Agrometeorology
University of Agricultural Sciences
G.K.V.K, Bangalore, Karnataka

Pranab Dutta
Programme Coordinator
Krishi Vigyan Kendra
West Tripura, Tripura – 799 207

Pritam Goswami
Aquaculture Research Unit
Department of Zoology
Tripura Central University
Suryamani Nagar – 799 130, Tripura (West)

Pulak Chaudhuri
Asstt. Director
HRC, Nagicherra
Tripura

R. Laha
Division of Veterinary Parasitology
ICAR Research Complex for NEH Region
Umroi Road, Umiam
Meghalaya

R. Bhagawati
Joint Director
ICAR Research Centre for NEH Region
Basar, West Siang District
Arunachal Pradesh

R.N. Chatterjee
Project Directorate on Poultry
Hyderabad – 500 030
Andhra Pradesh

Rajegowda, M.B.
AICRP on Agrometeorology
University of Agricultural Sciences
G.K.V.K, Bangalore, (Karnataka)

Rajib Nath
Department of Agronomy
Bidhan Chandra Krishi Viswavidyalaya
Mohanpur – 741 235, West Bengal

Ratan K Saha
Department of Fish Health & Environment
College of Fisheries
CAU, Lembucherra
Tripura

Roop Singh Maitry
ICAR Research Complex for NEH Region
Tripura Centre
Lembucherra – 799 210
Tripura

S. Banik
Aquaculture Research Unit
Department of Zoology
Tripura (Central) Univeristy
Suryamani Nagar – 799 130
Tripura

S. Baruah
ICAR Research Complex for NEH Region
Arunachal Pradesh Centre, Basar
West Siang District, (Arunachal Pradesh)

S. Chakraborty
Sericulture Development Officer
Grainage Unit
Indranagar, Agartala

S. Kalita
Subject Matter Specialists (Entomology) KVK
Assam Agricultural University
Kajalgaon – 783 385, Chirang, Assam

S.P. Singh
ICAR Research Complex for NEH Region
Arunachal Pradesh Centre, Basar
West Siang District
Arunachal Pradesh

S.S. Kundu
Dairy Cattle Nutrition Division
National Dairy Research Institute
Karnal-132 001, Haryana

S.V. Ngachan
ICAR Research Complex for NEH Region
Umiam, Meghalaya

Samir Malla
Aquaculture Research Unit
Department of Zoology
Tripura Central Univeristy
Suryamani Nagar – 799 130, Tripura

Santosh Kumar
ICAR-RC-NEH Region
Mizoram Centre
Kolasib – 796 081, Mizoram

Sarika Jena
Deparment of Agronomy
Bidhan Chandra Krishi Viswavidyalay
Mohanpur – 741235, West Bengal

Serajul Islam
Department of Horticulture
Calcutta University, Kolkata
West Bengal

Shilpa, C.N.
AICRP on Agrometeorology
University of Agricultural Sciences
G.K.V.K, Bangalore, Karnataka

Subrata Biswas
Sr Scientist (Plant Pathology)
ICAR Research Complex for NEH Region
Tripura Centre, Lembucherra – 799 210
West Tripura

Sukhen Chandra Das
College of Agriculture, Lembucherra
Agartala – 799 210, Tripura

Suseelendra Desai
Central Research Institute for Dryland Agriculture
Santoshnagar, Hyderabad – 500 059
Andhra Pradesh

Swaraj Kumar Dutta
Department of Agronomy
Bidhan Chandra Krishi Viswavidyalaya
Mohanpur – 741 235, West Bengal

T.K. Chakma
Director of Handloom Handicraft & Sericulture
Govt. of Tripura, Agartala, Tripura

Tapan Kr. Das
Dairy Cattle Nutrition Division
National Dairy Research Institute
Karnal-132 001, Haryana

U.C. Sharma
Vice President
International Commission on Water Quality (IAHS)
Jammu – 180 009, (J&K)

U.D. Singh
Indian Agricultural Research Institute
Pusa, New Delhi – 110 012

Vikas Sharma
S.K. University of Agricultural Sciences & Tech.
Chatha, Jammu – 180 009, (J&K)

Y. Ramakrishna
ICAR-RC-NEH Region
Mizoram Centre
Kolasib – 796 081, Mizoram

Contents

Section C: Climatic Influences Over Agricultural and Horticultural Crops

Section D: Climatic Influences Over Animal Husbandry and Fisheries

Section E: ICT Application in Agriculture

Section A: Weather and Agromet Services

1

Impact of Climate Change: Vulnerability, Mitigation and Adaptation – with Special Reference to Northeast India

U.C. Sharma and Vikas Sharma

Abstract

The north eastern region of India, consisting of the eight states of Arunachal Pradesh, Assam, Manipur, Meghalaya, Mizoram, Nagaland, Sikkim and Tripura, has an area of 255 090 km^2 with a vigorous monsoon dominated climatic and a dynamic hydro-geomorphic regimes. Endowed with a sub-tropical type of monsoonal climate, the region shows remarkable variability of weather and climate on a regional scale. It possesses about 30% of the total water resources potential and about 41 % of the total hydropower potential of India. The per capita and per hectare availability of water in this region is the highest in the country. The problems of environment degradation in the northeastern areas of India are, primarily impacted by demographic pressure, urbanization, natural disasters, soil erosion and land degradation. The region is visited by the southwest monsoons which make it one of the wettest areas in South Asia as well as highly frequented by the water induced disasters. The rapid melting of glaciers is initially expected to first contribute to excessive water flow and flooding in the region. Eventually with the full loss of glaciers, the shortage, when it comes, will likely arrive much more abruptly in time; with water systems going from plenty to want in perhaps a few decades or less.

Keywords: IPCC, Greenhouse Gases Glaciers, Soil Erosion, North East India.

Introduction

The Inter-Governmental Panel for Climate Change (IPCC 2007) has estimated that by 2100 the global average surface warming (surface air temperature change), will increase by 1.1-6.4 °C; the sea level will rise between 18 and 59 cm; the oceans will become more acidic and it is very likely that hot extremes, heat waves and heavy precipitation events will continue to become more frequent. There will be more precipitation at higher latitudes and less precipitation in most subtropical land areas. Many people are not resilient to extreme weather patterns and climate variability and they are unable to protect their families, livelihoods and food supply from negative impacts of seasonal rainfall leading to floods or water scarcity during extended droughts. IPCC (2007) identified that the severe impact, that may be associated with climate induced sea level rise, would be influenced not only by the rising waters but also the increased intensity of climatic forces such as increased waves and tidal activities, storm surges, typhoons, increased rainfalls, increase monsoon winds and flash floods. Sharma and Sharma (2009), mentioned about strong interaction among vegetation, livestock, soil and hydrology in the hinterland, causing soil erosion. Climate change is likely to have major implications for wetland ecosystems, which will include altered water level regimes due to modifications in local and catchment hydrology. However, substantial uncertainty exists in the precise impacts of clssimate change on wetlands partly due to uncertainty in GCM projections (Singh *et al.* 2010). Although elevated water levels may permit enhanced abstraction for irrigation and domestic uses, future increases in hydropower generation are limited by existing infrastructure. It makes an interesting learning of how lives and practices of ecosystem communities have very slowly but surely been undergoing changes under the clout of these new environmental situations in a river and water dominated landscape of Assam. Several studies have predicted significant changes in the weather-climate and water regimes of the Himalayan region in the coming decades because of impacts of global climatc change. The consequent social, economic, livelihood and environmental implications in the eastern Himalayan region are also of immediate concern to scientists and policy makers. The north-eastern region of India has already been experiencing rainfall extremes, more frequent and intense floods, moderate to intense drought like situations and flash flooding. The region suffers paradoxically both from water excess as well as water crises. Riparian people of a downstream state like Assam are always at the receiving end from fluctuating and changing character of rainfall and rivers.

Impact of Climate Change: In the IPCC's third assessment report, the anticipated increase in average global temperature over the next 100 years is between 1.4 and 5.8 °C (IPCC, 2001). This is increasing from 1 - 3.5 °C according to the panel's second assessment, which was released in 1995. The rise in temperature will affect all spheres of life (Table 1). Not only is climate change happening faster than previously predicted, but it may happen even faster than the latest predictions.

If we do not take action to stop global warming immediately, the damage will be irreversible. From the standpoint of hydrological applications and water resources management, it is necessary to determine the spatial and temporal structures of rainfall distribution to understand the relationship between rainfall and floods so that occurrence of flood events could be predicted well before the event to happen and take appropriate measures for evacuation of human and property to reduce or escape losses. At a global scale, higher temperatures of 4-7 degrees on land and 2 degrees in our oceans will lead to further ice-cap melting and thermal expansion of the world's oceans (Malcolm 2010). Not only will our oceans be warmer, but it is predicted that increased carbon uptake will raise acidity levels bringing further stresses to marine flora and fauna. Climate change has major impact on fisheries sector. The world is facing massive readjustment due to the forecast impacts from climate change and global warming. As ocean temperatures continue to rise, the level of the Earth's seas will also rise due to a combination of thermal expansion and ice caps melting (Malcolm, 2010). India being a developing nation has several problems typical of a transitional economy. There have been large scale flood events in Northeast India, carrying enormous load of sediments to the rivers and finally to the sea, causing clogging of sea shores and damaging ports and harbours. The sediment load in flood water contains 1500-30000, 6.4-25.8, 2.3-8.5, 15.4-33.8, 0.3-1.6, 0.8-2.4, 0.1-0.3 and 6.3-18.4 mg L^{-1} of soil, NO_3-N, P-PO_4, K_2O, Zn, Mn, Cu and Fe, respectively (Sharma and Sharma, 2005).

India's climate change policy is in an emergent stage. Under the global climate regime, India figures as a Non-Annex I nation and thus is not bound by quantitative emissions reduction targets. Even so, India acknowledges its substantial role in addressing the threats arising from climate change (Mahanta 2009). It was stated (Prime Minister's Council on Climate Change, 2008) that "There should be no doubt in anybody's mind," the prime minister recently stated, "that we fully recognise not just how important this issue is to India but also our obligation to address it." Yet it has also been criticized for not establishing hard targets, particularly in the crucial area of water management. Until now, India has addressed climate change predominantly in the energy sector—the water resource sector has not received the priority attention that it merits. There is a growing shortfall of water in the Indian Subcontinent and future threats imperil water supplies and ecosystems throughout the entire region. Indeed, the World Bank has warned that India is on the brink of severe water crises with most of its states living under the stress of water scarcity John, B. (2005).

Changes in climate may alter the distribution of important vector species (for example, malarial mosquitoes) and may increase the spread of such diseases to new areas. An increase of 3.8 °C in temperature and a 7% increase in relative humidity by 2050 over present level. Biodiversity is likely to be impacted under the projected climate scenarios due to the changes or shifts in forest or vegetation types in 57% to 60% of forested grids, forest dieback during the transient phase, and different species responding differently to climate change without change in forest type.

Table 1: Effect of rising temperature on water, food, health, land and environment

Temp. rise	Water	Food	Health	Land	Environment
1°C	Glaciers disappear completely, threatening water supplies	Modest increase in cereal yield in temperate regions	About 0.3 million people die from diseases	Permafrost thawing damages buildings and roads	At least 10 % of land species facing extinction
2 °C	20-30% decrease in water availability	Decline in crop yield in tropical regions	More people exposed to malaria	10 million more people exposed to flooding	15-40% species facing extinction
3 °C	1 to 4 billion people suffer water shortages, 1 to 5 b may gain water, flooding	150 - 550 additional millions at risk of hunger	1 – 3 million more people die from malnutrition	1–170 mil. more people affected by coastal flooding each year	20–50% of species facing extinction, including 25– 60% mammals,
4 °C	30 -50% decrease in water availability in some regions	Agriculture yield decline by 15-35%	80 million people exposed to malaria	7-300m people affected by coastal flooding	Loss of around half Arctic tundra.
5 °C	Disappearance of large glaciers in Himalayas	Seriously disrupting marine ecosystem		Sea level rise threatening low lying areas	

The mountain ecosystem of northeast India is highly fragile and bio-diverse. The vegetation varies from tropical rain forests to alpine conifers. North East India is biodiversity hotspot and renowned for its high species diversity. It is also recognized as one of the centres of origin of cultivated plants. In the northeastern region, the two important measures of climate change which have direct and significant impact on the biodiversity are the variation in precipitation and temperature. The increase in precipitation can change the nature of the forest in terms of the floral species dominance, canopy cover, forest dynamics etc. Irrigated rice in the Northeastern region is predicted to gain yields marginally due to warming as compared to rainfed crop. This is because irrigated rice tends to benefit from the CO_2 fertilization effect. The yield of wet rice may increase from 5 to 10%. The yield of maize and sorghum are likely to decline in NE India with climate change. A great concern for the Northeast is the projected threat of malaria in the region with climate change leading to more wetter and warmer periods. The northeastern region has a variable climate. The rapid changes in topography result in climate changes within short distance. Generally the daily temperature in the plains of Brahmaputra and the Barak Valley as well as in Tripura and in the western portion of Mizoram Hills is about 15 ^{0}C in January, whereas in other parts of the region, the temperature is lower. In temperate areas, the summer temperatures are relatively lower than the sub-tropical zones. The rapid melting of glaciers is initially expected to first contribute to excessive water flow and flooding in the region. Eventually with the full loss of glaciers, the shortage, when it comes, will likely arrive much more abruptly in time; with water systems going from plenty to want in perhaps a few decades or less. This would put an estimated 200 million mostly poor people in NE India and Bangladesh, who literally depend on mighty Ganges and Brahmaputra for fresh water and livelihoods, at risk. India produces nearly one third of the world's tea. Tea production is being impacted by climate change. In northeastern state, Assam, are carpeted with lush tea bushes whose leaves produce some of the world's finest tea. The rising temperatures may be affecting the tea plantations, resulting in declining productivity. Tea production in the Assam region has declined in recent years, although the area under cultivation has risen. Erratic rainfall patterns are of particular concern to planters because the tea plant is largely dependent on the weather. There have been some reports that the flavour of the Assam tea.

Vulnerability and adaptation strategies: With an economy closely tied to its natural-resource-base and climate-sensitive sectors such as agriculture, water, and forestry, the northeastern region of India faces a major threat because of the projected changes in climate. Crucial sectors in the region like agriculture, water resources, health, sanitation, and rural development are likely to be affected by climate change. India's large population primarily depends on climate-sensitive sectors like agriculture and forestry for livelihood. At the national level, increase of 0.4° C has been observed in surface air temperatures over the past century. A warming trend has been observed along the west coast, in central India, the interior peninsula, and north-eastern India.

Changes in key climate variables, namely temperature, precipitation, and humidity, may have significant long-term implications for the quality and quantity of water. The most significant changes are likely to occur in the glacier fed river systems, due to the melting of the Himalayan glaciers. Changes are likely in the long-term lean-season water flows of large snow- and glacier-fed river systems of the Brahmaputra, the Ganga, and the Indus. Wheat production for the country as a whole may decline after 2020 and rice production may be adversely impacted in the eastern states. Boundary changes in the growth of crops are also expected. Studies by the Indian Agricultural Research Institute (IARI) indicate the possibility of a loss of 4–5 million tonnes in annual wheat production with every 1 ^{0}C rise in temperature even after considering the carbon fertilization effect. However, yield losses associated with frost damage in north-western India may decrease. Small changes in temperature and rainfall have significant effects on the quality of fruits, vegetables, tea, coffee, aromatic and medicinal plants, and basmati rice. Pathogens and insect populations are strongly dependent upon temperature and humidity, and changes in these parameters may change their population dynamics. Other impacts on agricultural and related sectors include lower yields from dairy cattle and decline in fish breeding, migration, and harvests. Coral reefs may decline owing to rise in sea surface temperatures.

Mitigation is addressing the sources of climate change (increased carbon dioxide and equivalent gases in the atmosphere) and attempting to lower the source of this problem i.e. reducing the concentration of these gases in the atmosphere. Hence attempts at carbon trading schemes, carbon taxes and regulation of carbon emissions are all attempts at mitigation. Adaptation on the other hand is the action taken to adapt to the observed impacts of climate change – such as altering the location of buildings, building sea walls and other structures etc. There is a growing shortfall of water in the Indian subcontinent and future threats would affect water supplies and ecosystems. The World Bank has warned that India is on the brink of severe water crises with most of its states living under the stress of water scarcity (World Bank, 2005). The Intergovernmental Panel on Climate Change (IPCC), in its Second Assessment Report, defines vulnerability as "the extent to which climate change may damage or harm a system." It adds that vulnerability "depends not only on a system's sensitivity, but also on its ability to adapt to new climatic conditions" (Watson *et al.* 1996). Vulnerability Is the extent to which a natural or social system is susceptible to sustaining damage from climate change, and is a function of extent of climate change, the sensitivity of the system to changes in climate and the ability to adapt the system to changes in climate. Hence, a highly vulnerable system is one that is highly sensitive to modest changes in climate and one for which the ability to adapt is severely constrained. Adaptation to climate is the process through which people reduce the adverse effects of climate on their health and well-being, and take advantage of the opportunities that their climatic environment provides. Adaptation involves adjustments to enhance the viability of social and economic activities and to reduce their vulnerability to climate, including its current variability and extreme events as well as longer-term climate change. Adaptation can be spontaneous or planned, and can be carried out in response to or in anticipation of change in conditions (Watson *et al.* 1996). According to the IPCC Third Assessment Report, adaptation "has the potential to reduce adverse impacts of climate change and to enhance beneficial impacts, but will incur costs and will not prevent all damages." However, "options and incentives are greater for adaptation of human systems than for adaptation to protect natural systems" (IPCC 2001). It is generally accepted that there are three options for adaptation to climate change and associated rising sea levels: protect, accommodate or retreat (Taylor *et al.* 2010). 'Protect' relies on engineered solutions such as sea walls, raised road levels and construction of levees to provide protection from rising sea levels, tidal inundation and storm surges. Issues to be considered include the location of any proposed barriers, the design criteria and lifespan for the infrastructure, available resources and responsibility for construction and maintenance of any physical barriers. 'Accommodate' assumes that residents will adapt to and accept an increased frequency of inundation as lake levels continue to rise. New development would be directed to higher land, but the existing private and public infrastructure would remain in situ, and be subject to deeper flooding, more often. 'Retreat' requires strategic planning to relocate those parts of existing settlements that are subject to coastal inundation to land which is unlikely to be at risk from future sea level rise (Taylor *et al.* 2010).

Government Initiative: With a large population dependent on it, agriculture has always been vital to India's development policies. The National Policy on Agriculture desires to tap the vast growth potential of the agricultural sector, strengthen rural infrastructure to support faster agricultural development and promote value addition. Some of the programmes promoted by the government include (i) watersheds development in rainfed areas, (ii) drought-resistant varieties, (iii) crop diversification, (iv) on-farm water-efficient technologies, (v) promoting the National Agricultural Insurance Scheme and (vi) resource conserving technologies for crop production. Considering the possible impacts of climate change on Indian agriculture, the Indian Council of Agricultural Research (ICAR) had introduced a countrywide network on impacts, adaptation, and vulnerability of Indian agriculture to climate change. Several initiatives have been taken up for watershed development and management and water-harvesting measures including rejuvenation and revival of traditional water storage systems is high on the water conservation agenda. The National Water Policy (2002) stresses that non-conventional methods for utilization of water, including inter-basin transfers, artificial recharge of groundwater, and desalination of brackish or sea water, as well as traditional water conservation practices like rainwater harvesting, including roof-top rainwater harvesting, should be practised to increase the utilizable water resources. Programmes such as the Drought Prone Areas Development Programme have been started.

One of the major initiatives has been the introduction of the social forestry and participatory Forest Management Programme, which has been successful. Involvement of local stakeholders in the management and protection of local forest resources not only ensures their sustainable management but also prepares communities dependent on them to strengthen their coping practices in the context of uncertainties.

Gaps in the Regulatory Framework: The extent of gaps in regulatory framework has been given Table 2. The present regulations for GHGs are in scattered form in a number of laws. Further, the subject matter and objectives of these laws are incidental in reducing GHG emissions.

Table 2: Gaps in the Regulatory Framework for GHG Emissions

Framework	Present Status	Gaps	Improvements
General legal Framework	EPA, Air Act etc.	Scattered laws and regulations	Comprehensive and co herent legislation
National Environment Policy GHG	Emphasis on need for adaptation to climate change and the scope for incorporating it into programs with priority given to the right to development. Inclusion of environmental considerations in sectoral policy making. Emphasis on need for mechanisms that operationalize environmental concerns at all levels of Govt.	Sources of GHG emissions in various sectors are not given due consideration. Ignores need to reduce GHG emissions from various sources scattered across many sectors of the economy on the presumption that India's policies for sustainable development results in a relatively GHGs benign growth path.	Comprehensive environment policy on the need to reduce emissions from its various sources scattered all over the various sectors of the Indian economy
Institutional Framework	CPCB, SPCB, MoEF	Specific GHG regulatory body	GHG supervisory and regulatory body
Dispute Settlement	CPCB, SPCB, courts	Slow and not very efficient	GHG supervisory and regulatory body with powers to decide disputes
Sectoral	Some industry specific regulations	Does not cover all sources of GHG emissions across various sectors	Regulations Sector specific climate regulations integrated with overall regulations

This would in due course lead to a comprehensive legislation with the primary aim of regulating and limiting GHG emissions in various sectors of the economy. One of the drawbacks of the GHG regulatory framework in India is the absence of an effective institutional framework for GHG emissions. At present, this institutional framework is responsible for administering and supervising GHG-related regulations in India. There is also a need to develop sector-specific climate policies, measures and regulations that could mitigate GHG emissions as well as be integrated with the objective of sustainable development of the economy. The national strategy generally emphasizes cutting down fossil fuel use, but not on limiting the GHG emissions from these various sources of GHG emissions like factories, vehicles etc. Table 2 provides an overview of the gaps in the regulatory framework.

Conclusions

The people's participation is necessary for the success of implementation of the government policies. There is need to build up a reliable time-series database with trends for relevant hydrological variables. It is vital that this process include mechanisms for equitably sharing both the risks and benefits to be derived from climate change adaptation projects and focus not only on broad-based, long-term development, but also on those corresponding activities that have immediate impacts on water security at the community level. Institutions must be able to create and manage an environment of incentives and disincentives that encourage initiatives consistent with sound water policy objectives for the benefit of all stakeholders, including the most socially and environmentally vulnerable communities. There is need to address some issues like Climate Change such as level of understanding on climate change of the regional leaders, awareness of the actors involved and their willingness, mandate, opportunities and strengths, emerging climate issues and their possibility of mainstreaming and research needs to address climate change.

References

Intergovernmental Panel on Climate Change. 2007. Climate change 2007: Impacts, adaptation and vulnerability: contribution of working group II to the fourth assessment report of the IPCC, Cambridge: Cambridge University Press.

Intergovernmental Panel on Climate Change (IPCC). 2001. Climate Change 2001: Adaptation Vulnerability. Contribution of Working Group II to the Third Assessment Report of the Intergovernmental Panel on Climate Change. Geneva: UNEP/WMO.

John, B. 2005. India's Water Economy: Bracing for a Turbulent Future, World Bank Report No. 34750- IN (Washington, DC).

Mahanta, C. 2009. Climate change threats to India's water resources and emerging policy responses. In: Indian climate policy: choices and challenges (ed. by D. Michel and A. Pandya), Stimson Pragmatic Steps for Global Security.

Malcolm, A. D. 2010. Climate Change – The Need for Adaptation on The Gippsland Coast, In, Proceedings of the International Symposium on Coastal Zones and Climate Change: Assessing the Impacts and Developing Adaptation Strategie, D. Dutta and W. Wright (Ed), Monash University, Australia, pp 9-18.

Ministry of Environment and Forest. 1987. Mangroves in India – a status report, New Delhi, India.

Prime Minister's Council on Climate Change 2008. National Action Plan on Climate Change, Government of India, New Delhi.

Sharma U. C. and Sharma, V. 2005. "Impact of natural variability and anthropogenic factors on the flood events in NE region of India", In, Regional Hydrological Impacts of Climate Change. S.W. Franks (Ed). IAHS Publ. No. 296, IAHS Press, Wallingford, U.K., pp 265-272.

Sharma, U.C. and Sharma, V. 2009. "Plant-livestock-soil-hydrology interactions in the northeastern region of India", Biologia, 64: 460-464.

Singh, C. R., Thompson, J. R., French, J. R., Kingston, D. G., and Mackay, A. W. 2010. Modelling the impact of prescribed global warming on water resources of headwater catchments of the Irrawaddy River and their implications for Loktak Lake, northeast India, Hydrol. Earth Syst. Sci. Discuss., 7, 2781-2828.

Taylor, G., Neumann, P. and Dunn, A. 2010. "Loch sport – planning for climate change, In, Proceedings of the International Symposium on Coastal Zones and Climate Change: Assessing the Impacts and Developing Adaptation Strategie, D. Dutta and W. Wright (Ed), Monash University, Australia, pp 75-83.

Watson, R.T., Zinyoera, M.C., and Moss, R.H. 1996. Climate Change 1995: Impacts, Adaptations and Mitigation of Climate Change: Scientific-Technical Analysis. Contribution of Working Group II to the Second Assessment Report of the IPCC.

World Bank 2005. Briscoe, John. India's Water Economy: Bracing for a Turbulent Future, Report No. 34750- IN (Washington, DC).

❑❑❑

2

Resilient Agriculture in the Context of Climate Change

M. Datta, D. Daschaudhuri and S.V. Ngachan

Abstract

Climate change and its various forms of variability are a global phenomenon and are definitely a threat to our civilization. The change in climatic variables, being uncertain in nature has an impact on productivity of agriculture and other allied sectors thus producing a detrimental effect on the fragile ecosystem prevailed in the North East India. Low or high atmospheric CO_2 concentration, temperture fluctuations , rainfall and soil moisture variations, sunshine hours , soil temperature variability ,altogether considered as climatic complex could have impact on field crops, vegetables, fruit crops, animal and aquaculture productivity in a diversified way.Technological modalities in the domain of resilience in agriculture pertainining to natural resource management , crop production and livestock /fisheries need to implemented in a holistic manner so as to have institutional interventions to take the resilient measures in the doorstep of the farming community and such measures will be the weapon for the people to face the challenges of climatic uncertaintity

Keywords: Climate Change, Resilient Agriculture.

The predicted changes in climate, especially increased atmospheric CO_2, temperature and precipitation, associated with changes in nitrogen deposition, tropo- and stratospheric ozone levels, UV-B radiation, *etc*. can have great impacts on world agricultural production and supply patterns. In order for agricultural production to be sufficient to meet the demands of the ever-growing human population, the impact of the climate must be understood and integrated in any future planning. The risks associated with climate change lie in the interaction of several systems with many variables that must be collectively considered. Agriculture (including crop agriculture, animal husbandry, forestry and fisheries) can be defined as one of the systems, and climate the other. If these systems are treated independently, this would lead to an approach which is too fragmentary. The issue is more global. It is now held as likely that human activities can affect climate, one of the components of the environment. Climate in turn affects agriculture, the source of all food consumed by human beings and domestic animals. It must be further considered that not only climate may be changing, but that human societies and agriculture developmental trends and constraints of their own which climate change impact studies must take into consideration.

In general, global food production is variable in nature. There are, however, marked disparities between continents. For example, in Africa local production of cereals cannot keep pace with population increase, and production of root and tuber crops is growing faster than that of the nutritionally more valuable cereals. In contrast, arable land growth lags behind population growth, which indicates some intensification of production. In Asia, the upper limit of available land has been reached in several countries, resulting in very high cropping intensities and a dominant role for irrigation. In Latin America, the increase in arable land is achieved only at a high ecological cost (especially deforestation) which may have direct relevance to climate change. According to a recent FAO prospective study covering the years 1988/1990 to 2010 (*World Agriculture: Towards 2010 or AT-2010;* FAO, 1995a), 'the rate of growth of agricultural land will be further reduced during the next two decades. The pressures on *fresh* water *resources,* however, will be considerable, as will be those on the environment arising from the intensification of land use'. A significant reduction in world population growth rates is foreseen from 1.8% (1980/1990) to 1.4% (2000/2010), roughly equivalent to an increase of the population doubling time from 40 to 50 years. The projected continuation of the high population growth in Africa can be related to slow economic development during the coming two decades, since in general a reduction of poverty precedes reduction of population growth. The following significant trends are seen for the near future: (i) world production of cereals will continue to grow, but not in per capita terms; (ii) export cereals will undergo a modest growth in demand; (iii) the livestock sector in developing countries will continue to grow; (iv) root crops, tubers and plantains will retain their importance; (v) oil crops will undergo rapid growth in developing countries; and, most importantly, (vi) many developing countries will become net agricultural importers. As indicated above for population increase, large differences exist in the rate of agricultural development among the developing continents, while differences among the developed countries tend to level out. It is also likely that hunger and under-nutrition, which currently affects

800 million people (about 20% of the 4 000 million inhabitants of developing countries), will be reduced, but large pockets of malnutrition will persist. Pressure on environmental resources will continue to build up.

Climate variability is likely to increase under global warming (Katz and Brown, 1992), both in absolute and in relative terms. This is linked with thresholds which affect the occurrence of many meteorological phenomena. For instance, tropical cyclones are 'fed' by water vapour evaporating from oceans at a temperature above 26 or 27°C. Therefore, higher average sea surface temperatures are bound to result in a higher frequency of tropical cyclones. The rate of change itself is extremely important. For example, work (Lehman, 1993; Paillard and Labeyrie, 1994; Rahmsdorf, 1994; Holmes, 1995) on the saw-tooth temperature changes in the past as observed in the Arctic, raise concern that changes may occur abruptly, with average temperatures changing by 10 or 12°C in just a matter of decades. The mechanism of such changes is not clear as yet, but seems to involve the mechanical stability of ice sheets, and sudden changes in the hydrological phenomenon.

In addition to water vapour, important greenhouse gases are carbon dioxide (CO_2), methane (CH_4), nitrous oxide (N_2O), tropospheric ozone (O_3) and chlorofluorocarbons (CFCs). The basic characteristics of the first three gases are given in Table 1. The degree to which these greenhouse gases stem from agricultural sources is also given. The exact ratio between these land-use related emissions and those from natural ecosystems (swamps, tundra), fossil fuels (coal, oil, gas)

Table 1: An overview of agricultural greenhouse gases with the trends as currently envisaged (adapted from IPCC, 1992; Houghton *et al.*, 1995; Keeling *et al.*, 1995). ppmv and ppbv stand for parts per 10^6 and parts per 10^9, respectively, by volume.

	CO_2	CH_4	N_2O
Atmospheric lifetime (yr)	120	14.5	120
Direct GWP [1]	1	24.5 [2]	320
Pre-industrial concentration [3]	280 ppmv	0.8 ppmv	288 ppbv
Present-day levels	360 ppmv	1.72 ppmv	310 ppbv
Current annual increase (%)	0.5	0.9	0.25
Major agricultural sources [4]	deforestation	- wetland rice - ruminants - biomass burning	- synthetic N fertilizers - animal excreta - biological N fixation
Percentage of global source stemming from agriculture	30	40	25
Predicted change 1990-2020	-	+	+

[1]GWP, Global Warming Potential, is the direct warming effect in relation to CO_2 at a time horizon of 100 years.

[2] Includes indirect effects through chemistry.

[3] About year 1750-1800.

[4] Activities responsible for emissions are projected to increase by: rice (+ 10%); ruminant population (+30%); synthetic fertilizer use (+20%): animal excreta and biological N fixation increasing but rate not specified.

and geological sources (volcanoes) are a bone of contention among industrialized, oil producing and developing countries. Moral rights and duties in relation to the international flow of development and conservation funds are involved. Therefore, better estimates based on exact measurements of the net greenhouse gas emissions from agricultural practices in developing countries are urgently required to assess responsibilities properly. The required reductions of emissions to achieve stabilization of atmospheric concentrations of current levels are believed to be >60% for CO_2 15-20% for CH_4 and 70-80% for N_2O. This implies the need for up-to-date and complete information on land cover and land uses per country: not only the kind of crops grown but also their intensity, their rotation and the amount of inputs (energy, fertilizers).

The impact of the green revolution of the 1960s and 1970s in stimulating cereal productivity and ensuring near-global food security is well recognized. However, a current analysis of yield increases of eight key cereals relative to the increase in the human population since that period indicates that any green revolution-induced increases in cereal production, relative to population are now at the end (Ziska *et al* ,2012). If the gains of the green revolution relative to population have been negated, challenges for the agricultural sector remain. There is necessity for more: more cereals for bio-fuel, more grain for meat, more food for the additional 2 billion people expected by 2050; and as these demands mount, available resources are becoming strained, with less arable land, less water for irrigation and less energy for fertilizer production. Additional emerging constraints include the need to reduce net greenhouse gas emissions, desires to foster biodiverse landscapes and regions, ongoing degradation of soil and other natural resources and rising cost of the agricultural inputs.

Ecological and indirect climate effects

In qualitative terms, many indirect effects of climate change on agriculture can be conjectured. Most of them are estimated to be negative and they catch most of the attention of the media. These effects include:

- The overall predictability of weather and climate would decrease, making the day-to-day and medium-term planning of farm operations more difficult;
- Loss of biodiversity from some of the most fragile environments, such as tropical forests and mangroves;
- Sea-level rise (40 cm in the coming 100 years) would submerge some valuable coastal agricultural land;
- The incidence of diseases and pests, especially alien ones, could increase;
- Present (agro) ecological zones could shift in some cases over hundreds of kilometres horizontally, and hundreds of metres altitudinally, with the hazard that some plants, especially trees, and animal species cannot follow in time, and that farming systems cannot adjust themselves in time;

- Higher temperatures would allow seasonally longer plant growth and crop growing in cool and mountainous areas, allowing in some cases increased cropping and production.
- In contrast, in already warm areas climate change can cause reduced productivity;
- The current imbalance of food production between cool and temperate regions and tropical and subtropical regions could worsen.

General Description of North East India

Northeast India refers to the easternmost region of India consisting of the contiguous eight States [(Arunachal Pradesh, Assam, Manipur, Meghalaya, Mizoram, Nagaland, Tripura, Sikkim and parts of North Bengal (districts of Darjeeling, Jalpaiguri, and Kooch Behar)]. Northeast India is ethnically distinct from the rest of India and has strong ethnic and cultural ties with East Asia and Southeast Asia, that had escaped Sanskritization in the most groups. Linguistically the region is distinguished by a preponderance of Tibeto-Burman languages. The states are officially recognized under North Eastern Council constituted in 1971 as the acting agency for the development of the eight states. The North Eastern Development Finance Corporation Ltd was incorporated on August 9, 1995 and the Ministry of Development of North Eastern Region (DoNER) was set up in September 2001.

The Siliguri Corridor in West Bengal, with an average width of 21 km to 40 km, connects the North Eastern region with the mainland Indian sub-continent. The region shares more than 4500 kilometres of international border (about 90 per cent of its entire border area) with China (South Tibet) in the North, Mayanmar in the East, Bangladesh in the South-West, and Bhutan to the North-West. Earliest mention of North East India was by the Chinese explorer, Chang Kien, in 100 BC. Huen Tsang, the great Chinese traveller, visited the region in the 7th century. Throughout its history the region is known to have been an important physical and cultural bridge between India and Southeast Asia, as well as a potential location for the domestication of several important plants, and is thus an important focus area for archaeology.

The Northeast region can be physiographically categorized into the Eastern Himalayas, Northeast Hills (Patkai-Naga Hills and Lushai Hills) and the Brahmaputra and the Barak Valley Plains. Northeast India (at the confluence of Indo-Malayan, Indo-Chinese, and Indian biogeographical realms) has a predominantly humid sub-tropical climate with hot, humid summers, severe monsoons and mild winters. Along with the west coast of India, this region has some of the Indian sub-continent's last remaining rain forests which support diverse flora and fauna and several crop species. Similarly, reserves of petroleum and natural gas in the region constitute a fifth of India's total potential. The region is covered by the mighty Brahmaputra-Barak river systems and their tributaries. Geographically, apart from the Brahmaputra, Barak and Imphal valleys and some flat lands in between the hills of Meghalaya and Tripura, the remaining two-thirds of the area is hilly terrain interspersed with valleys and plains; the altitude varies from almost sea-level to over 7000 metres above MSL. The region's high rainfall averaging around 10000 millimetres and above creates problems of eco system, high seismic activity and floods. The states of Arunachal Pradesh and Sikkim have a montane climate with cold, snowy winters and mild summers. The following figures highlight the biodiversity significance of the region:

- 51 forest types are found in the region broadly classified into six major types - tropical moist deciduous forests, tropical semi evergreen forests, tropical wet evergreen forests, subtropical forests, temperate forests and alpine forests.
- Out of the nine important vegetation types of India, six are found in the North Eastern region.
- These forests harbour 80,000 out of 15,000 species of flowering plants. In terms of floral species richness the highest diversity is reported from the states of Arunachal Pradesh (5000 species) and Sikkim (4500 species) amongst the North Eastern States.
- According to the Indian Red data book published by the Botanical Survey of India, 10 per cent of the total flowering plants in the country are endangered. Of the 1500 endangered floral species, 800 are reported from North East India.
- Most of the North Eastern states have more than 60% of their geographical area under forest cover, a minimum suggested coverage for the hill states in the country.
- North East India is a part of Indo Burma 'hotspot'. The hotspot is, the second largest and next only to the Mediterranean basin with an area 2,20,60,000 Km among the 25 identified globally.
- The International Council for Bird Preservation, UK identified the Assam plains and the Eastern Himalaya as an Endemic Bird Area (EBA). The EBA has an area of 220,000 sq kilometre following the Himalayan range in the countries of Bangladesh, Bhutan, China, Nepal, Myanmar and the Indian states of Sikkim, northern West Bengal, Arunachal Pradesh, southern Assam, Nagaland, Manipur, Meghalaya and Mizoram. Because of a southward occurrence of this mountain range in comparison to other Himalayan ranges, this region has a distinctly different climate with warmer mean temperatures and fewer days with frost and have much higher rainfall. This has resulted in the occurrence of a rich array of restricted range bird species and more

than two critically endangered species, three endangered species and 14 vulnerable species of birds within this EBA. Stattersfield *et al.* (1998) identified 22 restricted range species out of which 19 are confined to this region and the remaining three are also present in other endemic and secondary areas. Eleven out of the 22 restricted range species found in this region are considered as threatened (Birdlife International 2001), a number greater than in any other EBA of India.

- WWF has identified the following priority ecoregions in North-East India: Brahmaputra Valley Semi Evergreen Forests ,The Eastern Himalayan Broadleaved Forests ,The Eastern Himalayan Sub-alpine Coniferous Forests and India–Myanmar Pine Forests

The 39 million people of Northeast India constitute only about 3.8 per cent of the total population of the country. Over 68 per cent of this population (26.64 million) lives in the state of Assam alone. The density of population varies from 13 per sq. kilometre in Arunachal Pradesh to 340 per sq. kilometre in Assam. The predominantly hilly terrain in all the states except Assam is host to an overwhelming proportion of tribal population ranging from 19.3 per cent in Assam to 94.5 per cent in Mizoram. The region has over 160 Scheduled Tribes and over 400 other sub-tribal communities and groups. It is predominantly rural with over 84 per cent of the population living in the countryside. According to 2011 census, the total literacy rate of the population in the region at 68.5 per cent, with female literacy rate at 61.5 per cent, is higher than the country's average of 64.8 per cent and 53.7 per cent respectively. Of course, there are significant variations in the literacy rates among different states with Assam, Arunachal Pradesh and Meghalaya below the national average. The economy in the entire Northeast India is agrarian in nature, although little land is available for settled agriculture. The inaccessible terrain has made rapid industrialization difficult in the region. Along with settled agriculture, *jhum*(shifting cultivation) is still practiced by a few indigenous groups of people.

Climate of North East India

The climatic conditions of North Eastern Region are very congenial for growth of lush green forest including rich variety of orchids. The upper most regions of Arunachal Pradesh being a part of the Himalayas have an alpine and tundra climate while middle part experiences a temperate climate. The lower plains and river valleys experience a humid, sub-tropical climate with hot summers and mild winters. During monsoon season temperature ranges between 22°C to 30°C whereas during winters the temperature drops down to 15°C to 21°C. There are heavy rains during monsoon. Arunachal Pradesh enjoys multitude of microclimatic niches differing from hill to hill. Assam experiences heavy rainfall and high humidity, a typical 'Tropical Monsoon Rainforest Climate'. Summer starts in mid May, accompanied by high humidity and rainfall. The maximum temperature during summers ranges between 35°C to 38°C and in winters it drops down to 6°C to 8°C. The climatic conditions of Meghalaya vary with altitudinal range. Khasi and Jaintia hills have a very soothing climate whereas Garo hills experience hot and humid in summer months. Plenty of rainfall coupled with moderate temperature helps to flourish lush and green vegetation including orchids. The climate of Manipur remains moderate throughout the year and ranges from tropical to temperate depending on the altitude .The temperature hardly reaches to the maximum 32°C during summer. It receives an annual rainfall of 1500 mm. This state enjoy salubrious climate round the year. The entire state covered with thick dense forest which depicts the abundance of forest resources as well as diversity in orchid species. Mizoram has a mild climate with maximum temperature of 30°C during summer months. It experiences heavy rainfall during monsoon period. Like other states of North Eastern Region, Nagaland has largely monsoon climate with high humidity. Temperature ranges from 21°C to 31°C in summer months. Similar to Arunachal Pradesh, Sikkim is a mountainous state with lush green forest harbouring variety of orchids. The northern part of the state have an alpine zone, rest of the state have subtropical to temperate climate with high rainfall and humidity. The most part of the state enjoy moderate temperature of about 30 °C during summer. This state is considered as one of the richest orchid diversity hot spot in the country. Tripura has overall a tropical savannah climate, although the undulating topography leads to slight variations in climatic conditions of the state. The maximum temperature in this state reaches to 35°C during summer. It receives heavy rains during monsoon season.

North Eastern Region of India are experiencing climatic change that would have impact on food production and security, sustained water supply, biodiversity of forests and other natural ecosystems, human health and settlements . This region is highly prone to the consequences to climate change due to its geo-ecological fragility, strategic location vis-à-vis the eastern Himalayan landscape and international borders, its trans-boundary river basins and socio-economic instabilities. The studies have shown that though there is no significant change in rainfall as a whole but in a part of north eastern region (South Assam Meteorological Subdivision) has recorded a significant change in seasonal rainfall. The summer monsoon rainfall is found to be decreasing over this region significantly during the last century at an approximate rate of 11 mm per decade. It is predicted that temperature of the region would rise to 2.0-3.5^0 C and the precipitation would increase 250 – 500 mm in the North Eastern region. Increase in rainfall may not have much impact on the forest resources of North East but change in temperature regime may cause severe impact on our natural resources including biodiversity. The cycles of climate change are driven by natural factors occurred over a period of centuries. It forces the species to migrate pole ward in an orderly manner (Darwin, 1859). The response of species or population in relation to climate change depends on the species biology and the geographic location of the population. In general populations in the flat terrain moved toward upward or pole ward during warming period (Jackson *et al.,* 1987). While those in mountainous area with mild slopes, migrated upward along the elevation gradient

(Thomson, 1990). However, in regions with complex habitat, steep and uneven slopes, rare species shrinking in population sizes and faced local extinction (Maschinski *et al.*, 2006). Apart changing distributional pattern, the plant species also show changes in timings of seasonal occurrences (phenology). Fluctuations in flowering time due to fluctuations in spring temperatures have been documented for a number of woody species and herbaceous species. Increase of temperature during winter and spring may hasten the flowering of orchids. The unusual flowering will fail to invite insect pollinator which would in long term decrease orchid population. Richness and abundance in plant species population another impact has been reported an impact of climate change.

There have been some conspicuous changes in temperature as well as rainfall pattern in northeast India over the past century which suggests that climate change is fast becoming an inconvenient reality of present rather than remaining a subject of future speculation only. The atmospheric temperature (annual mean temperature and annual mean maximum temperature) has been rising at a rate quite perceptible to even the common people of the region. According to a recently reported study by Deka *et al.* (2009), the annual maximum and mean temperature in northeast India during 1901-2003 has increased significantly by a rate of 1.02°C and 0.60°C/100 years, respectively (Table 2). Atmospheric temperature in the region is further projected to rise by approximately 3°C to 5°C during the latter third of this century (Table 3). It is worth noting that projected temperature rise in northeast India is greater than in other regions of the country. More pronounced still is the reduction in the annual as well as the monsoon rainfall over the years in north-eastern India. The most striking evidence of the declining rainfall in the region comes from the drastically reducing amount of annual rainfall in one of the wettest places in the world i.e. Cherrapunjee. According to a report, Cherrapunjee received less rain in the whole of 2001 – only 363 inches – than it got in just one month in 1861 (Terradaily, March 03, 2007). In July 1861 alone, Cherrapunji had 366 inches of rain. Between August 1860 and July 1861, Cherrapunjee got a record 1042 inches of rain – a world record. But now the annual rainfall there has sharply fallen to less than a third of that. The alarming deficits in annual as well as monsoon rainfall observed year after year in recent times which causes severe droughts across the various states further confirms the climate change-induced decline in rainfall in northeast India (Table 4).

Table 2: Trend rate of temperature in northeast India during 1901-2003

Seasons	Trend rate of temperature (°C/100 yr)		
	Maximum	Minimum	Mean
Pre-monsoon (March-May)	+0.78	+0.09	+0.44
Monsoon (June-September)	+0.75	-0.32	+0.21
Post-monsoon (October-November)	+1.54	+0.64	+1.09
Winter (December-February)	+1.22	+0.61	+0.92
Annual	+1.02	+0.18	+0.60

Table 3: Trend of deficit rainfall (%) over north-eastern India in recent times*

Assam	2005	2006	2009	2010
Assam	-23	-32	-34	-6
Arunachal Pradesh	0	-25	-29	-30
Nagaland	-22	-25	-63	-27
Manipur	-22	-25	-67	-46
Mizoram	-22	-25	-32	-28
Meghalaya	-23	-32	-56	-49
Tripura	-22	-25	-31	-16

**Source* Monoj –Kumar 2011

Table 4: Region-wise and season-wise projections of temperature and rainfall changes in India for the period 2070-2099 with reference to the base period 1960-1990*

Region	Jan.-March	April-June	July-Sep.	Oct.-Dec.
Temperature Change (°C)				
Northeast	4.95	4.11	2.88	4.05
Northwest	4.53	4.25	2.96	4.16
Southeast	4.16	3.21	2.53	3.29
Southwest	3.74	3.07	2.52	3.04
Precipitation Change (%)				
Northeast	-9.3	20.3	21.0	7.5
Northwest	7.2	7.1	27.2	57.0
Southeast	-32.9	29.7	10.9	0.7
Southwest	22.3	32.3	8.8	8.5

(Cline, 2007)
**Projections are average of predictions of six general circulation models including HadCM3, CSIROMk2, CGCM2, GFDL-R30, CCSR/NIES, and ECHAM4/OPYC3.*

Areawise Distribution of Agroclimatic Zones in NEH Region

Zones	Altitude	Approx. geographical area (sq. km)	Place
Alpine	More that 3500m	47068	Arunachal Pradesh, Sikkim
Temperate sub alpine	1500-3500m	33564	Meghalaya, Manipur, Sikkim, Mizoram, Nagaland
Sub-tropical hill	1000-1500m	29021	Arunachal Pradesh, Meghalaya Manipur, Sikkim, Tripura, Nagaland
Sub-tropical plain	400-1000m	812	Manipur, Nagaland, Meghalaya
Mild Tropical Hill	200-800m	26349	Arunachal Pradesh, Meghalaya, Sikkim, Mizoram, Nagaland
Mild tropical plain	0-200m	29333	Meghalaya, Manipur, Sikkim, Mizoram, Nagaland

Weather and Crops

Rice

- Pre monsoon rainfall for direct sown rice [AP-20% Rainfall, Assam- 8%, Manipur-7%, Meghalaya-7%, Mizoram-12%, Nagaland-11%, Sikkim-10% and Tripura -12%]
- 37% yield variability on pre-monsoon/monsoon rainfall.
- 750-2500mm water need of rice.
- Wet spells detrimental from flowering to maturity.
- Optimum temperature - 30°C (maximum) and 20°C (minimum).
- Sunshine hours about 1000 with 220-240 hours in the last 30 days.
- Strong winds inducing sterility.
- Flowering inhibited below 40% RH and optimum RH around 70-80% .

Pulses and oilseeds

- Heavy rains during kharif (July to Sept.) leading to seed rotting.
- Luxuriant vegetative growth but low pod formation.
- Pre and post kharif best season.
- Soil moisture stress noted for rabi pulse.
- Frost injury to rabi pulse viz.peas, lentil and gram in hills.
- Seed germination at maturity of groundnut in case of heavy rainfall and high humidity.
- Fungal growth during seed storage in groundnut.
- Low germination and cold injury of groundnut during winter in hills.
- Susceptibility of frost to mustard.
- Waterlogging or saturated moisture condition detrimental to sesamum.

Vegetables

- **Cabbage:** Can withstand extreme cold and frost, 12.8 to 15.6°C soil temperature required for seed germination.
- **Cauliflower:** Temperature (5-25°C), temperature affecting vegetative, curding and reproductive phase.
- **Tomato:** Temperature(15-27°C), low seed germination, poor plant growth and flower drop both at high and low temperature, excessive rain injurious to plant causing flower drop.

- **Capsicum:** Ideal temperature for flowering 26-28° C(day) and 16-18°C (night), low temperature in hills hamper growth unable to withstand heavy rains during flowering.

Tropical fruits

- **Cashew:** Unable to tolerate frost and extreme cold, temperature should not fall below 20° C, high temperature 39 to 42°C causing fruit drop, heavy rains and cloudy weather affecting yield.
- **Citrus:** Susceptible to cold injury, hot winds and excessive heat, low humidity (colour development) and high humidity juicy fruits
- **Banana:** Temperature(10-40°C) susceptible to high wind, low temperature(delay in ripening).
- **Litchi:** Prolonged rain not desirable at the time of flowering, fruit splitting caused by alternate spells of rain and dry heat.
- **Pomegranate:** Hot and dry climate during fruit development. Fruit quality affected by humidity.
- **Sweet orange:** Dry semiarid to subtropical climate.16-20°C.Rainfall 500-700 m
- **Coconut:** Optimum temperature 20-32°C, high temperature and heavy rain interferes with pollination, severe draught at flowering period causing inflorescence to abort.

Temperate fruits

- **Apple:** 1000-1500 hours of chilling 7°C. temperature below -2°C killing open blossoms, low temp.,rain and cloudy weather affecting cross pollination.
- **Strawberry**: Mainly temperate fruit but also grown in subtropical climate. Short day/long day/day neutral varieties.
- **Peach:** Moderately winter hardy, sensitive to low temperature, swelling buds injured at-6.5° C.
- **Plum:** Subtropical to temperate high hills, require 1000-1200 hr below 7° C during winter. Frost free spring, good drainage, adequate nutrients during summer.
- **Kiwis:** 700-800 chilling hr. below 7° C to break the rest period . High temp (>3° C) scorching of leaves.

Weather and Floriculture

Ornamental crops classified as temperate and tropical ones.

- **Crysanthamum:** 12 to 18°C, shows bronze colouration of leaves, pink colouration in white flowers in low temperature, Bleached inflorescence at high temperature.
- **Rose:** 16 to 25°C, high temperature causing yellow/brown colouration and low temperature killing flower buds and dieback of plants.
- **Anthurium:** 16-19°C, Gerbera-14-27°C, Carnation-10-24°C
- **Cacti/Succulants** need low humidity, **anthurium/orchids** require high humidity.
- Most foliage plants need more than 50% RH.

Weather and Livestock Productivity

Reducing impacts of climate on livestock production

For an animal to maintain homeothermy (no change in body temperature, other than normal circadian rhythms), the ambient environment and the animal must exchange heat at a rate that permits balancing the metabolic heat production and the energy gains/losses from the four transfer modes. Ruminant animals primarily adjust evaporative heat loss to maintain homeo-thermy during brief exposures to adverse weather, but will reduce feed intake to lower heat production during prolonged hot weather. Swine and poultry primarily adjust heat production to maintain homeothermy. Success in improving production and efficiency in most climates is possible if a rational approach is followed.

- Site selection
- Windbreaks
- Shades
- Partially or totally enclosed shelters
- Genetic improvement for adaptation

Many forms of environmental modification are available. In hot weather, water can serve as an effective cooling agent for farm animals, especially for species that maintain homeothermy primarily by regulating heat production (such as swine). Direct wetting of animals is often used as an emergency measure and can be a very effective protective device. Swine, as well as water buffaloes, are naturally wallowing animals and wallows for them have been shown to improve performance. The wetting by sprinklers has been used as a routine technique for swine (Nicholas *et al.*, 1982) and beef cattle (Morrison *et al.*, 1973) with measurable benefits, but not in other cases (Morrison *et al.*, 1981). Fogger nozzles, sometimes recommended for wetting animals, are a less effective method of cooling, as the fine droplets cling to the outer hair coat where the heat for evaporation comes from the air rather. than from the body. This is a minor problem for animals with very short, sleek hair coats, however.Air cooling using evaporative coolers designed to reduce ambient temperatures in livestock shelters can be quite effective. A correctly designed evaporative cooler will reduce the dry bulb temperature of air entering the cooler by 80 per cent of the wet bulb depression. A study carried by Igono *et al.* (1992) involved two dairy herds, one of which was exposed to evaporative coolers during the hottest period of the day, but not the other. The results showed that the average milk yield was almost the same for both herds during the cold months, but during the summer the production of the cows in the non-cooled herd was significantly lower than that of the cooled one. Shades, sprinklers and fans are very effective methods of improving thermal environment for dairy cows in hot, humid climates (McFarlane and Stevens, 1972; Bucklin *et al.*, 1991). Strickland *et al.* (1989) found that cows maintained in a shelter cooled with fans and sprinklers yielded 11.6 per cent more milk than control cows.

Because high water demand and wastewater runoff are a concern for dairy plants, however, a decrease in the use of water for sprinkling and fan cooling systems is desirable. Adequate cooling can be attained using the lowest water application rate of 313.4 L/h per nozzle or 215.9 L average daily water use per cow (Means *et al.*, 1992). This is a significant decrease in comparison with the amount of water used by Strickland *et al.* (1989). According to Means *et al.* (1992) one of the most inexpensive adjustments of a cooling system is reducing the size of the nozzle, thus saving significant amounts of water and reducing pumping costs.

Other options exist for hot environments, up to complete mechanical air conditioning. While air conditioning is technically feasible, high initial and operating costs preclude its use in almost all areas and situations. Cooling of roofs or other surrounding surfaces by evaporation of water (using water sprinklers on the roof, for instance) can effectively reduce the radiant heat load on animals. Theoretically, cooling floors beneath animals to increase conduction and radiation heat loss is also a means of microclimate modification. The condensation of moisture on the floor surface of dairy shelters would create unsanitary conditions, however. For pigs, cooled floors at air temperatures above 24°C provided increased conduction heat dissipation, with the increase being greater at colder floor temperatures (to 10°C), as pointed out by Restrepo *et al.* (1977). The increased heat dissipation by conduction was accompanied by a decrease in the evaporative heat loss at colder floor tempera-tures, however, so the overall benefits to the animals were almost unchanged. No performance benefits were measured in a separate field trial (Bond *et al.*, 1964).

For cold weather, the benefits of environmental modification beyond shelters or windbreaks to minimize the effects of weather extremes are less clear. Neonates of all species are vulnerable, and require some protection for survival. Growing and mature animals can survive relatively severe cold if they are adequately fed and disease prob-lems are absent. Production efficiency can be markedly reduced, however (National Research Council, 1981). Controlled ventilation systems in enclosed housing can use minimal sensible heat to buffer extremes of cold for improved efficiency, while added artificial heat is essential for survival or economically beneficial. The selection and use of a specific environmental modification practice or technique must be carefully evaluated, as not all will be cost-effective. Hahn and McQuigg (1970) have used probability techniques to establish the economic benefits that would result from environmental modification for dairy cows in hot weather. The work was based on the temperature–humidity index (THI), with values derived from hourly dry bulb and dewpoint temperatures. The distribution func-tion of THI allows the probability of a given summer line THI to be computed, together with the associated decline in milk production in naturally varying conditions based on a validated response function. It is estimated that , the total loss per 122-day summer season was approximately 90 kg for a cow producing 22.5 kg per day and 150 kg for a cow producing 45 kg per day. The technique used is applicable to any species, season and location for which an appropriate response function and climatological database exist, and provides a rational basis for estimating the benefits of environmental modification alternatives. Gates *et al*. (1991) also used the THI method to assess the feasibility of employing misting systems for growing-finishing hogs, and they observed that the potential improvements to the growing environment due to misting at minimum THI indicate that misting systems warrant further research as a cost-effective alternative method of cooling grow-ing-finishing hogs. For tropical regions subjected to intense solar radiation, the black globe humidity index (BGHI, proposed by Buffington *et al*., 1981) will probably be better than THI for evaluation of the livestock housing/ environment, if the black globe temperature is easily available.

Weather and Aquaculture

Climate change is modifying fish distribution and the productivity of marine and freshwater species. (Brander,2007).This has impacts on the sustainability of fisheries and aquaculture, on the livelihoods of the communities that depend on fisheries, and on the ability of the oceans to capture and store carbon (biological pump). The effect of sea level rise means that coastal fishing communities are in the front line of climate change, while changing rainfall patterns and water use impact on inland (freshwater) fisheries and aquaculture.

Aquaculture

The impacts of climate change can be addressed through adaptation and mitigation. The costs and benefits of adaptation are essentially local or national, while the costs of mitigation are essentially national whereas the benefits are global. Some activities generate both mitigation and adaptation benefits, for example, the restoration of mangrove forests can protect shorelines from erosion and provide breeding grounds for fish while also sequestering carbon.

Adaptation

Several international agencies, including the World Bank and the Food and Agriculture Organization have programs to help countries and communities adapt to global warming, for example by developing policies to improve the resilience of natural resources, through assessments of risk and vulnerability, by increasing awareness of climate change impacts and strengthening key institutions, such as for weather forecasting and early warning systems. The World Development Report 2010-Development and Climate Change, shows that reducing overcapacity in fishing fleets and rebuilding fish stocks can both improve resilience to climate change and increase economic returns from marine capture fisheries by US$50 billion per year, while also reducing GHG emissions by fishing fleets. Consequently removal of subsidies on fuel for fishing can have a double benefit by reducing emissions and promoting overfishing.Investment in sustainable aquaculture can buffer water use in agriculture while producing food and diversifying economic activities. Algal biofuels also show potential as algae can produce 15-300 times more oil per acre than conventional crops, such as rapeseed, soybeans, or jatropha and marine algae do not require scarce freshwater. Programs such as the GEF-funded Coral Reef Targeted Research provide advice on building resilience and conserving coral reef ecosystems,while six Pacific countries recently gave a formal undertaking to protect the reefs in a biodiversity hotspot – the Coral Triangle.

Mitigation

The oceans have removed 50% of the anthropogenic CO_2, so the oceans have absorbed much of the impact of climate change. The famous White Cliffs of Dover illustrate how the ocean captures and buries carbon. These limestone cliffs are formed from the skeletons of marine plankton called coccoliths. Similarly, petroleum formation is attributed largely to marine and aquatic plankton further illustrating the key role of the oceans in carbon sequestration. Exactly how the oceans capture and bury CO_2 is the subject of intense researc by scientists worldwide, such as the Carboocean ProjectThe current level of GHG emissions means that ocean acidity will continue to increase and aquatic ecosystems (Ficke *et al* 2007) will continue to degrade and change. There are feedback mechanisms involved here. For example, warmer waters can absorb less CO_2, so as ocean temperatures rise some dissolved CO_2 will be released back into the atmosphere. Warming also reduces nutrient levels in the mesopelagic zone (about 200 to 1000 m deep). This in turn limits the growth of diatoms in favour of smaller phytoplankton that are poorer biological pumps of carbon. This inhibits the ability of the ocean ecosystems to sequester carbon as the oceans warm.

What is clear, is that healthy ocean and coastal ecosystems are necessary to continue the vital role of the ocean carbon sinks, as indicated, for example, by the Blue Carbon assessment prepared by UNEP and the coastal carbon sinks repor of IUCN and growing evidence of the role of fish biomass in the transport of carbon from surface waters to the deep ocean.While the various carbon finance instruments include restoration of forests (REDD) and producing clean energy (emissions trading), few address the need to finance healthy ocean and aquatic ecosystems although these are essential for continued uptake of CO_2 and GHGs. The scientific basis for ocean fertilization – to produce more phytoplankton to increase the uptake of CO_2 – has been challenged, and proposals for burial of CO_2 in the deep ocean have come under criticism from environmentalists. The debate on these issues has underlined the need to increase scientific understanding of how the ocean sequesters carbon.

Modalities of Climate Resilient Agriculture

Farmers are being increasingly affected due to climate change & variability esp. extreme events making agriculture more risky and less profitable. Frequent droughts, delayed onset of monsoon, dryspells, water scarcity, rising temperature, high intensity rains, flood, cyclone, related disease & pest, sea inundation, salinity, *etc* The following are the strategy developed in National initiative on Climate Resilient agriculture launched by ICAR in 2010.

Module – I Natural Resource Management

- Conservation tillage where appropriate, residue incorporation, mulching, *etc*
- In-situ moisture conservation Conservation furrow, contouring, terracing, trench cum bunding, ridge & furrow system, *etc*
- Rainwater harvesting and recycling for supplemental irrigation Renovation of defunct rainwater harvesting structures, Farm pond, Nadi, *etc*
- Improved drainage in flood prone areas
- Ground water recharge: Percolation pond, injection wells, *etc*
- Water saving irrigation methods Micro irrigation system- sprinkler, drip Skip furrow irrigation, *etc*
- Soil system resilience by promoting soil health cards (encouraging INM) Representative soil samples collected from the selected 100 villages and analyzed
- Promoting multi- purpose trees
- Inventory of organic resources as benchmark
- Promoting mangroves and management of sea inundation/ salinity

Module – II Crop Production

- Introducing drought/temperature/flood tolerant varieties/ crops (combination of deep & shallow rooted crops, late & early maturing varieties)
- Advancement of planting dates of winter crops in areas with terminal heat stress
- Water saving paddy cultivation methods (SRI, aerobic, direct seeding)
- Community paddy nurseries in multiple dates for delayed onset of monsoon
- Location specific intercropping systems with high sustainable yield index
- Promote diversification towards high value crops (wherever possible)
- Farm machinery custom hiring centre for timely planting & harvesting, value addition (increase area sown & efficiency, reduce drudgery)

Module – III Livestock and Fisheries

- Augmenting fodder production through CPRs to mitigate scarcity during droughts/floods (managed by community)
- Bio-gas on community basis- by small holders
- Promoting cultivated fodder production (fodder crops & trees)
- Improved fodder/feed storage and conservation (hay, silage) methods
- Promoting stress tolerant high yielding breeds of livestock
- Promoting livestock & fishery as diversification strategy
- Improved shelters for reducing heat stress in livestock
- Management of fish ponds/tanks during water scarcity and excess water, *etc.*
- Prophylaxis & area specific mineral supplementation for livestock

Module – IV Institutional Interventions

- Farm machinery custom hiring centre- increased efficiency, reduced drudgery
- Seed bank/ seed village concept (crops for which seed storage and availability is a problem, *e.g.* groundnut)
- Community fodder bank- conservation and storage on community basis in fodder scarce area
- Institutionalizing agro-advisories by putting weather station in each KVK and selected village.

- Training young farmers as service providers (drip, sprinkler, livestock, farm machinery repair)
- Commodity groups/ collective marketing group,
- Facilitating introduction of weather index based insurance
- Enhancing climate literacy of the farmers: wall writings, pamphlets, talks, group meetings, *etc*.

Conclusions

Global climate change, if it occurs, will definitely affect agriculture. Most mechanisms, and two-way interactions between agriculture and climate, are known, even if not always well understood. It is evident that the relationship between climate change and agriculture is still very much a matter of conjecture with many uncertainties ; it remains largely a conundrum. Major uncertainties affect both the Global Circulation Models (GCMs) and the response of agriculture, as illustrated by differences among models, especially as regards effects at the national and subregional levels. In addition, many of the models do not take into consideration CO_2 fertilization and improved water-use efficiency, the effect of cloud cover (on both climate and photosynthesis), or the transient nature of climate change. The issues that give rise to uncertainty are also regarding temperature. It is assumed that the 4.0 to 5.2°C scenarios occur in 2060, but the most recent IPCC work suggests the mean estimate for 2060 is closer to 1.5°C and that the range of global temperature impacts by 2100 is likely to be between 1.0 and 5°C.

Given these uncertainties in both magnitude and direction of impact, a key issue is *vulnerability* to possible climate change. Vulnerability is used here to mean the *potential* for negative consequences that are difficult to ameliorate through adaptive measures given the range of possible climate changes that might reasonably occur. Defining an area or population as vulnerable is, thus, not a prediction of negative consequences of climate change; it is an indication that across the range of possible climate changes, there are some climatic outcomes that would lead to relatively more serious consequences for the region than for other regions.

Nearly all agricultural impact studies conducted over the years have considered some technological options for adapting to climate change. Among those that offer promise are:

Seasonal changes and sowing dates. For frost-limited growing areas (i.e., temperate and cold areas), warming could extend the season, allowing planting of longer maturity annual varieties that achieve higher yields . For short-season crops such as wheat, rice, barley, oats, and many vegetable crops, extension of the growing season may allow more crops per year, autumn planting, or, where warming leads to regular summer highs above critical thresholds, a split season with a short summer fallow may be possible. For subtropical and tropical areas where growing season is limited by precipitation or where the cropping already occurs throughout the year, the ability to extend the growing season may be more limited and depends on how precipitation patterns change.

Different crop variety or species. For most major crops, varieties exist with a wide range of maturities and climatic tolerances. Responses to climate change are strongly cultivar dependent Longer-season cultivars were shown to provide a steadier yield under more variable conditions . In general, such changes may lead to higher yields or may only partly offset losses in yields or profitability. Crop diversification has been identified as an adaptive response.

New crop varieties. The genetic base is broad for most crops but limited for some. Heat, drought and pest resistance; acid/ salt tolerance, and general improvements in crop yield and quality would be beneficial. Genetic engineering and gene mapping offer the potential for introducing a wider range of traits. Difficulty in assuring traits are efficaciously expressed in the full plant, consumer concerns, profitability and regulatory hurdles have slowed the introduction of genetically engineered varieties compared with early estimates.

Water supply and irrigation systems. Across studies, irrigated agriculture is in general less negatively affected than dryland agriculture but adding irrigation is costly and subject to the availability of water supplies. Climate change will also affect future water supplies. There is wide scope for enhancing irrigation efficiency through adoption of drip irrigation systems and other water-conserving technologies but successful adoption will require substantial changes in how irrigation systems are managed and how water resources are priced. Because inadequate water systems are responsible for current problems of land degradation, and because competition for water is likely to increase, there likely will be a need for changes in the management and pricing of water regardless of whether and how climate changes. Tillage method and incorporation of crop residues are other means of increasing the useful water supply for cropping.

Other inputs and management adjustments. Added nitrogen and other fertilizers would likely be necessary to take full advantage of the CO_2 effect. Where high levels of nitrogen are applied, nitrogen not used by the crop may be leached into the groundwater, runoff into surface water, or be released from the soil as nitrous oxide. Additional nitrogen in groundwater and surface water has been linked to health effects in humans and affects aquatic ecosystems.

Tillage. Minimum and reduced tillage technologies in combination with planting of cover crops and green manure crops offers substantial possibilities to reverse existing soil organic matter, soil erosion, and nutrient loss and to combat potential further losses due to climate change . Reduced and minimum tillage techniques have spread widely in some countries but are more limited in other regions. There is considerable current interest in transferring these techniques to other regions.

Improved short-term climate prediction. Linking agricultural management to seasonal climate predictions (currently largely based on the El Niño Southern Oscillation Phenomenon), where such predictions can be made with reliability, can allow management to adapt incrementally to climate change. Management/climate predictor links are an important and growing part of agricultural extension in both developed and developing countries .

References

Bond, T.E., C.F. Kelly and W.R. Garrett, 1964: Heat transfer from swine to a cold slab. Trans. ASAE, 7:34.

Brander, K.M. 2007. "Global fish production and climate change". Proc. Natl. Acad. Sci. U.S.A. 104 (50):19709-14.

Bucklin, R.A., L.W. Turner, D.K. Beede, D.R. Bray and R.W. Hemken, 1991: Methods to relieve heat stress for dairy cows in hot, humid climates. Appl. Eng. Agr., 7(2):241–247.

Buffington, D.E., A. Collazo-Arocho, G.H. Canton, D. Pitt, W.W. Thatcher and R.J. Collier, 1981: Black Globe-Humidity Index (BGHI) as a comfort equation for dairy cows. Trans. ASAE, 24:711–714.

Cline, W. 2007. Global Warming and Agriculture: Impact Estimates by Country, Washington D.C., Peterson Institute.

Darwin, C. 1859. On the origin of species. John Murray, London.

Deka, R.L., Mahanta, C., Nath, K.K. 2009. Trends and fluctuations of temperature regime of north east India, In: ISPRS Archives XXXVIII-8/W3 Workshop Proceedings: Impact of Climate Change on Agriculture.

FAO. 1995a. World Agriculture: Towards 2010, An FAO Study. N. Alexandratos (ed.). John Wiley, Chichester, UK, and FAO, Rome. 488 p.

Ficke, A.D., Myrick, C.A. and Hansen, L.J. 2007. "Potential impacts of global climate change on freshwater fisheries". Fish Biology and Fisheries 17 (4):581–613

Gates, R.S., J.L. Usry, J.A. Nienaber, L.W. Turner and T.C. Bridges, 1991: An optimal misting method for cooling livestock housing. Trans. ASAE, 34(5):2199–2206.

Hahn, G.L. and J.D. McQuigg, 1970: Evaluation of climatological records for rational planning of livestock shelters. Agric. Meteorol., 7:131–141.

Holmes, R. 1995. Arctic ice shows speed of climate 'flips'. New Scientist 145 (1967): 13.

Houghton, J.T., Meira Filho, L.G., Bruce, J., Lee, H., Callander, B.A., Haites, E., Harris, N. and Maskell, K. (eds.). 1995. Climate Change 1994. Radiative forcing of climate change; and an evaluation of the IPCC IS92 emission scenarios. Cambridge University Press, Cambridge, New York, Melbourne.

Igono, M.O., G. Bjovedt and H.T. Sanford-Crane, 1992: Environmental profile and critical temper-ature effects on milk production of Holstein cows in desert climate. Int. J. Biometeor., 36:77–87.

IPCC. 1992. Climate Change 1992. The supplementary report to the IPCC scientific assessment. J.T. Houghton, B.A. Callandar and S.K. Varney (eds.). Cambridge University Press. Cambridge.

Jackson, G., T. Webb III, E.C. Grimm, W.F. Ruddiman and H.E. Wright Jr. (eds.). 1987. North America and adjacent oceans during the last deglaciation. Geological Society of America. 3: 277-288.

Katz, R.W. and Brown, B.G. 1992. Extreme events in a changing climate: variability is more important than averages. Clim. Change 21: 289-302.

Lehman, S. 1993. Ice sheets, wayward winds and sea change. Nature 365: 108-110.

MacFarlane, J.S. and B.A. Stevens, 1972: The effect of natural shade and spraying with water on the productivity of dairy cows in the tropics. Trop. Anim. Health Prod., 4:249.

Manoj-Kumar 2011. Evidences, Projections and Potential Impacts of Climate Change on Food Production in Northeast India. Indian J Hill Frmg , 24(1&2): 1-10

Maschinski, J., J. Baggs, P.E. Quintana- Ascencio and E. Menges 2006. Using population viability analysis to predict the effects of climate change on the extinct risk of an endangered limestone endemic shrub,Arizona cliffrose. Conservation Biology. 20: 218-228.

Means, S.L., R.A. Bucklin, R.A. Nordstedt, D.K. Beede, D.R. Bray, C.J. Wilcox and W.K. Sanchez, 1992. Water application rates for a sprinkler and fan dairy cooling system in hot, humid climates. Appl. Eng. Agr., 8(3):375–379.

Morrison, S.R., R.L. Givens, and G.P. Lofgreen, 1973: Sprinkling cattle for relief from heat stress. J. Anim. Sci., 36:428.

Morrison, S.R., M. Prokop and G.P. Lofgreen, 1981: Sprinkling cattle for relief: activation tempera-ture, duration of sprinkling and pen area sprinkled. Trans. ASAE, 24:1299.

Paillard, D. and Labeyrie, L. 1994. Role of thermohaline circulation in the abrupt warming after Heinrich events. Nature 372: 162-164.

Restrepo, G., Shanklin, M.D. and Hahn, G.L. 1977: Heat dissipation from growing pigs as a func-tion of floor and ambient temperatures. Trans. ASAE, 20:145.

Rahmsdorf, S. 1994. Rapid climate transitions in a coupled ocean-atmosphere model. Nature 372: 82-85.

Strickland, J.T., R.A. Bucklin, R.A. Nordstedt, D.K. Beede and D.R. Bray, 1989: Sprinkler and fan cooling system for dairy cows in hot, humid climates. Appl. Eng. Agr., 5(2):231–233.

Thompson, R.S. 1990. Late quaternary vegetation and climate in the Great Basin. *In:* Betancourt, J.L., Van Devender, T.R. and Martin, P.S., Packrat middens: The last 40000 years of biotic change: Tucson, Arizona. The University of Arizona Press. pp. 200-209.

Ziska, Lewis H., James, A. Bunce, Hiroyuki Shimono, David, R. Gealy, Jeffrey T. Baker, Paul C.D. Newton, Matthew P. Reynolds, Krishna S.V. Jagadish, Chunwu Zhu, Mark Howden and Lloyd T. Wilson 2012. Food security and climate change: on the potential to adapt global crop production by active selection to rising atmospheric carbon dioxide. Proc. R. Soc. B 2012 279, 4097-4105.

❑❑❑

3

Probability Distribution Model of Weather Spells and Weather Cycles: A Case Study

Debashis Bhowmik, Banjul Bhattacharyya

Abstract

An attempt has been made in this paper to study probability model for modeling of spell distribution & Weather Cycles during the monsoon period (June to September) at Cochbehar, West Bengal. The daily rainfall data of 31 years (1970 to 2000), recorded at Cochbehar have been analyzed. In this study, it was found that daily rainfall during monsoon period follows Markov Chain model. Markov dependent truncated geometric distribution model is best fitted for studying the Dry spells, Wet spells & Weather Cycles during the monsoon period.

Keywords: Markov Chain, Transition Probability, Dry Spell, Wet Spell, Weather Cycles & Truncated Geometric Distribution.

Introduction

Quantification of variability of climatic factors has always been a matter of great importance for crop planning. Out of all climatic factors, rainfall has many facets. Mostly work in this direction is confined to the total rainfall, maximum and minimum amount of rainfall and its range etc. however, not only the inadequacy of rainfall but also excess or deficiency is harmful to a particular crop. It is also natural to imagine that for the crop production the total amount of rainfall is not of much importance but the pattern of its occurrence such as spells of rainy and dry days, expected number of dry days between the two rainy days and their repetitions are much use for the rain fed crops. The intermittent dry & wet spells can be useful for crop planning programme. By the knowledge of dry spell, wet spell & weather cycles farmer can plan sowing periods such a way that moisture sensitive stages do not fall during the prevailing dry spells. The knowledge of wet spell, dry spell & weather cycles can also have great deal in improving the efficiency of irrigation water utilization.

Among the earlier studies may be mentioned the work of Cochran (1938) who proposed a probability model based on the" theory of Runs" to study the "persistency " behavior of rainy days while others like Gabriel and Neuman (1957) suggested geometric distribution as a suitable model for wet and dry weather spells. Gabriel and Neuman (1962) considered empirical evidence to find out the suitability of the pattern of occurrence of rainfall with the help of Markov chain model. In this study, stochastic approach has been considered to predict the behaviour of spell distribution and weather cycles under the assumption that the probability of occurrence of precipitation of any day depends upon the previous observation and only on it. Bhargava *et al.*(1973) studied the pattern of occurrence of rainfall using a two state Markov Chain Model. Hann *et al.*(1976) presented a Markov Chain Model for studying the behavior of daily rainfall. Manohar and Siddappa (1984) conducted a study of weather spell and weather cycles at Raichur using Markov Chain Model.

The aim of this study is to fit probability distribution model for studying the pattern of dry spell, wet spell & weather cycles event in Cochbehar, W.B.

Materials and Methods

Thirty years rainfall data (1970 -2000) during monsoon period (from May 31st to September 30th) of Cochbehar meteorological center, W.B utilized. The daily rainfall data are collected from the India Meteorological Department, Pune, India.

The basic discrete distributions in truncated form are considered for probability model of wet spells, dry spells and weather cycles. The truncated geometric probability model based on two state Markov chain model is considered to fit the frequency of dry spell wet spell and weather cycles.

Markov chain model of daily rainfall: A sequence of days with nil or trace (< 2.5 mm) amount of daily rainfall is called a period of dry spell. A day receiving at least 2.5 mm of rainfall from 08.30 hrs to 08.30 hrs the next day, as wet day, otherwise a dry day. This classification gives a sequence of wet and dry days which can be regarded as a two – state Markov chain with wet and dry days. Each day of this sequence is classified as one of the following possibilities.

i) A dry day preceded by a dry day

ii) A wet day preceded by a dry day

iii) A dry day preceded by a wet day

iv) A wet day preceded by a wet day

Thus for each year, 31st May which is regarded as the initial day may be considered for classifying 1st June for each year. Thus, for each year, the nature of a day is classified as one of the four possibilities depending on the previous day is dry or wet. Repeating the process each cell frequencies for the above four possibilities are obtained.

Let these frequencies be f (D/D), f (W/D), f (D/W) and f (W/W) respectively with f (D/D) + f (W/D) = n_1 and f (D/W) + f (W/W) = n_2

The various transition probabilities are to be estimated by the corresponding relative frequencies which are the maximum likelihood estimates (Anderson and Goodman). The transition probability matrix P, which describes the two states Markov Chain model is represented by

$$P = \begin{bmatrix} p_{00} \; p_{01} \\ p_{10} \; p_{11} \end{bmatrix}$$

where,

p_{00} = the conditional probability that a day is a dry day given that the preceding day was a dry day. p_{11} = the conditional probability that a days is a wet day given that preceding day was a wet day. p_{01} = the conditional probability that a day is a wet day given that the preceding day was a dry day. p_{10}= the conditional probability that a day is a dry day given that the preceding day was a wet day.

The conditional probabilities can be estimated as below:

$$p_{01} = \text{f (W/D)}/\ n_1 \quad \text{and p11} = \quad \text{f (W/W)}/\ n_2$$

So, under the two state Markov chain model the other two conditional probabilities are easily obtained as p_{00} = 1- p_{01} and p_{10} = 1- $p_{11.}$

The P matrix enables the estimates of the equilibrium or steady state probabilities p_0 and p_1 corresponding dry and wet days from

$$p_0 = p_{10}/(p_{01}+p_{10)}$$

$$p_1 = p_{01}/(p_{01}+p_{10)}$$

and the expected number of dry and wet days in a month as well as over the season (pooled), i.e. E(D) and E(W), are then obtained as E(D) = n p_0 and E (W) = n p_1 , where, n is the days consisting the month under study.

Tests of Independence: For determining weather, the occurrence of a wet day and a dry day depends only on the immediately preceding day's weather; we test the hypothesis of independence against Markovian dependence using the test statistic is given by

$$\sum_{i}^{n}\sum_{j}^{n}\frac{(n_{ij}-\frac{n_i n_j}{n})^2}{n_i n_j / n},$$

Where,

n_{ij} = number of frequency corresponding to a day is j-th state given that preceding day was in i-th state.

n_i = total number of frequency in the i- th state.

n_j = total number of frequency in the j- th state.

n = total number of frequency

which is follow limiting Chi square with $(m-1)^2$ degree of freedom.

Spell and weather cycle distribution: A rainfall event is characterized by two sets of meteorological phases such as wet day and dry day. In a sequence of wet and dry days, there are possibilities of the occurrence of wet days and dry day in succession. A number of subsequent wet days preceded and followed by a dry day are regarded as the length of wet spell. Subsequently a dry spell is defined in the similar way. A weather cycle is defined as the combination of wet spell with the immediate successive dry spell, recognized as wet-dry weather cycle and a dry spell with the immediate successive wet spell referred as dry-wet cycle.

The probabilities of obtaining these three events, a wet spell of length w – days, a dry spell of length d days and weather cycle of length n – days can be obtained from the Markovian dependent truncated geometric model. The probability function for wet spell length (X), dry spell length (Y) and the weather cycle length (Z) are given below:

$$P(X=w) = (p_{11})^{w-1}\,(1- p_{11})$$

$$P(Y=d) = (p_{00})^{d-1}\,(1- p_{00})$$

$$\text{And } P(Z=n)= (p_{01} \times p_{10}) \times \left[\frac{p_{11}^{\,n-1} - p_{00}^{\,n-1}}{p_{11} - p_{00}}\right]$$

Where, P (X=w) = probability of a wet spell of length w days ,

P (Y=d) = probability of a dry spell of length d days,

P (Z=n) = probability of a weather cycle of length n days

p_{00}, p_{01}, p_{10} and p_{11} are the conditional probability of a day is dry day preceded by a dry day, a wet day preceded by a dry day , a dry day preceded by a wet day and a wet day preceded by a wet day respectively.

Length of Spell and weather cycle: Using the parameters of two state Markov chain model on the daily rainfall & considering the distribution, expected length of dry spell of length "d" days is given by

$$E(d) = \frac{1}{p_{01}}$$

The expected length of wet spell of w days i.e.," w " successive wet days followed by a dry day

then is given by $E(w) = \frac{1}{(1-p_{11})}$

The expected length of a weather cycle i.e dry spell followed by wet spells and vice- versa is then given by $E(n) \frac{1}{p_{01}} = + \frac{1}{(1-p_{11})}$

where p_{01} and p_{11} are the usual notations given in earlier sub section.

Results and Discussion

From the analysis of daily rainfall data, it is revealed that the hypothesis of independence of weather of consecutive days is to be rejected and therefore, the occurrence of wet day or dry day is influenced by the previous days weather. This indicates that the Markov chain model is very much effective on the determination weather pattern in Coochbehar, during the monsoon months as well as over the season. It is also found that, the expected frequencies for dry spells, wet spells and as well as dry-wet wet-dry weather cycles of varying length are the good agreement with the observed frequencies, as Chi-square statistics value of goodness of fit test found non significant at 5 % level of significance during all the monsoon months and over the season. Thus, Markov dependent truncated geometric distribution is well fitted to the frequency distribution of dry spells, wet spells and as well as dry-wet, wet-dry weather cycles of varying length during the monsoon months and season.

In Cochbehar, expected length of dry spells varies from 1.82 to 2.59, expected length of wet spells varies from 2.50 to 3.57 and expected length of weather cycle varies from 5.06 to 5.39. It is also observed that the expected length obtained through Markov dependent truncated geometric model is almost same as observed length of dry spell, wet spell as well as weather cycle during all the monsoon months and season.

Table 1: Transition probability matrix and steady state probability p_0 and p_1 in Cochbehar during monsoon months and season.

State	June		July		August		September		Season	
	Dry	Wet	Dry	Wet	Dry	Wet	Dry	Wet	Dry	Wet
Dry	0.53	0.47	0.45	0.55	0.61	0.39	0.61	0.39	0.56	0.44
Wet	0.31	0.69	0.28	0.72	0.40	0.60	0.37	0.63	0.33	0.67
π_0	0.40	0.33	0.50	0.49	0.43					
π_1	0.60	0.67	0.50	0.51	0.57					

Table 2: Chi-square statistic values of test for Markovian independence

Months	Chi- Square Value (Cal)
June	36.68
July	24.11
August	30.0
Sept.	40.90
Season	155.55

Table.3: Observed and expected frequency of dry spell in Cochbehar during monsoon months and season.

Length	June		July		August		September		Season	
	Obs	Exp	Obs	Exp	Obs	Exp	Obs	Exp	Obs	Exp
1.	85	80.88	95	92.66	75	73.84	77	68.32	332	309.35
2.	36	42.63	40	41.25	49	44.37	37	41.80	162	172.44
3.	21	22.46	19	18.36	22	26.66	21	25.57	83	96.13
4.	16	11.84	6	8.17	14	16.02	13	15.65	49	53.58
5.	5	6.24	2	3.64	11	9.62	12	9.57	30	29.87
6.	3	3.29	2	1.62	5	5.778	5	5.86	15	16.65
7.	4	1.73	2	0.72	1	3.47	5	3.58	12	9.28
8.	0	0.91	0	0.32	3	2.09	2	2.19	5	5.17
9.	0	0.48	1	0.14	2	1.25	0	1.34	3	2.88
10.	0	0.25			1	0.75	0	0.82	1	1.61
11.	1	0.13			1	0.45	1	0.50	3	0.90
12.	85				0	0.27	2	0.31	2	0.50
13.	36				0	0.16	1	0.19	1	0.28
14.	21				0	0.10			0	0.16
15.	16				1	0.06			332	0.09
Chi-square	3.49 (NS)		10.97 (NS)		12.19 (NS)		7.94 (NS)		18.16 (NS)	

Table 4: Observed and expected frequency of wet spell in Coochbehar during monsoon months and season.

Length	June		July		August		September		Season	
	Obs	Exp	Obs	Exp	Obs	Exp	Obs	Exp	Obs	Exp
1.	60	54.38	60	49.64	91	75.22	72	64.98	270	239.45
2.	36	37.48	36	35.95	41	45.28	27	40.85	143	159.70
3.	21	25.83	27	26.04	21	27.26	31	25.68	92	106.52
4.	19	17.81	12	18.86	13	16.41	20	16.15	64	71.04
5.	11	12.27	10	13.66	6	9.88	9	10.15	32	47.38
6.	8	8.46	7	9.89	4	5.95	5	6.38	24	31.60
7.	4	5.83	10	7.16	5	3.58	2	4.01	21	21.08
8.	5	4.02	5	5.19	4	2.16	4	2.52	19	14.06
9.	5	2.77	4	3.76	2	1.30	2	1.59	15	9.38
10.	1	1.91	2	2.72	1	.78	1	1.00	10	6.25
11.	1	1.32	3	1.97	1	.47	0	0.63	7	4.17
12.	2	0.91	1	1.43			1	0.39	8	2.78
13.	0	0.63	1	1.03			1	0.25	2	1.86
14.	1	0.43	1	0.75					2	1.24
15.	1	0.30	0	0.54					1	0.83
16.			1	0.39					1	0.55
Chi-square	3.49 (NS)		10.97 (NS)		12.19 (NS)		7.94 (NS)		18.16 (NS)	

Table 5: Observed and expected frequency of Dry-wet weather cycle in Cochbehar during monsoon months and season

Length	June		July		August		September		Season	
	Obs	Exp	Obs	Exp	Obs	Exp	Obs	Exp	Obs	Exp
2.	30	24.04	30	25.41	35	27.61	28	23.52	116	97.28
3.	34	29.33	35	29.73	37	33.41	25	29.17	135	119.66
4.	24	26.99	29	26.55	24	30.32	25	27.13	110	110.68
5.	19	22.20	15	21.43	30	24.46	23	22.43	82	91.24
6.	11	17.22	10	16.47	13	18.50	18	17.39	57	70.70
7.	12	12.88	8	12.33	10	13.43	13	12.94	46	52.73
8.	10	9.42	5	9.09	5	9.48	12	9.36	32	38.33
9.	2	6.78	8	6.64	5	6.56	3	6.64	29	27.36
10.	8	4.83	8	4.82	8	4.46	4	4.63	18	19.27
11.	5	3.41	5	3.49	2	3.00	3	3.19	18	13.44
12.	3	2.40	1	2.52	5	2.00	4	2.18	13	9.30
13.	2	1.68	4	1.82	1	1.32	0	1.48	7	6.40
14.	0	1.17	3	1.31	1	.86	2	.99	5	4.38
15.	3	.81	2	.95	0	.56	0	.66	4	2.98
16.	0	.56	0	.68	0	.37	2	.44	2	2.03
17.	1	.39	1	.49	0	.24	1	.29	3	1.38
18.	0	.27	0	.35	0	.15			0	.93
19.	0	.19	1	.25	1	.10			2	.63
Chi-square	11.96 (NS)		14.24 (NS)		11.80 (NS)		4.71 (NS)		15.74 (NS)	

Table.6: Observed and expected frequency of wet-dry weather cycle in Coochbehar during monsoon months and season.

Length	June		July		August		September		Season	
	Obs	Exp	Obs	Exp	Obs	Exp	Obs	Exp	Obs	Exp
1	25	24.04	35	25.26	34	27.61	30	23.38	114	96.99
2	33	29.33	30	29.55	36	33.41	23	28.99	122	119.30
3	29	26.99	28	26.39	31	30.32	30	26.96	118	110.35
4	19	22.20	17	21.30	21	24.46	20	22.29	78	90.97
5	15	17.22	11	16.37	13	18.50	14	17.28	58	70.49
6	11	12.88	7	12.25	11	13.43	14	12.86	43	52.57
7	8	9.42	6	9.03	9	9.48	11	9.31	34	38.21
8	7	6.78	8	6.60	3	6.56	2	6.60	20	27.28
9	7	4.83	7	4.79	0	4.46	3	4.60	22	19.21
10	3	3.41	6	3.47	6	3.00	7	3.17	17	13.40
11	2	2.40	3	2.51	3	2.00	2	2.17	10	9.27
12	2	1.68	3	1.81	6	1.32	2	1.47	10	6.38
13	0	1.17	1	1.30	0	.86	1	.99	4	4.36
14	1	.81	1	.94	1	.56	1	.66	4	2.98
15	1	.56	0	.68	1	.37	0	.44	2	2.02
16	1	.39	1	.49	1	.24	1	.29	4	1.37
17	1	.27			1	.15	0	.19	2	.93
18							0	.13	0	.63
19							0	.08	2	.42
20								0	.05	0
21										
22								0	.04	1
23										
24							0	.02	2	.13
25							1	.01	1	.09
Chi-square value	2.28 (NS)		12.32 (NS)		9.02 (NS)		8.87 (NS)		20.78 (NS)	

Table 7: Observed and expected frequency length of Dry spell, wet spell and weather cycle in Cochbehar during monsoon months and season

Particular	June	July	August	September	Season
Observed length of Dry spell	2.14	1.80	2.51	2.59	2.27
Expected length of Dry spell	2.13	1.82	2.56	2.56	2.27
Observed length of Wet spell	3.21	3.66	2.58	2.70	3.03
Expected length of Wet spell	3.23	3.57	2.50	2.70	3.03
Observed length of weather cycle	5.31	5.29	5.00	5.32	5.31
Expected length of weather cycle	5.35	5.39	5.06	5.27	5.30

References

Bhargava P.N., P. Narin, K.G. Aneja and Asha Pradhan. 1973. A study of occurrence of rainfall in Raichur district with the help of Markov chain model. J. Ind. Soc. Agril. Stat, 25:197-204

Gabriel K. R. and J. Neuman. 1962. A Markov chain model for daily rainfall occurrence at Tel Aviv. Quart J. R. Met. Soc. 88:90-95

Hann C. T. , D. M. Allen and J. O. Street. 1976. A Markov chain model of daily rainfall. Water resource research . 12:443-449

Manohar N and B. Siddappa. 1984. A study for weather spells and weather cycles at Raichur using Markov Chain model . Indian J. Agric. Sci. 54:55-60

Pragyan Kumari, C. V. S. Sastri, A. K. Srivastava and P. Kumari. 2000. Probability of occurrence of dry and wet spells during rice growing season in Biohar. Ann. Agric. Res. 21:355-359.

Sundararaj N and S. Ramachandra. 1975. Markov dependent geometric model for weather spells and weather cycles a study. Indian Met. Hydro. Geophys. 26:221-226.

❑❑❑

4

Weather Based Agromet Advisory Services: An Improved Tool to Mitigate Climate Change Impact on Crops

Shilpa, C.N., Pavithra, B.V., Padmashri, H.S., Rajegowda, M.B and N.A. Janardhana Gowda

Abstract

Under Ministry of Earth Sciences, India Meteorological Department has launched District level Agro-meteorological Advisory Services in June 2008. The agro-advisory bulletins containing the forecast on rainfall cloud cover, wind direction and speed aid minimum and maximum temperature with their likely impact on crops are in use twice in a week.

Keywords: Agromet Advisory Services, NCMRWF, IMD, Multipurpose Community Telecenters, Decision support system.

Introduction

The Government of India has been focusing on strengthening farmers' knowledge on sustainable farm practices in their overall efforts to augment food security of the nation. In spite of successful research on agricultural practices and technologies, Indian farmers are facing a multitude of problems to sustain crop productivity. Many of these problems are linked to variability in weather and climate. To address this issue effectively, the Government launched District-level Agro meteorological Advisory Service (DAAS) in June 2008 as one of the flagship programme of

Ministry of Earth Sciences. The DAAS aims to generate Agro meteorological information (weather forecast and Agromet-advisories) and develop suitable dissemination system, to the farming community in order to improve crop/livestock productivity. It enables farmers to take advantage of benevolent weather and minimize the adverse impact of malevolent weather on crops. Presently, the weather based agro-advisories are disseminated to the farming community at district level through mass media (Radio, Print and TV). But there exists a wide information gap between information generator and user. The outreach of IAAS system to deliver the information at Block and Panchayat (village) level, in a timely manner, needs to be stepped up.

The NCMRWF in collaboration with IMD, Indian Council of Agricultural Research (ICAR) and State Agricultural Universities (SAUs) is providing Agrometereological Advisory Services (AAS) at the scale of Agroclimatic Zone to the farming community based on location-specific Medium Range Weather Forecast (MRWF). The NCMRWF provides AAS in the form of Agromet Advisory Bulletins, which contain expert; advise on crop, soils and weather to the farming community. Thus, the AAS set up exhibits a multi-institutional, multi-disciplinary synergy to render an operational service for the use of farming community. Over the past decade, the NCMRWF has established an impressive infrastructure and also developed suitable methodologies for giving quantitative medium range weather forecasting services. These units are located within SAUs headquarters, their regional research stations and ICAR institutes. Location-specific weather forecast for six parameters, viz. rainfall, cloud cover, wind direction and speed, and minimum and maximum temperature are obtained twice a week. Weather forecasts are given by NCMRWF to AAS Units biweekly (Tuesday and Friday). The forecast, in quantitative terms, is issued for above-mentioned six parameters. The AAS Units assisted by an advisory board consisting of agricultural scientists representing wide spectrum of agricultural disciplines discuss the forecast of their zone to prepare the AAS Bulletins. These advisories contain location specific and crop specific farm level advisories prepared in local language containing description of prevailing weather, soil & crop condition, and suggestions for taking appropriate measures to minimize the loss and also, optimize input in the form of irrigation, fertilizer or pesticides. The Bulletin contains the forecast on six parameters along with their likely impact on crops. The Bulletin also provides the advice to farmers on the measures to be taken to maintain crop yields and to minimize weather-induced losses. The AAS Bulletin is disseminated to farmers by mass media (TV, radio, newspapers) in vernacular language. The Bulletin is also disseminated to the contact farmers in several villages by phone, post, or by hand.

Method of Disseminating Agromet Advisory Services

Indian Meteorological Department forecast weather data to different departments; it involves all stakeholders such as State Agricultural Universities (SAUs), Indian Council for Agriculture Research (ICAR), Krishi Vigyan Kendras (KVK), Department of Agriculture and Cooperation, State Departments of Agriculture/ Horticulture/ Animal Husbandry/ Forestry (up to District level offices), NGOs, Media Agencies, *etc*. It includes meteorological (weather observing and forecasting), agricultural (identifying weather sensitive stress and preparing suitable advisory using weather forecast), extension (two way communication with user) and information dissemination (Media, Information Technology, Telecom) agencies. After receiving the forecast from IMD Nodal officers of AICRPAM will prepare the AAS Bulletin then it send through Mass Medias like TV, Radio *etc*. (Fig. 1).ultimately disseminated to end users/farmers.

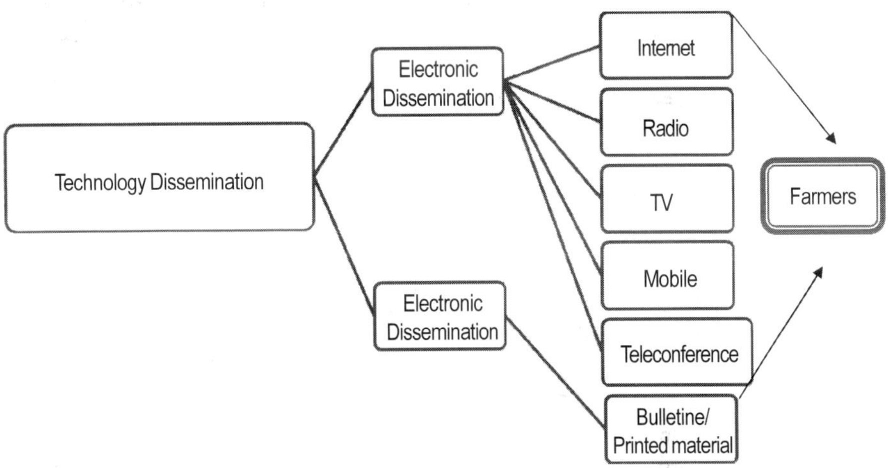

Fig.1: Agromet Advisory Dissemination Technologies

IMD giving forecasts for 6 weather parameters viz., rainfall, maximum and minimum temperatures, wind speed and direction, relative humidity and cloudiness, besides weekly cumulative rainfall forecast. The products are disseminated to Regional Meteorological Centres and Meteorological Centres of IMD located in different states. These offices undertake value addition to these products twice a week on Tuesday and Friday and communicate to 130 AgroMet Field Units (AMFUs) located with State Agriculture Universities (SAUs), ICAR *etc*.

The District level Agro-advisory based on the above forecast products and the crop information available from districts, the AMFU prepares district-wise agro-advisories. These are operated at State Agriculture Universities (SAUs), Indian Council of Agricultural Research institutions (ICAR), Indian Institute of Technology (IIT) by providing grant-in-aid from IMD. These units are responsible for recording agro-meteorological observations, preparing medium range weather forecast-based Agromet advisories for the districts falling under precinct of concerned agroclimatic zone and dissemination of the same. Concerned universities/institutes have appointed Nodal Officer and Technical Officers, who prepare the advisory bulletins in consultation with the panel of experts already created at these units. The Agromet bulletins include specific advice on field crops, horticultural crops and livestock *etc.*, which farmers need to act upon. Its frequency is twice a week i.e. Tuesday and Friday.

Advisory Dissemination Mechanism the weather based advisories are being disseminated to the farmers through mass media dissemination, Internet *etc.* A mechanism has also been developed to obtain feedback from the farmers on quality of weather forecast, relevance and content of agromet advisory and effectiveness of information dissemination system. The dissemination mechanism needs to be scaled up to communicate advisories through multi-channel dissemination system including Radio, TV, and print media, Cell Phone-SMS, Internet and Common Service Centers. District level Agromet Advisory services as shown in the Fig 2.

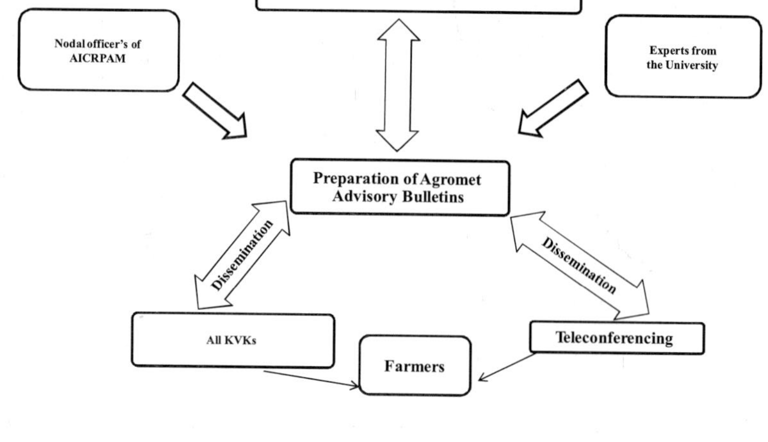

Fig. 2: Distric wise Agromet Advisory Services through Teleconferencing

Conclusion

Weather based Agromet Advisory Service information used for decision making, represents part of a continuum; at the other end is scientific knowledge and understanding. Other components of this continuum are the collection of data and transforming data into useful information. Information has value when it is disseminated in such a way that the end-users get the maximum benefit in applying its content. The Internet will play an important role in the collection and transfer of information. In developing countries, Multi-Purpose Community Telecentres (MCTs) will be the equivalent of an information supermarket. Radio can be used to transfer information from MCTs to rural areas. Using response farming as an example, a prototype information system that can have wide applicability is suggested. Future concerns about the information needs of diverse end-users, information on a fee basis, and the training needs of end-users and intermediaries are discussed. Although modern technology has improved weather based AAS information and increased the number of end-users, continued improvements are necessary to ensure that the content of the information is adequate to fulfill the requirements of the farming community's. Finally we can say that AAS play a important role to mitigate Climate change effect on crops.

References

Kushwaha H.S., Phool Chand and Manisha Rani 2008. Economic impact of MRWF based agromet advisory services on farmers field of humid sub-tropics of Uttarkhand, India. Journal of Agrometeorology (Special issue – Part I): 235–239.

Sharma SK, Kothari AK, Sharma RK and Jain PM 2008. Capitalizing on agro-advisoty for higher productivity in rainfed agroecosystem – A case study. Journal of Agrometeorology (Special issue- Part 1): 219–224.

Ravindra babu, B.T, Janardhanagowda, N.A., Jagadeesha, N., K.R., Rajashekhar, Rajegowda, M.B. 2007. Application of weather based agroadvisories in eastern dry zone of Karnataka, Journal of Agrometeorology 9(2): 259-264.

Venkataraman 2004. Climatic characterisation of crop productivity and input needs for agrometeorological advisory services, Journal of Agrometeorology 6(11):98-105.

❑❑❑

5

Climate Change Threatens Food Crops Across the World: An Overview

Md. Hedayetullah, C.K. Kundu, and Serajul Islam

Abstract

Climate is changing naturally at its own pace, since the beginning of evolution of earth, 4–5 billion years ago, but presently it has gained momentum due to inadvertent anthropogenic disturbances. Global climate is changing and it can have serious implications for our food security through its direct and indirect effects on crops, soils, live stocks, fisheries and pests. In the developing countries including India there has been relatively less attention paid to this topic in an integrated manner. Agriculture is the largest human activity in the world which depends on climatic parameters. We can manipulate soil, water and nutrient management practices but manipulation of climate is beyond our control. It influences crop-pest equilibrium. More than 50% differences in yield are due to climatic variation. Different causes of climate change are: A) Natural cause: Earth orbit, Global warming, Mountains, Sun, and Volcano. B) Anthropogenic cause: Global warming, Deforestation, Change in land use pattern. Climate change is a present reality and a major risk for future generations. Implementing energy conservation methods and utilizing renewable resource such as green power (power from moon, hydrogen fuel, nuclear power, sewage power, geothermal power and power of the mighty sun). Government should also encourage the use of mass transit; provide tax rebates for people who buy low and no pollution vehicles and subsidies to fossil fuels and the nuclear industries. According to Dr. M. S. Swaminathan our present Indian agriculture is vulnerable to climate change and climatic hazards due to dependence on less no. of crops. Therefore we have to go for crop diversification.

Keywords: Climate change, Cause, Food crops, Challenge, Recommendation.

Introduction

The atmosphere surrounding the earth is made up of nitrogen (78%), oxygen (21%) and the remaining 1%, is made up of trace gases (called so because they are present in very small quantities) that include carbon dioxide, methane and nitrous oxide. These gases also called greenhouse gases act as a blanket and trap heat radiating from the earth and make the atmosphere warm. Beginning with the industrial revolution global atmospheric concentrations of these greenhouse gases have increased markedly as a result of human activities. The global increases in carbon dioxide concentration are due primarily to fossil fuel use and land use change, while those of methane and nitrous oxide are primarily due to agriculture. According to one estimate, industrialized countries are responsible for about 83% of the rise in cumulative fossil fuel related CO_2 emissions (Loske, 1996) since 1800. As a result we are witnessing global warming.

The increasing green-house gases (GHG) resulted in global warming by 0.74B0C over past 100 years and 11 of the 12 warmest years were recorded during 1995-2006. The Intergovernmental Panel for Climate Change (IPCC) projections on temperature predicts an increase of 1.8 to 4.0B°C, by the end of this century. Some changes will affect agriculture through their direct and indirect affects on crops, soils, livestock, fisheries and pests. Tropical countries are likely to be affected more compared the countries situated in temperate regions. The brunt of environmental changes is expected to be very high in India due to greater dependence on agriculture, limited natural resources, alarming increase in human and livestock population, changing pattern in land use and socio-economic factors that pose a great threat in meeting the food, fibre, fuel and fodder requirement. There is a likelihood of a considerable impact on agricultural land-use due to snow melt, availability of irrigation, frequency and intensity of inter- and intra-seasonal droughts and floods, soil organic transformation matters, soil erosion and availability of energy as a consequence of global warming, impacting agricultural production and hence, the nations food security (Okammoto, 2004). Global warming due to greenhouse effect is expected to impact hydrological cycle viz. precipitation, evapo-transpiration, soil moisture *etc*, which would pose new challenges for agriculture.

What is Climate Change?

Climatic change is the overall increase or decrease in average annual mean climatic parameters from previous years to the succeeding years. Some evidences such as rise in temperature and CO_2 level; depletion in rainfall; shifting and shrinking of cooling period; changing pattern of monsoon; occurance of natural disaster; early blossoming of trees; appearance of grasses in Antarctica; changing cropping pattern; high latitudinal shifting of temperate crops are the real life projections of climate change (Piyasiri, 2002).

Global and Indian Scenario of Climate Change: The global atmospheric concentration of carbon dioxide, a greenhouse gas (GHG) largely responsible for global warming, has increased from a pre-industrial value of about 280 ppm to 379 ppm in 2005. Similarly, the global atmospheric concentration of methane and nitrous oxides, other important GHGs, has also increased considerably. The increase in GHGs was 70% between 1970 and 2004. Eleven of the last twelve years rank among the 12 warmest years in the instrumental record of global surface temperature since 1850. The mean earth temperature has changed by 0.74B°C during 1906-2005. Most of the observed increase in global average temperatures since the mid-20th century is very likely due to the observed increase in anthropogenic greenhouse gas concentrations. During last 50 years, cold days, cold nights and frost have become less frequent, while hot days, hot nights, and heat waves have become more frequent. The frequency of heavy precipitation events has increased over most land areas. Global average sea level rose at an average rate of 1.8 mm per year over 1961 to 2003. This rate was faster over 1993 to 2003, about 3.1 mm per year. Climate change is likely to impact all the natural ecosystems as well as socio-economic systems as shown by the National Communications Report of India to the UNFCCC (UNFCCC, 2004).

Analyses done by the Indian Meteorology Department and the Indian Institute of Tropical Meteorology, Pune (MS), generally show the same trends for temperature, heat waves, glaciers, droughts and floods, and sea level rise as by the Intergovernmental Panel on Climate Change of United Nations. Magnitude of the change varies in some cases. At all India level, there is no trend in monsoon rainfall during last 100 years, but there are some regional patterns Areas of increasing trend in monsoon rainfall are found along the west coast north Andhra Pradesh and north-west India, and those of decreasing trend over east Madhya Pradesh and adjoining areas, north-east India and parts of Gujarat and Kerala (-6 to -8% of normal over 100 years). Surface air temperature for the period 1901 96 2000 indicates a significant warming of 0.4B°C for 100 years (Piyasiri *et al*., 2004).

Causes of Climate Change

Earth's Orbit: The Earth orbits the Sun once every year. This orbit is highly regular, but there do exist tiny variations in it that affect the amount of heat received from the Earth by the Sun, and therefore the Earth's climate. These variations occur over tens and hundreds of thousands of years, and are believed to explain the switching of the Earth's climate between Ice Ages and warmer periods. The last Ice Age occurred about 18,000 years ago. Currently the Earth's climate lies in a warm period, but some scientists believe that the next Ice Ages are not too far away.

Global Warming: Scientists now know that there are many causes of climate changes, and most of these are natural. However, perhaps the most well-known cause, at least amongst the public, is the man-made pollution of the air that may be contributing to global warming. Air naturally contains greenhouse gases which trap heat in the atmosphere and keep the Earth warmer than it should be, and suitable for life. This is called the natural greenhouse effect. In the last 200 years, mankind has increased the amount of greenhouse gases in the atmosphere by burning fossil fuels like coal, oil and gas. This is trapping more heat, which may be causing a change in the Earth's climate. Since the beginning of the 20th century the Earth has warmed up by over 0.5°C. This does not sound like much, but during the last Ice Age, the Earth was on average only 5°C colder than today. Global warming, as it is known, is predicted to continue in the 21st century unless we take action to reduce air pollution. By 2100, the Earth may be another 2 or 3°C warmer.

Greenhouse Effect: The Sun is the only source of external heat for the Earth. Sunlight heats the Earth's surface during the day. As the Earth's surface warms up it also releases heat, but we cannot see this as light like sunlight from the Sun. Sometimes, on hot days we can see the heat shimmers rising from a hot black road. This shows us that a lot of heat is being given off by the surface. At night, when the surface is no longer heated by the Sun, the Earth continues to give off heat in this way. In fact, at the scale of the whole Earth, as much heat is given off by the Earth as is received by the Sun. This keeps the average temperature of the Earth fairly stable, currently at about 15°C. If the Earth didn't have an atmosphere, its surface would on average be as cold as the surface of the Moon, about -18°C. The reason that it is much warmer is because of greenhouse gases in the air. Greenhouse gases exist naturally in air, heating it by trapping energy that has originally come from the Sun. This process is called the greenhouse effect. The combustion of fossil fuels and other human activities are the primary reasons for increased concentrations of CO_2 and other greenhouse gases. Between 1990 and 1999, an estimated 6.3 GtC/year was released due to the combustion of fossil fuels, and another 1.6 GtC/year was released due to the burning of forest vegetation. This was offset by the absorption of 2.3 GtC/year each by growing vegetation and the oceans. This left a balance of 3.3 GtC/per year in the atmosphere (Climate change, 2001).

Mountains: Mountains are the largest features found on the Earth's continents. The highest ones in the world rise over 5 miles into the sky. Mountains are usually grouped together in ranges, which can be several hundred miles wide and many thousands of miles long. Mountain ranges are built by the same processes that cause the continents to drift over million or tens of millions of years. When continents collide, the rock in the middle is crunched upwards to form mountains. Most of the world's weather occurs in the lowest 5 to 8 miles of the atmosphere.

Sun: The Sun is our main source of light and warmth. It is the driving force behind the world's weather. We are all aware that a cloudy day is usually colder than a sunny one, and that nighttime is almost always colder than daytime. It is likely then that changes in the amount of energy coming from the Sun could well affect the Earth's climate.

Volcanoes: Large volcanoes can inject huge quantities of dust and other gaseous pollution into the air. The most explosive volcanic eruptions are able to send material many miles up into the atmosphere, above the level of most rain clouds. Here the pollution remains for up to two years, blown around the globe by winds. This volcanic pollution blocks out some of the sunlight. At the Earth's surface this causes a cooling of the climate. If the pollution spreads all around the Earth, a global cooling of as much as 0.3°C is possible which can last for 2 years. An example of this happened in 1991 when Mount Pinatubo in the Philippines erupted.

Role of Global Warming in Climate Change

Global warming is when the earth heats up. It happens when the green house gases (CO_2, water vapour, nitrous oxide& CH_4) trap heat and lights from the sun in the earth's atmosphere which increases the temperature. Global warming is also known as "green house effect". The green house effect is when the temperature rises because the sun's heat and light is trapped in the earth's atmosphere.

What cause Global Warming?

Electrical pollution:One thing that causes global warming is electrical pollution. In most cases fossil fuels are burned to create electricity. Fossil fuels are made of dead plants and petroleum. One example of fossil fuels is oil and petroleum. Many pollutants are sent to the air, when fossil fuels are burned. Some of these chemicals are called green house gases. Some other examples of polluting the air are – turning on a light, watching T.V, listening to a stereo, washing or drying clothes, using a hair dryer, riding a car, heating a meal in micro wave, using an air conditioner, playing a video game *etc*. When we do these things we are causing more GHG to send into the air.

Forest destruction creates more heat: Another thing that makes global warming worse is when people cut down trees. Trees and other plants collect CO_2, sinks of GHG. The gas (CO_2) collects light and heat, produced by the sun, and this makes the earth warmer. However when trees are burned the carbon locked in the structure is released into the air in the form of CO_2. Today the shrinking world forests are not able to absorb all the CO_2 created by us while burning fossil fuels. Everyday over 5,500 acres of rain forest are destroyed, every year 50 million acres are destroyed. Global CO_2 levels rise approximately 0.4 % each year to levels not experienced on this planet for millions of years (Odjugo and Isi, 2003).

Population growth contributes to global warming: The intellectual powers the we enjoy has enabled us to make effective use of technology and there by changed the environment. Technology is partly responsible for explosive population growth and responsible for the resulting damage to the earth resources. The industrial revolution caused a rapid increase in the population growth, as oil and gas fuels were exploited for our use. Presently we have too many people on earth who are using technologies that are destructive for the earth (Kumar *et al.*, 2001).

Impacts of Climate Change on Agriculture

Destabilization of local climate: This may lead to complete change in cropping pattern.

Glacier retreat and disappear: According to UN climate report the Himalayan glaciers that are the source of Asia's biggest rivers (Ganga, Indus, Brahmaputra, Yangtze, Mekong, Salween, Yellow) would disappear by 2035 as temperature rise.

Ocean becomes acidified: It acts as a sink of CO_2 and due to increased level of CO_2 in environment it has lead to ocean acidification.

Sea level rise and flood defense: due to rise in sea level some countries will be much affected which are low lying such as Bangladesh and Netherlands, which would be most heated by sea level rise.

Crop yield: Increased CO_2 concentration will increase branching leaf area of the crops that will also reduce soil moisture. Decrease in potential yield is likely to be caused by shortening of the growing period decrease in water availability and poor verbalization (Peng *et al.*, 2004). As increase in temperature beyond 22% although has no effect on photosynthesis, cause sterility in rice resulting in reduced grain yield. The combination of decrease in soil moisture and increase in heat stress is likely to result in lower wheat yield. In North India warming would offset some losses in yield by early pod set in winter grain legumes like chickpea and lentil and by arresting forced damage in oilseed Brassicas (Isik and Devadoss, 2006).

Duration of the crop: Duration of crop growth cycles are reduced in high temperature so there will be reduction in yields in some of the areas of the world in case of major food crops. Annual crops the duration between sowing and harvesting will shorten the cycle and would rather have adverse effect on productivity because of senescence occurring sooner. Switch over the winter sown cereals viz., wheat, barley and oat would take advantage of longer growing season and reduce damage by high evaporation.

Soil fertility: Soil fertility would probably be modified, because soil degradation is more likely to occur. A soil constant is its C/N ratio. A doubling of carbon is likely to imply a higher storage of nitrogen in soil, thus providing higher fertilizing elements for plants. The increase in the temperature would induce a greater mineralization and the atmospheric CO_2 concentration would tend to increase it.

Weeds: Weeds would undergo the same acceleration of cycle than cultivated crops, and would also benefit of carbonaceous fertilization. Most weeds being C_3 plants, they are likely to compete even more than now against crops such as corns.

Disease and insect pest: The increase in rainfall is likely to lead to an increase of atmospheric humidity and maybe to the duration of moistening. Combined with higher temperatures and humidity, this could be an increased pressure from insect and disease vectors. Global warming may increase the rate of development of pest and number of generations produced round the years, extend geographical range. There will be increase in density of *Helicoverpa armmigera* that leads to damage to cotton resulting in significant economic losses (Scherm, 2004).

Management practices: extreme climatic conditions become more frequent, particularly in terms of intense rainfall, droughts, and heat spells. There will be substantial increase in the need for irrigation which will probably lead to higher cost of production and possible shift towards less water demanding uses. In areas where greater rainfall is likely increase in fertilizer use may be required.

Cropping pattern: the combination of the effects of climatic change on agricultural crops is likely to bring about a spatial shift of crop potential. The area which at present is judged to be most suited to a given crop or combinations of crops may no longer remain as such after a climatic shift. Wheat belts may be replaced by barley by maize by soybean and so on. The reduction in the length of growing season of cotton will make this crop even more suitable in double cropping system in Indo-Gangetic plains of India (Shibasaki and Tan, 2003).

Future Impacts of Climate Change

Freshwater Resources

Global climate change is likely to affect freshwater resources by altering the availability of water. In some areas water levels may increase, while in others it may decrease. According to the IPCC, by mid-century, annual average river runoff and water availability are projected to increase by 10-40% at high latitudes and in some wet tropical areas, and decrease by 10-30% over some dry regions at mid-latitudes and the dry tropics. Water supplies stored in glaciers and snow cover also are projected to decline, further reducing water availability in regions supplied by melt water from major mountain ranges. Reduced water availability means drought-affected areas will likely increase in extent; heavy precipitation events are likely to increase in frequency, thereby augmenting flood risk.

Food and Agriculture

According to the IPCC, a warmer climate will result in long-term regional changes to precipitation, length of the growing season, and the growth range of certain crops. While some of these changes may be beneficial in the short-term (like increased global food production), these benefits will eventually diminish as global average temperatures continue to rise (Parry *et al.*, 2004). The IPCC predicts that crop productivity will increase slightly at mid- to high latitudes for local mean temperature increase of up to 1-3 °C, depending on the crop. At lower latitudes, though, especially seasonally dry and tropical regions, crop productivity is projected to decrease for even small local temperature increases (Darwin, 2004), thereby increasing the risk of hunger in certain parts of the world. Globally, the potential for food production is projected to increases with increases in local average temperature over a range of 1-3 °C. Above this temperature range, however, global production is projected to decrease (Calanca *et al.*, 2005). Increases in the frequency of droughts and floods also are projected to affect local crop production negatively. The worst-case scenario is stark: Africa's Sahel region will produce fewer cereals, rice cultivation in Asia will be under threat, there will be fewer vegetables — with potatoes and beans potentially wiped out — and livestock and fisheries will be severely stressed (Smith, 2004).

Climate change is making crop scientists review their research agenda. Until now, their main focus was on improving yields. But with successive International Panel on Climate Change (IPCC) reports warning that increased droughts and floods will shift crop systems, 'climate-proofing' of crops has become crucial. The Consultative Group on International Agricultural Research (CGIAR) institutes are now investigating how to make crops' more resilient to environment stresses (Bosello and Zhang, 2006).

Rice crops most vulnerable

About 90 per cent of the world's rice is grown and consumed in Asia (where 70 per cent of the world's poor live), and sub-Saharan Africa is the world's fastest growing rice consumer. The most vulnerable (Sukla *et al.*, 2003) agricultural systems are the rain-fed uplands and lowlands that form almost 80 per cent of total rice land in Africa. Reiner Wassman, coordinator of the Rice and Climate Change Consortium at the International Rice Research Institute (IRRI) in the Philippines, says IRRI strategies should include breeding rice that can survive climate change. He wants to see plants that can tolerate higher temperatures and flooding, that flower in the mornings before temperatures rise, and that transpire (lose water through evaporation from leaves) more efficiently to cool the air around them (Nakagawa *et al.*, 2003). His hopes are buoyed by IRRI's latest research into the rice line 'sub1', which survived submersion for 17 days (see Scientists create flood-resistant rice). The line could provide genes for flood tolerance. In Africa, the Africa Rice Centre (WARDA) is focusing on its NERICA (New Rice for Africa) varieties. These combine traits of Africa's *Oryza glaberrima* — such as drought and local disease tolerance — with the high yields of Asia's *Oryza sativa* (Matthews and Wassmann, 2003).

Looming disaster for wheat

Drought is also a big concern for the International Maize and Wheat Improvement Centre (CIMMYT) in El Batan, Mexico. The IPCC's predictions of increasing droughts spell disaster for half of the developing world's wheat growing areas. The problem is particularly acute in central and West Africa, where the poor depend on wheat but get an annual rainfall of less than 350 mm, says CIMMYT scientist Rodomiro Ortiz. CIMMYT has launched a hunt for drought tolerance in wild wheats and 'landraces' — traditional crops that have adapted to local conditions over centuries. The centre is also teaming up with the Japan International Research Centre for Agricultural Sciences to map drought-tolerant genes in wheat and maize. CIMMYT is using its findings in both traditional breeding and genetic engineering programmes. For example, researchers are working on genetically engineered wheat containing the DREB gene of *Arabidopsis thaliana* — a relative of mustard plants — that may confer tolerance to drought, saline soils and low temperatures. CIMMYT is testing yields of genetically engineered plants with the DREB gene under varying water stress. However, Ortiz cautions that the plant is still experimental. Most published studies simulated drought conditions in greenhouses more rapidly than would occur naturally (Attri and Rathore, 2003).

Shrinking Diversity

Scientists look for useful genes in plants grown only locally, and CIMMYT already has maize breeding programmes that work with local communities. But researchers fear many useful wild species could disappear. "Climate change is leading to significant losses of genetic resources in several regions of the world," says Atta-Krah. He says diversity among crop species must be effectively conserved, managed, and used to improve crops and adapt to climate change. One striking example of shrinking diversity is Latin America's beans. Peter Jones, a scientist at the International Centre for Tropical Agriculture (CIAT) in Columbia, says that of the 17 wild species of the *Arachis* genus — the pea family that includes the peanut — 12 will be extinct by 2055 due to climate change.

Recommendation

Implementing energy conservation methods and utilizing renewable resource such as green power (power from moon, hydrogen fuel, nuclear power, sewage power, geothermal power and power of the mighty sun). Trees play a unique role in global carbon cycle. They are largest land – based mechanism for removing CO_2 from the air. An acre of forest will absorb about ten times the CO_2 amount absorbed by an acre of crop land or grass land. So planting more trees and reducing timber cut worldwide will help to restore the ecological balance. Government should also encourage the use of mass transit; provide tax rebates for people who buy low and no pollution vehicles and subsidies to fossil fuels and the nuclear industries.

Try to follow the following environmental policies of"Reduce.... Reuse....Recycle". Reuse of anything is the easiest and best way to recycle. Save containers, bags, everything that can be used in the future. Also we use cloth; towels and napkins instead of paper ones, and use rechargeable batteries instead of disposable ones. Check population growth rate thereby decrease in pressure on agricultural growth rate and give a chance to explore the biodiversity. Tropical deforestation is an international issue, requiring urgent international action to the medium to long term; forests should be integrated fully into global carbon trading. All countries will need to adapt to a changing climate- for example through more resilient infrastructure, buildings, agriculture and enhanced social protection.

Conclusion

Agriculture is most vulnerable and uncertain to climate change compare to other enterprise. Agriculture in India and other developing countries is at present undergoing rapid transformation due to changing demands, market and agricultural technologies. Agricultural intelligence system needs to be strengthened.

The overall global policy framework should be designed to satisfy the following three key principles. It must be: effective - the frameworks must involve action that can affordably keep risks from climate change at acceptable levels; efficient - mitigation should be undertaken where it is cheapest, with carbon pricing and markets playing a central role in determining type and origin of mitigation; equitable - commitments must be perceived as equitable which requires rich countries taking a lead - this is a shared problem with differential responsibilities.

References

Abler-D, Shortle, J., Carmichael J., Horan R. 2002. Climate change, agriculture, and water quality in the Chesapeake Bay Region. Climatic-Change., 55(3): 339-359.

Alexandrov, V.A., Eitzinger, J. 2005. the potential effect of climate change and elevated air carbon dioxide on agricultural crop production in Central and Southeastern Europe. Journal-of-Crop-Improvement. 13(1/2):291-331.

Attri, S.D., Rathore, L.S. 2003. Simulation of impact of projected climate change on wheat in India. International Journal of Climatology, 23:(6)693-705.

Bosello, F., Zhang Jian. 2006. The effects of climate change on agriculture. Questione-Agraria. (1):97-124.

Calanca, P., Fuhrer, J., Jasper, K., Torriani, D., Keller, F., Dueri, S. 2005. Climate change and agricultural production. Agrarforschung. 12(9):392-397.

Climate Change. 2001. The Scientific Basis, Summary for Policy Makers and Technical Summary of the Working Group I Report, Intergovernmental Panel on Climate Change, Geneva, Switzerland, 2001.

Darwin, R. 2004. Effects of greenhouse gas emissions on world agriculture, food consumption, and economic welfare. Climatic Change. 66(1/2):191-238.

Faisal, I., M., Saila Parveen. 2004. Food security in the face of climate change, population growth, and resource constraints: implications for Bangladesh. Environmental Management. 34(4):487-498.

Gregory, P.J., Ingram, J.S.I., Brklacich, M. 2005. Climate change and food security. Philosophical Transactions of the Royal Society of London Series B, Biological Sciences. 360(1463):2139-2148.

India's Initial National Communications to the United Nations Framework Convention on Climate Change, Ministry of Environment and Forests, New Delhi, 2004.

Isik, M., Devadoss, S. 2006. An analysis of the impact of climate change on crop yields and yield variability. Applied-Economics. 38(7):835-844.

Kumar, K.S.K., Jyoti Parikh; Parikh, J. 2001. Socio-economic impacts of climate change on Indian agriculture. International Review for Environmental Strategies., 2:(2)277-293.

Loske, R., Scope of the Report: Setting the Stage: Climate Change and Sustainable Development, Third Assessment Report of Working Group III to Intergovernmental Panel on Climate Change, 1996.

Matthews, R., Wassmann, R. 2003. Modelling the impacts of climate change and methane emission reductions on rice production: a review. European Journal of Agronomy. 19(4):573-598.

Nakagawa, H., Horie, T., Matsui, T. 2003. Effects of climate change on rice production and adaptive technologies: Rice science: innovations and impact for livelihood Proceedings of the International Rice Research Conference, Beijing, China, 16-19 September 2002 pp. 635-658.

Odjugo, A.O.P., Isi, A.I. 2003. The impact of climate change and anthropogenic factors on desertification in the semi-arid region of Nigeria. Global Journal of Environmental Sciences. 2(2):118-127.

Okammoto, K., Shindo, J., Toda, H., Kawashima, H. 2004. Environmental changes and food production in Asia. JIRCAS International Symposium Series. (12):55-62.

Parry, M.L., Rosenzweig, C., Iglesias, A., Livermore, M., Fischer, G. 2004. Effects of climate change on global food production under SRES emissions and socio-economic scenarios. Global Environmental Change. 14(1): 53-67.

Parry-ML. 2002. Turning up the heat: how will agriculture weather global climate change? Sustainable food security for all by 2020: Proceedings of-an International Conference, Bonn, Germany, 4-6-September, 2001. pp. 116 -123.

Peng ShaoBing; Huang Jian Liang; Sheehy, J.E., Laza, R.C., Visperas, R.M., Zhong, X.H., Centeno, G.S., Khush, G.S., Cassman, K.G. 2004. Rice yields decline with higher night temperature from global warming. Proceedings of the National Academy of Sciences of the United States of America. 101(27):9971-9975.

Piyasiri, C.H., Peiris, T.S.G., Samita, S. 2004. Change in rainfall and temperature of low country wet intermediate (IL1) region of Sri Lanka. Tropical Agricultural Research. 16:271-281.

Scherm, H. 2004. Climate change: can we predict the impacts on plant pathology and pest management? Canadian Journal of Plant Pathology. 26(3):267-273.

Shibasaki, R. and Tan, G.X. 2003. Global estimation of crop productivity and the impacts of global warming by GIS and EPIC integration. Ecological Modelling. 168(3):357-370.

Shukla, P.R., Sharma, S.K., Ravindranath, N.H., Bhattacharya, S. and Garg, A. (eds), Climate Change and India: Vulnerability Assessment and Adaptation, Universities Press, Hyderabad, 2003.

Smith, D.L., Almaraz, J.J. 2004. Climate change and crop production: contributions, impacts, and adaptations. Canadian Journal of Plant Pathology. 26(3): 253-266.

❑❑❑

Section B: Pest and Diseases

6

Climate Change and Integrated Pest Management

M. Premjit Singh

Abstract

Climate change or elements of variability is a matter of great concern production positive, negative or natural impact on pest dynamics because of the specific nature of interactions of host, the pest and the environment. New pest such as Rice Whorl maggot, Flea beetal in rice/soybean and bark eating beetle in citrus are emerging with the change in climate pattern. Pest outbreaks in epemic form such as Rice hispa and rice ear cutting caterpillar are also noted. Location specific IPM packages with an emphasis on biological control are the need of the hour considering the threat of environmental pollution. The ecosystem diversification along with the study on natural enemies, spiders and coccinellids may be brought under purview.

Keywords: Climate change, New pest emergence, Biological control; IPM Strategy.

Introduction

Climate change is slow and definite change in climate of a place. In the recent past the atmospheric concentration of greenhouse gases (GHG_S) like carbon dioxide (CO_2), methane (CH_4) nitrous oxide (N_2O) and chlorofluorocarbons (CFC_S) have been increasing at an alarming rate due to anthropogenic activities, which in turn are causing global warming. During the past century global-average surface temperature has increased by approximately 0.6°C. The expected rate of increase is now 0.1°C per decade. The third Intergovernmental Panel on Climate Change (IPCC) report predicts that global average surface will increase by 1.4-5.8°C by 2100 with atmospheric CO_2 concentrations expected to rise between 540 and 970 ppm over the same period. Mahlman (1997) predicted that a doubling of atmosphere CO_2 (from a current concentration of 360 ppm) would result in a warming between 1.5 and 4.5°C. The rate of evaporation would increase in a warmer climate, which would lead to an increase in global precipitation of 2±0.5% per 1°C warming. Decrease in soil moisture because of increased temperature has also been predicted, although this could be offset by simultaneous increase in precipitation. Ultra-violet B (UVB) penetration and extreme events (*e.g.*, flood, storm and drought) are also predicted to increase but there is less certainty about the magnitude of these changes. Climate change can have positive, negative or neutral impact on individual pest systems because of the specific nature of interactions of host, the pest and the environment. (Table 1)

Impact of Extreme Climate/Weather on Pests

Emergence of new pests

Spreading of insect pests to the newer areas due to climate change is reported, which may be due to their ability to survive in the changed climatic conditions.

- Pigeonpea crop was heavily damaged by pod borer weevil, *Apion clavipes* with 59.73% in NEH region which was recorded for the first time in 1977. (Pathak, 1985).
- *Parasa bicolor* was reported for the first time infesting rice in Ukhrul district of Manipur during 1980 (Singh,1981).
- Flea beetle, *Chaetocnema basalis* was first observed infesting rice during transplanting and at the tillering stages in Assam during 1980 (Krishnasamy, *et al.*, 1982)
- During September-November, 1986, four species of bugs, were recorded in major rice growing districts of Manipur for the first time. These were Pentatomids, *Dolycoris indicus, Menida histrio, Scotinophara coarctata* and coreid, *Cletus signatus.* The stink, bug, *C. signatus* was predominant with 10-40% grain infestation (Singh and Singh, 1987).
- Root aphid, *Rhopalosiphum rufiabdominalis* which was earlier known to occur only in upland rice crops during rainy season as minor pest became serious pest in winter nursery in Manipur (Barwal, 1985).
- Rusty plum aphid, *Hysteroneura setariae* was reported for the first time in Assam during 1989-91 and it has the potential to become a pest of economic importance (Phukan *et al.*, 1995).

Table 1: Components of changing climate and their effect on host pants, insects and diseases

Variable	Mode of action	Effect-/+	Impact
Increased CO_2	Increased C:N ratio (Ball, 1997) resulting	-	Increase feeding of herbivore to derive more amino acids.
	increased plant size and canopy density	-	Greater fungal spore production, favour pathogens; Physiological change in host plant increases host resistant to pathogen(Coakley *et al*., 1990)
	More carbon storage in roots	+	Diminished losses from soil borne pests.
	Lower plant decomposition rate could increase the crop residue (Ball, 1997)	-	Disease organism can overwinter, resulting in higher inoculum levels at the beginning of the growing season
Enhanced	Increase in food quality due to increase in nitrogen under stress	-	Sudden resurgence of insect pests
Temperature	Influence the distribution of pests and diseases	-	Increase in 1°C temperature would enable species to spread 200 k northwards or 140 km upwards in altitude (Parry *et al*., 1989).
		+/	Species richness of insects tends to increase with temperature (Turner et al., 1987), thus more species will be gained than lost.
		+/-	Unfavorable areas become suitable, thereby, undoing migration; overwinter at higher altitude causing pole ward extension; activity of certain pests such as cabbage butterfly may decline in plains.
	Onset of hibernation may be delayed	-	Increasing period of pest activity resulting in more damage
	Faster development rate (Multivoltine spp. from Aphididae and Lepidoptera)	-	Additional generation within a year (Pollard and Yates, 1993)
	Predators of aphids on rabi crops may become more active	+	May check aphid population
	Natural enemies may enter new areas	+	Aiding in biological control of pests
	The triptrophic interaction between plants, herbivores insect and parasitoids results	-	Disruption of biological control process, lethal and sub lethal damage to parasitoids
	from a long coevolution process are more likely to be affected at higher trophic levels	--	The endosymbiont bacteria associated with either parasitoids or their hosts can be suppressed by a short exposure to high temperature
	Plant defense system is lowered thereby plant become more susceptible to pest attack	-	Wheat and oat become more susceptible to rust disease
	Increased lignification in forage spp.	+	Many forage spp. become more resistant to fungi with increase temperature
Pollutants(O_3, N_2O)	Adverse effect on plant growth	-	Necrotrophic pathogens may colonize plants weakened by ozone
		+	Obligate biotrophic infection may be lessened
		-	Directly or indirectly affect the process of water and nutrient uptake
Precipitation frequent /extreme	Creating conducive conditions for pathogens	-	More pathogen and disease development

- Rice whorl maggot, *Hydrellia philippina* which was previously recorded in plains and low lying areas of north-eastern region has spreaded in the upland rice fields (Gupta, 1991).
- Damage of white backed planthopper to early rice was noticed in Manipur during 1983 (Barwal, 1984) and it was attributed due to unusual heavy rains and flooding in the Imphal valley at that time. During 2000 and 2001 mixed population of brown planthopper and white backed planthopper was observed on *pre-kharif* rice and has now become serious pests of *kharif* rice in some pockets of Manipur (Devi, 2003).
- Flea beetles, *Chaetocnema basalis and Monolepta signata* were noticed from 1984 onwards in the north-east region on soybean causing minor to medium damage at early vegetative stage which were not noticed earlier (Azad Thakur, 1985).
- Brown scale, *Saissetia coffeae* was recorded for the first time in Manipur on brinjal and the pest now occurs regularly (Rao and Barwal, 1983).
- Green bug, *Rhynccocoris humeralis* which was not known to occur earlier has emerged as the most serious pest of citrus in Mizoram (Sachan and Gangwar, 1982).
- Bark eating beetle, *Odontolabis alces* which was not known earlier has been reported to damage tree trunks of citrus in Arunachal Pradesh in serious form (Padmanabhan, 1988).
- Recently, the papaya mealy bug, *Paracoccus marginatus* has invaded papaya in Tripura. If prompt action is not taken, the pest is likely to spread to all the north-eastern states and cause heavy damage not only to papaya but several other alternate crop plants like pigeonpea, okra, tomato, tapioca, guava, mango, mulberry, teak, rubber *etc.*

Pest Outbreaks

The north-eastern region is facing changing pattern of annual precipitation with the increase in temperature. Although many pest out-breaks occur in the region due to change in climatic conditions, only a few are recorded.

- Outbreak of rice ear cutting caterpillar/armyworm, *Mythimna separata* was initially recorded during 1982 in Assam and then spread to Manipur, Arunachal Pradesh, Meghalaya and Tripura.
- Two outbreaks of rice armyworm occurred in October, 1982 and November, 1986 in the Imphal valley of Manipur following unusual heavy rains between February and September that caused repeated flooding (Barwal, 1983; Singh and Singh, 1987).

- During April-June, 1982 scarabaeid beetle, *Alissonotum simile* emerged in epidemic form and 90% of the local tall rice variety, "Nagaland Special" was killed due to cutting of the young plants just below the soil surface by the adults, and larvae feeding on the roots (Sharma, 1985).
- Rice hispa, *Dicladispa armigera* appeared in epidemic form during 1978 in Tripura and found 100% infestation of the crop. Such epidemics are a recurrent phenomenon in Tripura, Meghalaya, Assam and foot-hills of Arunachal Pradesh (Das, 1980), Rice hispa outbreaks were recorded during 1987 in Meghalaya and Arunachal Pradesh.
- Asian long horned beetle, *Anoplophora glabripennis* emerged in epidemic form and in association with *Fusarium sp.* killed tree been on large scale throughout the State of Manipur.

Changing Pest Status

The north-eastern region is also experiencing increase in the average temperature, shortening cold winters, and erratic rainfall. Increase in temperature and shortening the winter period may increase the number of generations of insect pests, reducing or eliminating hibernation and remaining active throughout. In the north-eastern region, there are certain reports on the changing pest status due to climate change which are discussed herein.

- In the upland rice, flea beetle, *Chaetocnema basalis* which was earlier noticed to damage seedlings, has now established as a pest of regular occurrence in the early vegetative stage.
- Since 1986, brown planthopper, *Nilaparvata lugens* posed a threat in Meghalaya and Arunachal Pradesh.
- Leaf roller, leafhoppers and rice gundhi bugs became major pests of rice throughout the north-eastern region.
- Rice green horned caterpillar has become a pest of rice in Assam and Manipur.
- Based on the reaction patterns noted in differential rice varieties and DNA fingerprinting of gall midge population of Manipur by Amplified Length Polymorphism (AFLP) analysis confirmed the occurrence of a new virulent "Biotype 6" in Manipur, which may have arisen by migration from Myanmar/ South China, rather than by mutation of "biotype 3", which was earlier prevalent in this region (Singh, 1996; Katiyar *et al.*, 2000).
- Rice stem borers are serious pests in Tripura and Meghalaya, but are now surpassed by other pests like hispa, leaf folder, WBPH and BPH.
- Cabbage butterfly, *Pieris rapae montana* appeared as a major pest outnumbering *P. brassicae* at mid-altitude hills.

- Diamond back moth, *Plutella xylostella* and *Helulla undalis* which were not important earlier, now cause heavy loss to cole crops.
- In Arunachal Pradesh, bark eating beetle, *Odontolabis alces* has been reported to damage tree trunks of citrus in serious form (Padmanabhan, 1988)
- Serpentine leaf miner, *Liriomyza trifolii* has become serious pest on tomato and several vegitable crops in NE region.
- Thrips appeared as a major pest on certain crops like Chilli, groundnut, citrus and pomegranate.
- Some alien and new pests could become invasive and some secondary pests could assume serious proportions. For example, Sugarcane woolly aphid, *Ceratovacuna lanigera*, Coconut eriophyiid mite, *Aceria guerreronis*, spiralling whitefly, *Aleurodicus dispersus* and serpentine leaf miner, *Liriomyza trifolii* are some examples.

IPM Strategy: A flow diagram of the impact of climate change and the need of IPM reorientation are presented in Fig.1.

Breeding for Host resistance

Development of new genotypes tolerant to multiples stresses: drought, floods, heat, salinity, pests and diseases will help increase the food production. Cultivar resistant to pathogens may become more effective because of increased static and dynamic defenses from changes in physiology, nutritional status, and water availability. Durability of resistance may be threatened, however, if the number of infection cycles within a growing season increases (Coakley *et al.*, 1999). Increasingly, plant breeders will need to combine pest resistance with tolerance to factors such as heat and drought when they develop new varieties. The search for such tolerance has made good progress in a few dryland cereals, such as pearl millet, sorghum and durum wheat, but has been less successful-or hardly attempted at all – in most other crops. Flexible global arrangement will be needed to facilitate the speedy exchange of germplasm with the necessary resistance characteristics.

Biological Control

Changes in temperature may have highly nonlinear effects on tri-trophic interaction of host, pathogen and biocontrol agents. A rise in temperature from 17- 22° C in wheat resulted in an increase in aphid (*Sitobion aveanue*) reproduction by 10% and predatory activity of lady beetle (*Coccinella septumpunctata*) by 250% (Triltsch *et al.*, 1996).

Parasitoids depend on a series of adaptations to the ecology and physiology of their host and host plants for survival and are thus likely to be highly susceptible to changes in environmental conditions. A careful analysis is needed so that researchers may predict and manage the consequences of global changes at the

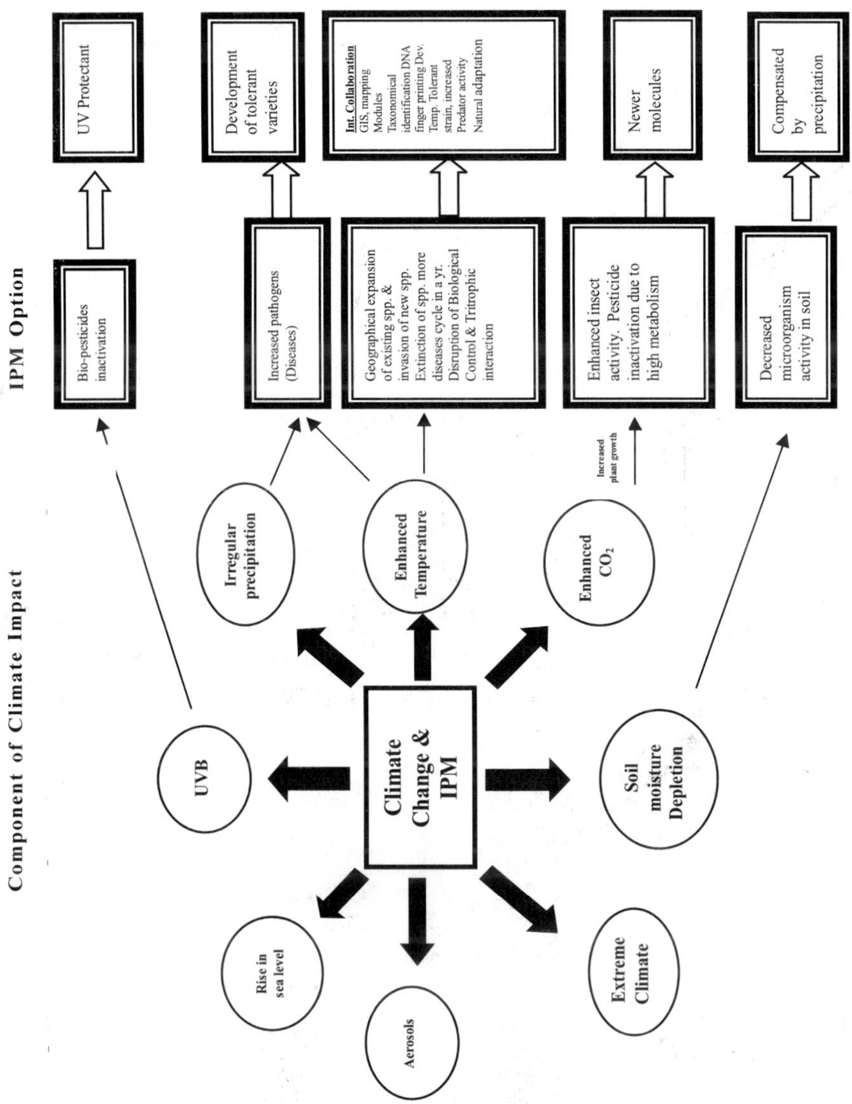

Fig.1: Impact of climate components and reorientation of IPM strategy in agriculture

ecosystem level (Hance *et al.,* 2007). Development of temperature tolerant strain may play important role. Moreover, elevation in the temperature would be a slow process so there is all the possibility that some of the parasitoids or predators may become tolerant to the higher temperature. it is being further supported by work carried out by PDBC. They developed a temperature tolerant strain of *Trichogramma* by gradually exposing them to higher temperature.

The shift away from chemical pesticides towards bio-insecticides will doubtless continue as an IPM strategy. Some bio-insecticides, however, are highly susceptible to rises in temperature and ultra-violet radiation. For example, the half-life of commercial preparations based on the granulosis virus-widely applied as a dust to stored potatoes to destroy the potato tuber moth and other pests-falls by 60 per cent when temperatures rise from 25° C to 30° C.

Climate change may alter the composition and dynamics of microbial communities in aerial and soil environments sufficiently to influence the health of plant organs (Ayres *et al*., 1996; Gunasekera, 1997). Changed microbial population in the phyllosphere and rhizosphere may influence plant disease through natural and augmental biological control agents.

Chemical Pesticides

Efficacy of crop protection through chemicals could be affected by climate change in two ways.

a) Change in temperature and precipitation may alter the dynamics of pesticide residue on the crop canopy. As the climate change model (Fowler and Hennessy, 1995) project an increase in the frequency of intense rainfall events, therefore, there would be increase in wash off of the pesticide applied and ultimately reduced control.

b) Elevated CO_2 would bring morphological and physiological changes in crop plants and could affect uptake, translocation and metabolism of systemic pesticides. For example, increased thickness of the epicuticular wax layer on leaves (Bowes, 1993) could result in slower and/or reduced uptake of pesticide by the host, whereas the increased canopy size could negatively affect spray coverage and lead to dilution of the active ingredients in the host tissue.

IPM-Case Studies

Development and validation of location specific IPM packages

During 2001-2003, under National Agricultural Technology Project, on farm adaptive trials were carried out in the northeastern region to quantify the yield gap on the farmers' fields and realize potential yield through IPM. The trials were undertaken in Manipur. Five villages were indentified at each location and in each village, two sites (replications) were selected. At each site, there were three treatments viz., i) Farmers practice (FP) involving conventional practices of the farmers, ii) Scheduled treatment (ST) involving application of insecticides based on a regular schedule irrespective of pest incidence and iii) IPM treatment (IPM) consisting of a set of optimized location specific control components. A plot size of 1 ha was maintained per treatment. The treatments specific to each location were worked out based on the historical data collection, survey information as well as the experience of the scientists organizing the trial. The details of the location specific treatments including IPM module components in the on farm trial are given in Table 2

Table 2: Development and validation of location specific IPM packages: Case studies in north-eastern region: Imphal, Manipur

		Treatment particulars
Farmers' practice (FP)	:	• One application of either Monocrotophos 36 WSC or Chlorpyriphos 20 EC @ 500 g a.i./ha at 35-65 DAT
Scheduled treatment (ST)	:	• Seed treatment with Carbendazim + Mancozeb @ 1 g/kg of seed • Application of Monocrotophos 36 WSC @ 500 g a.i./ha twice at 15 and 35 DAT followed by one application of Chlorpyriphos 20 EC @ 500 g a.i./ha at 65 DAT
IPM treatment (IPM)	:	• Use of CAU Selection-1, a variety resistant to gall midge and blast • Early planting during June in *kharif* season • Clean cultivation • Balanced and split application of nitrogenous fertilizer (N,P,K, @ 60:40:30 kg/ha – 30:40:30 kg N,P,K as basal + 15 kg N at maximum tillering stage + 15 kg N at panicle initiation stage) • Regular pest monitoring in nursery and main field by the use of pheromone traps @ 8 traps/ha in case of yellow stem borer • Release of *Trichogramma chilonis* @ 70,000 parasitized eggs/ha, twice starting from 30 DAT

Salient Findings

Incidence of insect pest and natural enemies

The pest incidence was moderate (Table-3); incidence of silver shoots was 2.3% in IPM significantly lower than that of ST (5.5%) and FP (5.7%); gundhi bug populations were also lowest in IPM (3.5 per 20 hills) compared to FP (4.9 per 20 hills) and ST (5.2 per 20 hills); GLH populations were similar across the treatments (5 hoppers per 20 hills); defoliators like leaf folder, case worm, whorl maggot and grass hopper as well as stem borer damage at harvest was low and below 5.0%. The mean infestation data of these pests revealed no discernible differences among the treatments. Overall, the insect pest incidence was less in the IPM and ST treatments than FP.

Among the natural enemies, spiders and coccinellids, were predominant; population of coccinellids was higher in IPM (3.2) compared to FP treatment (2.7). However, no discernible differences were observed in spiders (Fig-2)

Grain yield and economics

The grain yield was the highest in ST (4.6 t/ha) followed by IPM plots (4.2 t/ha) and FP (4.0 t/ha); however, the cost benefit ratio significantly higher (Fig.3) in IPM treatment (1:2.0) than ST (1:1.6)and FP (1:1.4)

Table 3: Insect pest incidence in different treatments, Imphal (2001-2003)

Treatment	Gall Midge (%SS)	Leaf Folder (%DL)	Case worm (%DL)	Whorlmaggot (%DL)	Grass hopper (%DL)	Gundhi Bug (No./20 hills)	GLH (No./ 20 hills)
IPM	2.3	1.7	1.2	1.5	3.3	3.5	5.3
ST	5.5	1.3	1.2	1.3	3.1	5.2	5.2
FP	5.7	2.4	1.5	1.6	3.4	4.9	5.3

Mean of observations in 10 sites, GLH-Green leafhopper, SS – silver shoots, DL- damaged leaves

Awareness on Location Specific IPM-whole Village Approach

IPM technologies were demonstrated through whole village approach at Imphal (Manipur) in 2004. Five villages were identified as IPM villages for adoption of IPM technology and their performance was compared with conventional farmers practices. The results are summarized herein.

Salient Findings

Incidence of insect pest and natural enemies

Eight insect pests viz., gall midge, stem borer, leaf folder, case worm, whorl maggot, green leaf hopper, grasshopper and gundhi bug were observed, and the pest incidence was moderate; silver shoots was 0.3% in IPM significantly lower than that FP (3.4%); gundhi bug populations lower in IPM (3.2 per 20 hills) compared to FP (5.8 per 20 hills); GLH populations were 4.4 hoppers per 20 hills in IPM lower than that of FP(5.4 per 20 hills); defoliators like leaf folder, case worm whorl maggot and grasshopper as well as stem borer damage at harvest was low and below 3.0%. The mean infestation data revealed no discernible differences among the treatments. Overall pest incidence was less in the IPM treatment than FP (Table-4).

Among the natural enemies, spiders and coccinellids were predominant (Fig-4). The spiders higher in IPM treatment (5.2 per 20 hills) compared to FP treatment (3.9 per 20 hills); coccinellids higher in IPM treatment (3.1 per 20 hills) than FP treatment (1.9 per 20 hills).

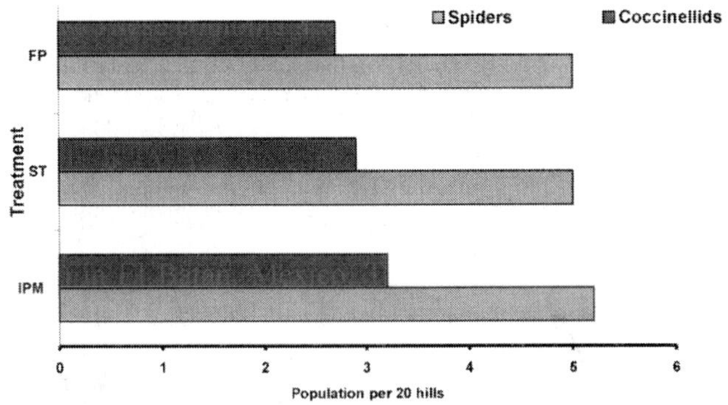

Fig: 2: Natural enemy populations in different treatments, Imphal

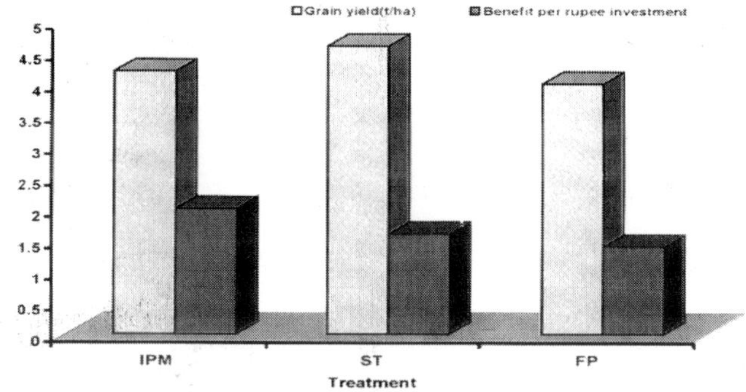

Fig. 3: Grain yield and economics of different treatments, Imphal

Treatment	Grain Yield (t/ha)	Cost benefit ratio
IPM	4.2	1:2.0
ST	4.6	1:1.6
FP	4.0	1:1.4

Grain yield and economics

The grain yield was higher in IPM plots (4.5 t/ha) and FP (4.1 t/ha). The net return

(Rs 15,390/ha) and cost-benefit ratio (1:2.1) were also higher in IPM treatment (Table 5) compared to FP (Rs. 8815/ ha and 1:1.5).

Table 4: Insect pest incidence in demonstration trail, *kharif* 2004, Imphal

Treat-ment	Gall midge (%SS)	Leaf folder (% DL)	Case worm (% DL)	Whorl maggot (% DL)	Grass hopper (% DL)	Gundhi bug (No./20 hills)	GLH (No./20 hills)	Stem borer (% WE)
IPM	0.3	2.3	0.0	0.8	1.9	3.2	4.4	0.6
FP	3.4	2.8	0.4	0.8	2.0	4.8	5.4	1.7

* Mean of observations in 10 sites, GLH – Green leafhopper, SS – Silver shoots, WE – White wars, DL – Damaged leaves

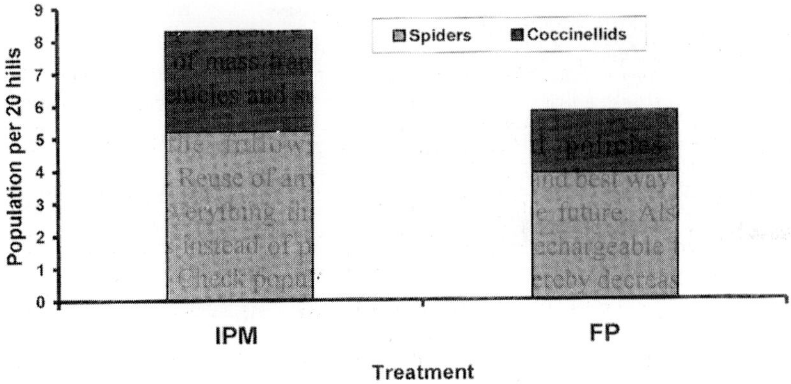

Fig. 4: Predatory populations in IPM and FP treatments, *kharif* 2004

Table 5: Grain yield and economics of IPM and FP treatments, *kharif* 2004

Treatment	Yield (kg/ha)	Cost of cultivation (Rs/ha)	Gross returns (Rs/ha)	Net return (Rs/ha)	Cost benefit ratio (CBR)
IPM	4,540	14,120	29,510	15,390	1:2.1
FP	4,110	17,900	26,715	8,815	1:1.5

Conclusion

To establish an integrated pest management program, the following basic guidelines should be followed.

- Understand the biology of the crop and identify the key pests, know their biology, recognize the damage they inflict and evaluate their economic status.
- Identify the key environmental factors that affect pest and potential pest species in the ecosystem.
- Consider concepts, methods and materials that individually or in combination, will help to permanently suppress pest.
- Anticipate unforeseen developments and move with caution to face the changes that can occur within it.
- Seek the weak links in the lifecycle of the key pest species and direct control practices as narrowly as possible at these weak links. Avoid broad impact on the resource ecosystem and consider and develop methods, which preserve, complement and augment the biotic and physical mortality factors that characterize the ecosystem, and
- Diversify the ecosystem.

References

Ayres, P.G., Gunasekera, T.S., Rasanayagam, M.S. and Paul, N.D. 1996. Effects of UV-B radiation (280-320 nm) on foliar saprotrophs and pathogens. In: Fungi and environmental change.(eds. Frankland, I.C., Magan, N., Gadd, G.M.). Cambridge University Press, Cambridge. pp.32-50.

Azad Thakur, N.S. 1985. Yield losses caused by major soybean insect pests. Agriculture Science Digest 5(4):213-14.

Ball A.S. 1997. Microbial decomposition at elevated CO_2 levels-effect of litter quality. Global Change Biology, 3:379-86.

Barwal, R.N. 1983. Outbreak of rice ear-cutting caterpillar, Mythimna separata (Walker) in Manipur, India; International Rice Research Newsletter 8(5):11-12.

Barwal, R.N. 1984. Changing insect pest status in the Imphal valley. International Rice Research Newsletter 9(4):12-13

Barwal, R.N. 1985. Insect pest of early season rice nurseries in the Imphal valley. International Rice Research Newsletter, 10(6):21-22.

Bowes, G. 1993. Facing the inevitable: Plants and increasing atmospheric CO_2. Annual Review of Plant Physiology and Plant Molecular Biology, 44:309-32.

Coakley, S.M., Scherm, H. and Chakraborty, S. 1999. Climate change and plant disease management. Annual Review in Phytopathology, 37:399-426.

Das, B. B. 1980. A note on hispa epidemic in rice field of Tripura, Indian Journal of Entomology, 42(3):530-531.

Devi, S.B. 2003. Studies on the seasonal incidence and management of plant hoppers, Nilaparvata lugens (Stal.) and Sogatella furcifera (Horv.) infesting pre-Kharif rice crop in Manipur. M. Sc. (Agri) Thesis, CAU, Imphal, 89 pp.

Fowler, A.M. and Hennessy, K. J. 1995. Potential impacts of global warming on the frequency and magnitude of heavy precipitation. Natural Hazards, 11:283-303.

Gunasekera, T. S., Paul, N.D. and Ayres, P.G. 1997. Responses of phylloplance yeasts to UV-B (290-320) radiation: interspecific differences in sensitivity, Mycological Research, 101:779-85.

Gupta M. K. 1991 Rice whorl maggot (RWM) incidence in Assam. International Rice Research Newsletter, 16(3):23.

Hance, T., Joan van, B., Vernon, P. and Biovin, G. 2007. Impact of extreme temperatures on parasitoids in a climatic change perspective. Annual Review of Entomology, 52:107-26.

Katiyar, S. K. Chandel, G, Tan, Y., Zhang, Y, Huang, B., Nugailiyadde, L., Fernando, K., Bentur, J.S., Inthavong, S., Constantino, S. and Benett J. 200 Biodiversity of Asian rice gall midge (*Orseolia oryzae* Wood - Mason) from five countries examined by AFLP analysis. Genome. 43:322-332.

Krishnasamy, N., Chauhan, D.P. and Das, R.K. 1982. The flea beetle as a rice pest in Assam India. International Rice Research Newsletter, 7(6):16.

Mahlman, J D. 1997. Uncertainties in projections of human-caused climate warming. Science 278:1416-1417.

Padmanabhan, B. 1988. Insect pest of citrus in Arunachal Pradesh. Annual Report. Division of Entomology, ICAR Research Complex, Shillong.

Parry, M.L., Carter, T.R. and Porter, J.H. 1989. The greenhouse effect and the future of UK agriculture. Journal of Royal Agricultural Society, 150:120-121.

Pathak, K. A. 1985. Insect pest of pigeonpea. Annual Report. Division of Entomology, ICAR Research Complex , Shillong.

Pollard, E., and Yates, T.J. 1993. Monitoring butterflies for ecology and conservation. Chapman and Hall, London.

Pukhan, E, Baruah, B. P. and Bora, D. K. 1995. Occurrence of rusty plum aphid, Hysteroneura setariae (Thomas) on rice in Titabar, Assam, Journal of the Agricultural Science Society of North East India, 8(1):92-93.

Rao,N.S. and Barwal, R. N. 1983. Brown scale, Saissetia coffeae (Walker) (Coccidae: Homoptera), a new pest of brinjal and its parasites in Manipur. Science & Culture, 49:255.

Sachan, J. N. and Gangwar, S. K. 1982. Insect pest of citrus. Mandarin Oranges in NEH Region and their Control. Technical Bull. No. 16. ICAR Research Complex, Shillong, 33 pp

Sharma S. 1985 Scarabaeid beetle outbreak in upland rice. International Rice Research Newsletter, 10(1):24.

Singh, M.P. 1996. A virulent rice gall midge biotype in Manipur. International Rice Research Newsletter, 21(1):31-32.

Singh, M.P. and Singh N.I. 1987. First recorded incidence of rice bugs in Manipur, India. International Rice Research Newsletter, 12(2):31.

Singh, S.A. 1981. Parasa bicolor (Walk): a new pest of rice in Manipur. International Rice Research Newsletter, 6(4):13.

Singh, S.A. and Singh, N. O.,1987. Some observations on the incidence of armyworm in rice in Manipur. Plant Protection Bulletins, India, 9(3):10-12.

Triltsch, H., Freier, B., and Rossberg, D. 1996. Temperatur-Schltisselfaktor fur Nutz-lingsleistungen im Winterweizen? Mitteilengen. biologische Bundesanst land Torstwirtschngen, 321:447 (Abstr.).

Turner, J.R.G., Gatehouse, C.M. and Corey, C.A. 1998. Does sloar energy control organic diversity-butterflies, moths and British climate? Oikos, 195-205.

❑❑❑

7

Effects of Weather Parameters on the Population Dynamics of Insect Pest of Rice: A Case Study

K. A. Pathak, Boopathi, T., Santosh Kumar, Y. Ramakrishna and Nitin Singh

Abstract

The present experiment was conducted to study the influence of weather parameters on the population dynamics of insect pests of rice during 2009-10 at ICAR-RC for NEH Region, Mizoram Centre, Kolasib, Mizoram. The result of the study indicated that, the mean populations of peak abundance of major insect pests rice crops during August were stem borer, flea beetle and gundhi bug. The Correlation studies showed that stem borer population a significant positively correlation was observed with maximum and minimum relative humidity. Nevertheless non significant negative correlation was obtained with maximum and minimum temperature. The finding of the present study indicated a positive non significant correlation with maximum relative humidity, maximum and minimum temperature where as negative insignificant correlation with rainfall and minimum relative humidity with flea beetle population. The highest minimum temperature showed a positive significant correlation with the Monolepta. Chaetocnema beetle had positive correlation with maximum temperature and humidity where as negative correlation with minimum temperature, minimum humidity and rainfall. Gundhi bug showed negative significant correlation with minimum temperature nevertheless non significant positive correlation was only obtained with maximum temperature, remaining weather parameters showed negative correlation with Gundhi bug population. A sifnificant positively correlation with minimum relative humidity moreover negatively significant correlation was obtained with maximum temperature in respect of orange fly population.

Keywords: Rice, Stem borer, Flea beetle, *Monolepta, Chaetocnema* beetle, Orange fly, Weather factors, Correlation.

Introduction

Rice is the most important cereal food crop of India. It occupies about 23.3% of gross cropped area of the country. It plays vital role in the national food grain supply. Rice contributes 43% of total food grain production and 46% of the total cereal production of the country. Rice is the staple food of more than 60% of the world's population especially for most of the people of South-East Asia. It is the grain with the second-highest worldwide production, after maize (corn). Among the rice growing countries in the world, India has the largest area under rice crop and ranks second in production next to China. The staple food of mizo people is primarily rice. The state's rice production declined average production in rice during 10th plan period was 1.64Mt/Ha, there has been a sharp decline in rice production during 2006-2007 with an average yield at 0.55 Mt/Ha. Successful cultivation is hampered due to several abiotic and biotic factors. Among biotic factor, insect pest is one of the major factors in minimizing the productivity. Climatic factor affect insect pest directly by limiting or expanding their distribution, growth, reproduction, diapauses and dispersal and indirectly through plant, mechanism and natural enemies that regulate insect population. Moreover, rice is the second in consumption of chemical pesticides particularly the insecticides. Their indiscriminate use has affected the ecology adversely. In addition to environmental pollution and health hazards this has led to resurgence of pests and pesticide resistance leading to failure of chemical protection and heavy loss to crop.

The seven insect pests and one mite invaded in the upland rice ecosystem. Among them, small black flea beetle and stem borer were found economically important, most dominant pest and it categorized as major pest. Weather plays an important role on the population dynamics and distribution of temperature, rainfall and relative humidity are the chief weather parameters influencing the pest incidence. Hence, knowledge of influence of weather factors not only pest but also on their natural enemies is useful. Interaction between crops phenology and insect phenology studies by degree day concept so that the phenomenon of peak pest attacking coinciding with the critical phenological phases of the crops can be avoided by adjusted by agricultural practices and period of optimum weather condition for pesticide application.

Methods and Materials

A monitoring of insect pest in rice crops was under taken during kharif season of 2009 from 25.06.09 to 14.09.09 in upland filed at ICAR -RC for NEH Region, Kolasib, Mizoram. The fields are ploughed and harrowed in summer for achieving the required-tilth Farmyard manure is uniformly distributed 2-3 weeks before sowing. The seed is sown directly with the onset of the monsoon showers. The crop was fed with NPK @ 120:60:60 kg ha-^{1} which was supplied by urea, single super phosphate and muriate of potash, respectively. The whole amount of phosphorus and potassium and half amount of nitrogen were applied as basal where as remaining half dose applied in two equal split doses at the tillering and panicle initiation stages. Insecticide was not applied during any of the seasons. The experiment was laid out in a randomized block design (RBD).With the other normal packages of practices for healthy crop growth. Very intensive monitoring of 5 hills from each fixed plot was observed at random for the population of insect pests in rice crops and record data appearance of insect pest population. Five weather parameters *viz.*, maximum temperature (T max), minimum temperature (T min), maximum relative humidity (RH1), minimum relative humidity (RH2) and rainfall (RF) were recorded from the weather station located at ICAR-RC Kolasib, Mizoram.. The data were subjected to suitable statically analysis by using simple correlation between the five weather parameters and insect pest population of rice crops.

Result and Discussion

The result of the study indicated that, the seasonal abundance of major insect pest's complex of rice in upland ecosystem during 2009-2010 (Fig. 1). Stem borer appeared during reproductive phase and attained the peak level of infestation on third week of August. Moreover small black flea beetle population started appearing from vegetative phase and continued till the harvest of the crops attained its peak activity on first week of August. *Monolepta* sp. it also categorized as major insect pests of rice and it was noticed through out the cropping period but attained the peak level of infestation on second week of August. The gundhi bug was more abundant during last week of June. The non-insect pest, red spider mite was more dominant pest and it could cause severe damage to the crop.

Correlation between the 5 weather parameters and insect pest of rice crops population in respect of different growth parameters furnished in table.1. The correlation study the relationship of 5 weather parameters with rice insect population indicated that the increase in relative humidity there would be increase of stem borer population a significant positively correlation was observed with maximum and minimum relative humidity. Nevertheless non significant negative correlation was obtained with maximum and minimum temperature. The present finding in respect of relative humidity and its correlation in rice stem borer population is in agreement with finding of Bala Subramanian *et al.* (1981).

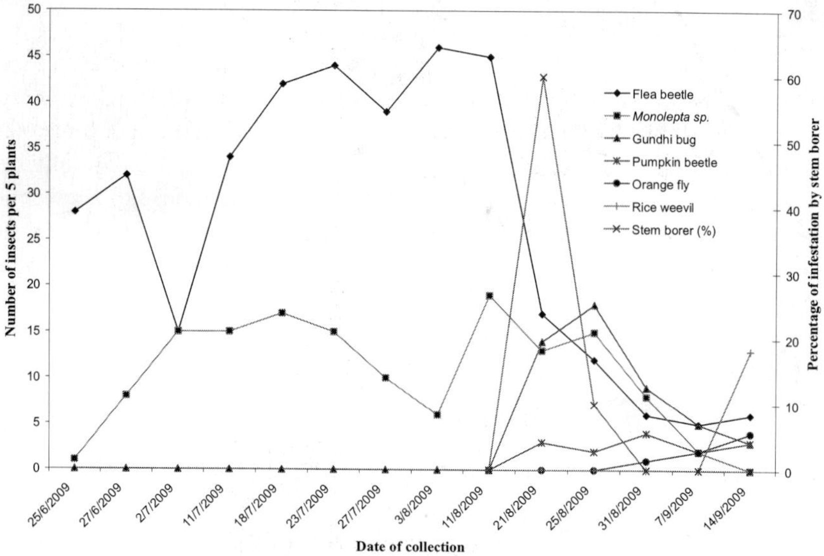

Fig.1: Seasonal abundance of major insect pests of upland rice during 2009-2010

Flea beetle population had a positive non significant correlation with maximum relative humidity, maximum and minimum temperature where as negative insignificant correlation with rainfall and minimum relative humidity which is not agreement with the finding of David (1995). The highest minimum temperature showed a positive significant correlation with the *Monolepta*. According to Mavi and Tupper (2005). If there were substantial increase in minimum temperature and evening humidity and an appreciable decrease in sun shine hours with occasional rain *monolepta* population at high level in this week. The present finding supported in respect of minimum temperature. *Chaetocnema* beetle had positive correlation with maximum temperature and humidity where as negative correlation with minimum temperature, minimum humidity and rainfall. Gundhi bug showed negative significant correlation with minimum temperature nevertheless non significant positive correlation was only obtained with maximum temperature,

Table 1: Correlation coefficient of rice insect pest population with the weather parameters

Sl.No.	Rice Insect	Mini. Temp	Max. Temp	Max. RH	Mini. RH	Rainfall
1.	Stem borer	-0.76**	-0.44	0.82**	0.73**	0.20
2.	Flea beetle	0.39**	0.39**	0.01	0.25	0.15
3.	*Monolepta*	-0.12	0.35	-0.25	-0.19	-0.44
4.	*Chaetocnema*	-0.37	0.33	0.30	-0.25	-0.02
5.	Gundhi bug	-0.87**	-0.44	0.48	-0.01	-0.08
6.	Orange Fly	0.60	-0.90**	-0.59	0.92**	0.07

*Significant at 0.05 level ** Significant at 0.01 level

remaining weather parameters showed negative correlation with Gundhi bug population. Orange fly recorded positively significant correlation with minimum relative humidity moreover negatively significant correlation was obtained with maximum temperature which is in agreement with finding of Kuwayama (1954) they pointed out that, when lower temperatures occur in summer, severe damage to the rice crops occur due to higher number of orange fly and extended adult longevity.

References

Adiroubane D and Raja. K. 2007. Influence of Weather Parameters on the Occurrence of Rice Yellow Stem Borer, Scirpophaga incertulus (Walker). Journal of Rice Research 3(1):4-9.

Balasubramanian, G., M. Gopalan., M. Balasubramanian and R. Kulandaivelu. (1982). Influence of weather factors on the damage potential of leaf folder in rice. Madras Agric. J., 69(6) :392-393.

Cammell M.E. and J.D. Knight 1992. Effects of climatic change on the population dynamics of croppests. Adv. Ecol. Res. 22:117-162.

Chelliah, S. and Heinrichs, E.A. 1980. Factors affecting insecticide induced resurgence of brown plant hopper (*Nilaparvata lugens*) on rice. Environmental Entomology, 5:773-777.

Harpal S. Mavi and Graeme, J. Tupper 2005. Agrometerology Principal and Applicatications of climate studies in agriculture. International Book Distributing Co. Lucknow, UP. pp. 130-131.

Kumar, Vijay and Patil, B.V. 2004. Activity of Brown Plant Hopper, Nilaparvata lugens Stal. (Homoptera: Delphacidae) and its Natural Enemies in Tungabhadra Project Area of Karnataka. Karnataka J.Agric.Sci., 17(4):717-720.

Kuwayama, S. 1954. Investigations on the insect pests of rice plants in the Hokkadio and their control. In Japanese with English summary Hokkadio Natl. Agric. Exp.Stn. 46:1-106.

Ngoan, N.O. 1972. Recent progress in rice insect research in Vietnam. Proceedings of Symposium on tropical Agricultural Research, Japan, 5:133-141.

❑❑❑

8

Climatological Influences in Occurrence of Pest with Special Emphasis to NE India

Subrata Biswas

Abstract

Pest damage is an age old problem of Indian agriculture. The increased accumulation of green house gases always changes the climatic conditions that have potential effects in increasing pest population. The insect pests are cold-blooded animals and their body temperature varies with the surrounding temperature. Thus, any change in the climate and weather is bound to influence the activity of pests. The pests are directly influenced by temperature, precipitation, humidity, wind speed and other environmental factors in terms of their occurrence, development, reproduction, distribution, migration and adaptation. In many cases the insect out breaks which classified as gradient out breaks are not self-driven but are entirely dependent on external environmental conditions. For example, the number of aphids has frequently been shown to be correlated with several environmental factors. Such correlations indicate an understanding of aphid out breaks. The increased temperature as a result of climate change has several implications for agricultural pests. These are as follows: (i) changes in population growth rates, (ii) increased number of generations, (iii) extension of development season, (iv) extension of geographical range, (v) increased over wintering, (vi) changes in crop-pest synchrony, (vii) changes in interspecific interactions, (vii) increased risk of invasion by migrant pests, (ix) introduction of alternate hosts, and (x) availability of over wintering hosts.

Keywords: Pest population in Crops, Environment, Pests.

In North East India, the change in climatic conditions is occurring very fast, due to simultaneous effects of global warming and deforestation. A large number of pests, like aphids, white fly, leaf miner, pod borer, shoot borer, fruit borer leaf folder, jassids, beetles, hispa, gall midge, scale insects and mealy bugs are very common in various crop plants in this region. Certain insect pests which were considered as minor during early 1970s have started appearing in serious proportions during the recent years. For example, brown plant hopper (BPH, Nilaparavata lugens), white backed plant hopper (WBPH, Sogatella furcifera), green leaf hopper (GLH, Nephotettix virescens), gall midge (Orseolia oryzae) and leaf-folder (Cnaphalocrocis medinalis) are found to cause considerable damage of rice crops in different rice growing areas of this region. The details of climatological influences in pest occurrences are discussed here.

Introduction

The effects of climate change on agricultural production are the growing concerns in all most all areas of the World including India. In North East region, also, the changes in climatic conditions are occurring very fast day by day due to human activities, like, fossil fuel combustion, deforestation, biomass burning, etc. Thus, time has already come to take adequate precautionary measures to combat the changing climatic situation in agricultural crop production. The greenhouse gases affect heat balance of the earth by absorbing long-wave radiation which would otherwise escape to the space. The earth becomes warmer day by day which has direct impact in pest population. In the current treatise some effects climate change in pest population is discussed.

Influence of Environment on Pests

Pest damage is an age old problem of Indian agriculture. The increased accumulation of green house gases always changes the climatic conditions that have potential effects in increasing pest population. The insect pests are cold blooded animal. The temperature of their body is more or less same to that of environment where they live. Thus temperature plays an important role in influencing insect behavior, distribution, development, survival and reproduction. In most cases the effect of temperature overwhelms the effects of other environmental factors. It is found that for a rise of 2° C in temperate zone the insect experiences 1-5 additional life cycles per season (Yamamura and Kirtani, 1998). As regards the impact of other environmental factors, moisture and CO_2 affect largely on insect population (Coviella and Trumble, 1999; Hunter, 2001; Hamilton *et al.*, 2005). In many cases the insect out breaks, which are classified as gradient out breaks, are not self-driven but entirely dependent on external environmental gradients in time and space (Berryman, 1987). For example, the number of aphids has frequently been shown to be correlated with several environmental factors. Such correlations indicate an understanding of aphid out breaks.

Temperature: The rising temperature affects insect populations in several ways. Although, in certain cases increased temperature in climate depresses insect population, but in cooler places, particularly in temperate zones it causes to develop more types of insect pests in higher populations of insects. The increased temperature as a result of climate change has several implications for agricultural pests, such as, (i) changes in population growth rates, (ii) increased number of generations, (iii) extension of development season, (iv) extension of geographical range, (v) increased over wintering, (vi) changes in crop-pest synchrony, (vii) changes in interspecific interactions, (viii) increased risk of invasion by migrant pests, (ix) introduction of alternate hosts, and (x) availability of over wintering hosts (Porter *et al.*, 1991). Increased temperature affects potentially on insect survival, development, geographic range and population size. It can impact insect physiology and development directly through the physiology or existence of hosts. Certain insects (cicada, arctic moths, *etc.*) which required several years to complete one life cycle in temperate climatic condition behave variability in moderate temperature (Bale, 2002; Petzoldt and Seaman, 2011). The crop pests develop more rapidly during periods of time with suitable temperatures (Fig.1). Increasing in temperature accelerates development of these insects, thus, causes more generations per year.

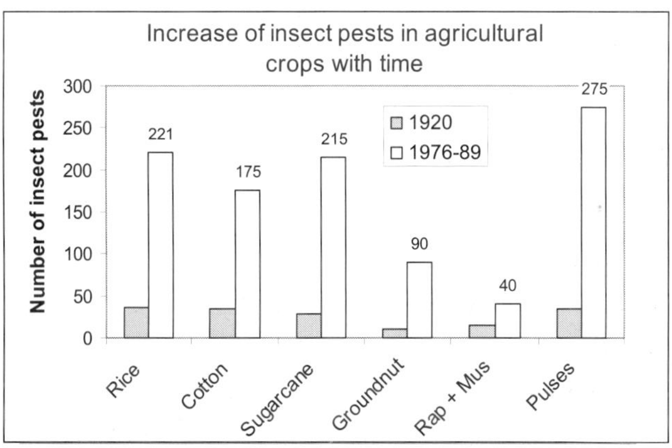

Fig 1: Increase of insect pests in agricultural crops from the base year 1920 to present time (based on data of 1976-1989)

In the year of 1920 there were 35 pests in rice crop and that now increased to 221 pests (Fletcher, 1920; Nayar *et al.*, 1976). In other crops also the pest populations were increased considerably with time. For example, during the year 1920 there were 34, 28, 10 (approx), 15 (approx) and 36 different pests in cotton, sugar cane, groundnut, rapeseed- mustard, and pulses, respectively and now the pest numbers in the respective crops are 175, 215, 90, 40 and 275 (Fletcher, 1920,

Singh *et al.*, 1988; CICR, 1989; David and Nandagopal, 1986; Reddy and Ghewande, 1986, Bakhetia and Arora, 1992; Singh *et al.*, 1989).

The influence of rising temperature on migratory insect pests (corn earworm) revealed with the expansion of overwintering areas and early immergence of the pests. The vulnerable life stages of the pests pass more quickly in raised temperature. So there exists little chance for the natural enemies to be hyperparasitized on the crop pests. Temperature also affects gender ratios of some pest species, such as, thrips, thus, affecting reproduction rates (Lewis, 1997). Winter mortality of insects due to warmer climatic condition becomes less, thus, increasing insect pest population (Harrington, 2001). Further, more and more areas in temperate zones which earlier were unsuitable for sowing crops are now becoming suitable for sowing crops due to hot climatic condition. This caused expanded crop areas for acting as hosts of increased pest population. Based on the evidence from fossil study, it is found that insect species and intensity of feeding increased historically with increase in temperature (Bale, 2002).

Carbon dioxide: In general, the effects of CO_2 on insects are considered as indirect. It is evident that increased CO_2 concentration in atmosphere causes more simple sugars in leaves which attract more insect pest attacks (Hamilton *et al.*, 2005). Further, insect feed more leaves that have lower nitrogen content in order to get sufficient nitrogen for their metabolism (Coviella and Trumble, 1999; Hunter, 2001). Recently, free air gas concentration enrichment (FACE) technology was used to create a required atmosphere with CO_2 and O_2 concentrations (Petzoldt and Seaman, 2011). In that study it was observed that the atmosphere with elevated CO_2 caused 57% more damage of soybeans from insect pests (primarily from Japanese beetle, potato leaf hopper, western corn rootworm and Mexican bean beetle). On the other hand, increased carbon to nitrogen ratios in plant tissue resulting from increased CO_2 levels may retard insect development and increase the life stages vulnerable to attack by parasitoids (Coviella and Trumble, 1999).

Rainfall: Comparatively less works have been carried out to study the effects of rainfall on insects. However, rainfall has enormous effects on insect pest population. The insect population increases during rainy season. Particularly, pea aphids can not tolerate draught condition (McVean and Dixon, 2001). In contrast, some insects, such as onion thrips, are very sensitive to precipitation and are killed or removed from crops by heavy rains (Reiners and Petzoldt, 2005). The insects, such as cranberry fruit worm and other insect pests of the crop, which overwinter in soil, are killed by flooding the soil (Vincent *et al.*, 2003). Rainfall also increases the fungal pathogens of insects due to high atmospheric humidity.

Impact of Climate Change

The insect pests are directly influenced by temperature, precipitation, humidity, wind speed and other environmental factors in terms of their occurrence, development, reproduction, distribution, migration and adaptation. In addition, these are indirectly influenced by the host plant conditions, natural enemies and interspecific interactions with other insects. The potential climate changes have significant role on the development, distribution and population density of insect pests (Parry *et al.*, 1990; Porter *et al.*, 1991). The growth of pest population is strongly influenced by the rate of development. The maximum rate of development is occurred when temperature and humidity are optimal. Hence higher temperature (below the lethal limit) causes faster rates of development and shorter developmental periods that resulted more number of generations of multivoltine insects and mites, and eventually, causes higher pest population (Parry, 1990). The climate changes are expected to affect distribution of pests which are at margins of their distribution during the present time. The increasing temperature due to climate change may result in greater ability to overwinter the species which are presently limited by low temperature. For example, in north India, overwintering ability and population of *Helicoverpa armigera* (Hubner) will increase to damage cotton. The extent of pest damage is determined by the timing of pest activity in relation to crop growth. The increased developmental rate in warmer condition will lead early establishment of pest populations while crops are at young and relatively susceptible growth stages (Dhaliwal and Bedi, 1993).

Climate Change and Pests in NEH Region

In North East India, the change in climatic conditions is occurring very fast, due to simultaneous effects of global warming and deforestation. The climatic conditions of this region are becoming warmer day by day. Several insect pests which were minor pests during early 1970, are now causing considerable damage of various crops (Table 1). In addition to the minor pests, several other pests are causing great crop losses in north east hill (NEH) region. The pests irrespective of earlier status of occurrence which causing great loss in this region are listed in Table 2.

Conclusion

The climate change is likely to be a gradual process that will give some opportunity to adapt. It is likely that the crops in changed situation will be attacked more frequently by certain pests. On the other hand, a few pests may be less in population to attack crops as change occurred. So, the precise impact of climate change on insects is somewhat uncertain because some climate changes may favour insects while others may inhibit. However, the preponderance of evidence indicates that there will be an overall increase of pest outbreaks in agricultural crops. The use of pesticides will also be increased simultaneously to combat the higher pest attacks

Table 1: Common occurrence of minor pests of 1970 during recent years in NEH region

Crop	Pest		Region	References
	Common name	Scientific Name		
Rice	Brown plant hopper (BPH)	*Nilaparavata lugens*	Tripura, Meghalaya	Personal observation, AR-06-07
	White back plant hopper (WBPH)	*Sogatella furcifera*	Meghalaya	AR2005-06
	Green leaf hopper (GLH),	*Nephotettix virescens*	Nagaland, Mizorum, Meghalaya	AR04-05, AR-2005-06AR-06-07
	Leaf folder	*Cnaphalocrocis medinalis*	Meghalaya, Tripura, Sikkim	AR 2004-05 Personal observation FAO 1989
Groundnut	Leaf miner	*Aproacerema modicella*	Tripura	Personal observation
Toria	Aphid	*Myzus persicae*	Tripura	Personal observation
Pulses, tomato, Chilli	White fly	*Bemisia tabaci*	Tripura	Biswas, 2003; Biswas & Singh, 2008

and that have negative effects on environment due to its hazards. Thus, close monitoring of insect pest attacks, keeping records pest attacks and management systems over time, and adoption of integrated pest management system are essential to manage the higher pest attacks in the changed situation.

Table 2: Major pests of crop plants in North East Hill region

Crop	Pest	Region	References
Rice	Leaf folder	Tripura, Nagaland	AR-04-05
	Gundhi bug	Tripura, Meghalaya	AR-04-05
		Nagaland, Mizoram	AR05-06
	Shoot borers	Tripura, Nagaland	AR-04-05,
	(Yellow stem borer)	Mizoram	6AR 05-0
	Hispa	Tripura	
	White back plant hopper	Meghalaya	AR 05-06
	Leaf hopper	Mizoram, Meghalaya	AR 05-06, AR 07-08
Maize	Cob borer	Meghalaya, Mizoram	AR-04-05AR 05-06
	Stem borer	Meghalaya, Nagaland	AR-04-05
		Mizoram, Sikkim	AR 05-06
	Aphids	Mizoram	AR 05-06
	Armyworm	Sikkim	AR 04-05
Mustard	Aphid	Tripura, Meghalaya	AR 05-06
Gram/Pea/Arhar	Pod borer	Tripura	Personal observation
	Weevil	Mizoram	AR 05-06
Blackgrm/Greengram	White fly	Tripura	Biswas, 2003
Soybean	Semi looper	Manipur	AR 05-06
	Blister beetle	Manipur, Tripura	AR 05-06
	Epicauta beetle	Manipur	AR 05-06
Groundnut	Leaf eating caterpillar	Manipur	AR 05-06
	Blister beetle	Manipur, Tripura	AR 05-06
	Aphid	Manipur	AR 05-06
Tomato	Fruit borer	Tripura	Personal observation
	White fly	Tripura	Biswas & Singh, 2008
	Leaf miner	Tripura	Personal observation
Brinjal	Shoot & fruit borer	Tripura	Personal observation
	Aphid	Tripura	Personal observation
	Leah hopper	Tripura	Personal observation
	Mealy bug	Tripura	Personal observation
Citrus	Trunk borer	Meghalaya	AR 04-05
	Aphids	Meghalaya	AR 05-06
Banana	Pseudo stem weevil	Mizoram	AR 04-05
	Rhizome weevil	Mizoram	AR 04-05
	Banana beetle	Mizoram	AR 04-05
	Banana aphid	Mizoram	AR 05-06

References

AR 2004-05. Annual Report, 2004-05. ICAR Research Complex for NEH Region, Umiam, Meghalaya.

AR 2005-06. Annual Report, 2005-06. ICAR Research Complex for NEH Region, Umiam, Meghalaya.

AR 2006-07. Annual Report, 2006-07. ICAR Research Complex for NEH Region, Umiam, Meghalaya.

Bakhetia, D.R.C. and Arora, R. 1992. Changing scenario of pests and pest management in rapeseed-mustard in India. In: National Seminar on Changing Scenario in pest and pest management in India, January 31-February 1, 1992, Hyderabad, Abstr. P-3.

Bale, J.S., Masters, G.J., Hodkinson, I.D., Awmack, C., Bezemer, T.M., Brown, V.K., Butterfield, J., Buse, A., Coulson, J.C., Farrar, J., Good, J.E.G., Harrington, R., Hartley, S., Jones, T.H., Lindroth, R.L., Press, M.C., Symrnioudis, I., Watt, A.D. and Whittaker, J.B. 2002. Herbivory in global climate change research: direct effects of rising temperatures on insect herbivores. Global Change Biology, 8:1-16.

Berryman, A.A. 1987. The theory and classification of outbreaks, pp. 3-30. In Eds, Barbosa, P. and Schultz, J.C. Insect outbreaks. Academic Press Inc., new York.

Biswas, S. 2003.Yellow mosaic disease on black gram and green gram in Tripura. Agrinews (ICAR, Tripura):1(4):1

Biswas, S. and Singh, N.P. 2008. Influence of host genotypes on the incidence of major diseases of tomato during winter season in Tripura. Indian J. Hort. 65(3): 356-357.

CICR 1989. Technology for increasing cotton production in India. Technology bulletin, Central Institute for Cotton Research, Regional Station, Coimbatore, Tamil Nadu.

Coviella, C. and Trumble, J. 1999. Effects of elevated atmospheric carbon dioxide on insect interactions. Conserv. Biol. 13: 700-712.

David, H. and Nandagopal, V. 1986. Pests of sugarcane distribution, Symptomology of attack and identification. In: Eds, David, H, Easwaramoorthy, S. and Jayanthi, R. Sugarcane entomology in India, Sugarcane Breeding Institute, Coimbatore, pp. 4-29.

Dhaliwal, G.S. and Bedi, S. 1993. Climate change and Indian agriculture: Ecological perspective, pp. 321-343. In: Eds, Gill, K.S., Dhaliwal, G.S. and Hansra, B.S. (1993). Changing scenario of Indian agriculture. Commonwealth Publishers, New Delhi.

FAO 1989. Quarterly Newsletter, Asia and Pacific Plant protection Commission, Food and Agriculture Organization, Bangkok.

Fletcher, T.B. 1920. The annotated list of Indian crop pests. In: Proc. 3rd Ent. Soc. Meeting, Pusa, PP-33-313.

Hamilton, J.G., Dermody, O., Aldea, M., Zangeri, A.R., Rogers, A., Berenbaum, M.R. and Delucia, E. 2005. Anthropogenic changes in Tropospheric Composition Increase susceptibility of soybean to insect herbivory. Environ. Entomol. 34 (2): 479-485.

Harrington, R., Fleming, R. and Woiwood, I.P. 2001. Climate change impacts on insect management and conservation in temperate regions: can they be predicted? Agricultural and forest Entomology, 3: 233-240.

Hunter, M.D. 2001. Effects of elevated atmospheric carbon dioxide on insect-plant interactions. Ag. Forest. Entomol. 3: 153-159.

Lewis, T. 1997. Thrips as crop pests. CAB International, Cambridge: University Press. P-740.

McVean, R. and Dixon, A.F.G. 2001. The effects of plant draught stress on populations of the pea aphid Acyrthosiphon pisum. Ecol. Entomol. 26: 440-443.

Nayar, KK., Ananthakrishnan, T.N. and David, B.V. 1976. General and applied entomology, Tata McGraw-Hill Publishing Co Ltd., New Delhi.

Parry, M.L. 1990. Climate change and world agriculture, Earthscan Publications Ltd., London.

Parry, M.L., Porter, J.H. and Carter, T.R. 1990. Climate change and its implications for agriculture, Outlook Agric., 19(1):9-15.

Petzoldt, C. and Seaman, A. 2011. Climate change effects on insects and pathogens. In Climate change and agriculture: promoting, practical and profitable responses, internet document.

Porter, J.H., Parry, M.L. and Carter, T.R. 1991. The potential effects of climate change on agricultural insect pests. Agric. For. Meteorol., 57:221-240.

Reddy, P.S. and Ghewande, M.P. 1986. Major insect pests of groundnut and their management. Pesticides, 20(5):52-56.

Reiners, S. and Petzoldt (eds) 2005. Integrated crop and pest management guidelines for commercial vegetable production. Cornell Cooperative Extension Publication. 124VG http://www.nysaes.cornell.edu/recomends/

Singh, J., Arora, R., Sohi, A.S. and Sidhu, A.S. 1988. New record of some insect species attacking American cotton in the Punjab. J. Insect Sci., 1(1):106-107.

Singh, O.P., Verma, S.N. and Nema, K.K. 1989. Insect pests of soybean in India. International Book Distributors, Dehradun.

Vincent, C., Hallman, G., Panneton, B. and Fleurat-Lessardu, F. 2003. Management of agricultural insects with physical control methods. Ann. Rev. Entomol. 48: 261-281.

Yamamura, K. and Kirtiani, K. 1998. A simple method to estimate the potential increase in the number of generations under global worming in temperate zones. Appl. Ent. and Zool. 33:289-298.

❑❑❑

9

Integrated Disease/Pest Management in Plantation Crops

Amar Bahadur and Sukhen Chandra Das

Abstract

Plantation crops like tea, Arecanut, Coconut, Cashewnut and Rubber are significant sources of farm income and also suffer variable diseases and pest attack thus decreasing th productivity tangibly. Pest identification followed by suitable recommendation needs to be implemented.

Keywords: Plantation crops, Diseases and insect attack on plantation crops.

Introduction

India is believed to be the native of several economic plant species which have spread to various part of the world. While agriculture continuous to stagnate in underdevelopment and insufficiency in food production. India enjoys variable climatic conditions and multiplicity of topographic feature, elevation and rainfall. The climatic conditions and soil type are so diversified that the various combinations of these factors favor unlimited production of a wide variety of crop plants. Now started move and rapid changes are taking place largely as results of bio-technology to advance scientific agriculture. Bio-technological approaches introduction of genetically modified in plant protection. A small percentage of former are well educated on the advantages of scientific forming and are adopting newer technique at limited extent. The highly yield variety has revolution to adopting intensive forming practices. Study of meteorological factors influencing the relationship and the microclimate factors affecting the development and spread of the diseases. It is well known that is several plant species nutritional deficiency lead to microbial diseases. The natural balance between beneficial and harmful pest changes with cropping patterns, pest management practices and variability in environment. Variability of microorganisms by heterocaryosis, mutation, hybridization and some other method induce variability in genetic characters of pathogen species and thus help in development of new stains or race. Pathogen often develop the capacity to perform biochemical reaction which were not present in them earlier, in this way they can utilize protoplasm of unfavorable host. This is known as adaptation to new cytoplasm. Global climate changes resulting changes in temperature rise and precipitation patterns could accentuate the vulnerability of agriculture. Global worming way increase average water vapor and evaporation increase in precipitation in high-altitude regions significantly alter the monsoon pattern which could influence the pathogens. Climate change leads to shifts in the pest incidence migration and viability thresholds. Increase UV-radiation is likely due to depletion of ozone layer resulting that might change the host–pathogen relationship and the status of the individual pest. Disturbance in the ecological balance among different component of biotic and a biotic makes certain organism reach pest status. Building farmers knowledge and skills in making best use of local resources and natural processes and community action. Pest management that such eco approaches can bring in both ecological and economical benefits to the formers. The beneficial and eco-friendly micro-organisms like fungi, bacteria, virus and protozoan capable of killing specific diseases causing microbes, nematode and insect pest and also those promoting plant growths are being considered as potential biological attention in eco-friendly Agriculture. Use of bio-control agent and bio-pesticides are increasingly gaining acceptance with formers. Integrated management of pest and diseases in crop, cultural practices such as selection of healthy and certified seed of registrants or tolerant variety, adopting proper crop rotation, seed treatment to control soil and seed born disease and sucking pest, timely sowing and irrigation, fertilization and weed management *etc*. play a crucial role. Bio-based pesticide and synthetic chemical control measures must be judiciously integrated. The use of botanical, microbial pesticides, pheromones and GM crop are to be integrated appropriately and exploited. The use of chemical pesticide and other agro-chemicals are getting reduce and being banned globally because of their toxic effect on humane being and his live stock. Residual toxicity environment problem, pest out breaks and drastic effect on beneficial microorganism and insects. Therefore, it is imperative to develop a holistic system of tacking pest to make it more eco-friendly, economically viable and socially acceptable for the formers.

The pathogen interactions under the influence of global climate change have the potential to disrupt both agriculture and natural ecosystems severely emerging of new pest and pathogen. Atmospheric CO_2 a major green house gas. The changes in global climate have started showing effects on general plant health as well as plant- pathogen system. Temperature can also bring in some desirable effect of some microorganism like isolates of bio-control potential *Trichoderma harzianum* can withstand high temperature and survival of more difficult for susceptible pathogen, moisture and precipitation favor some pathogen for rapid multiplication and dispersal. Climate change has a very big impact on insect vector and thereby influencing spread of some pathogens. Diseases management strategies thus will require adjustment. The induce host resistance may also be overcome quickly by more rapid diseases cycles of pathogens. Climate change and its probable impact of population dynamics of pathogens and design management practices before loss of crop. The role of beneficial endophytic mycorrhizal and rhizospheric microbe will become significant in sustaining agriculture. Plantation crop suffering different diseases and pest they can manage by integrated approaches to break their registrant and race changing the microbes and severity also. In this regard to tackle the pest and diseases of major crops integrated management practices have been adopted. In plantation crops like Tea (*Camellia sinensis*) Areca nut (*Areca catechu*) Coconut (*Cocos nucifera*) Cashew nut (*Anacardium occidentale*) Rubber (*Hevea brasiliensis*) can boost their production of farm income.

Diagnosis of Pathogens

Before going to manage crop diseases should know about correct identification of diseases. In tripura climate is very variable warm and humid, several micro organism survive in this climate some are useful and other harm full. In this aspect the some micro organism is antagonistic effect for their survival. In this aspect the useful micro organism should Isolation and characterization for diseases management those are better surviving and compatible in this climate. In tripura several medicinal plants are also growing in this climate should also exploits to diseases management for ecological, economical and environmental safe.

The diagnosis is done by percept and experience. The visual observations based on experience, precept and intuitive judgment is still the most widely used method for identification of diseases. In the field the symptoms provide the causes for possibilities but confirmation comes only after performing diagnostic test. Other methods such as host range, morphology of pathogen, isolation of pathogen on selective media and molecular diagnosis. Several sophisticated techniques are being used for observations which include microscopy, isolation and identification of biotic agents associated besides serology, immunological, biochemical and physiological and genomic analysis.

Symptoms are the visible expression of host pathogen interaction and presence of pathogen structure from the characteristic sign the basis for preliminary diagnosis. The morphological features such as spores, fructifications or sporophores is important part of diagnostic of pathogen. If diagnostic structures of the fungus are not readily visible on the infected host surface the parasite may be induce to sporulation of the infected tissue (Tuite 1969).

Nematode extraction of plant tissue or soil can be identified using various illustrations description and keys as describe (Mai and Lyou 1975). Identification of viral diseases have been described and reviewed (Bos 1970, Ball 1974, Mathews 1991). These methods include symptomatology, mode of transmission, host range, particle morphology, antigenicity and electrophoresis mobility in gels. Characteristic of virus and description of symptomatology are available in several sources (CMI 1970, Smith 1972).

Today photographs can be scanned digitalized and sent around the world within seconds (Holmes *et al.* 2000) . Modern telecommunication system was first used in agricultural Sciences in early 1990 as Digitally Assisted Diagnosis (DAD). The equipment used are computer, digital camera, dissecting and compound microscope with digital camera attachment flatbed scanner and a colour inkjet printer (Briere and France 1998, Holmen *et al*, 2000). Some symptoms of herbicide injuries on various plants have been described by Lockerman *et al* (1975). Besides routine culture media, several selective nad semi-selective culture media have been developed, which should be employed for identification of the pathogens. Selective media have been developed for *Pythium, Phytophthora, Fusarium* (Tsao 1970, Papavizas 1967) *Phoma betae* (Bugbeen 1974), *Macrophomina phaseolina* (Papavizas and Klay 1975) *Rhizoctonia solanacearum* (Ko and Homa 1971), *Sclerotium rolfsii* (Black man and Rodriguez - Kabana 1976) have been described. The range of choice is relatrively cheap, easy to use diagnostic kits now being developed should allow use to monitor even low level of diseases on the spot under field conditions (Klausner 1987, Miller *et al.* 1988,1990, 1992). Immunological and nucleic acid hybridization technique are increasingly becoming popular for the rapid detection of many of these pathogens which cannot be easily identified by other routine ways. In addition nucleic acid hybridization technique can detect uncultivable pathogen.

Diseases and Pests on Plantation Crops

1. Cashewnut (*Anacardium occidentale*)

In India cashew is largest producer of raw Cashew (*Anacardium occidentale*) nut in the world. India is also the largest exporter of cashew kernels. Among the Agricultural commodities exported from India. Cashew plant suffering different diseases some important diseases are given below.

Die-back pink diseases caused by *Botryobasidium salmonicolor* is prevalent in cashew plantation during rainy season. Main symptoms of this attack are appearance of white patches on branches followed by drying of twigs from the tip.

It can control affected part of apply Bordeaux mixture 1% during may-june and October.

Tea mosquito bug (*Helopeltis antonii*) is the most serious pest of cashew in

India. Both adult and nymphs suck sap from tender portions causing necrotic lesion which results in drying up of young leaves and inflorescence. Spray endosulfan (0.05%), carbaryl (0.1%), spray should be done thrice- When emergence of new vegetative flushes in October-november , commencement of penical emergence in December –January and fruit set in January- February.

Stem and Root borer (*Plocaederus ferrugincus*) is a serious pest of cashew destroying its tree. It lays egg in the crevices of loose bark in the trunk. The grub that hatches out bores in to the trunk and feeds on sub –epidermal and vascular tissues and tunnels are formed. The affected trees show yellowing of leaves drying of twings and presence of holes at the base of the stem with exuding sap and frass, control the pest, swab the exposed roots and trunk up to 1 m height with a suspension of carbaryl (0.2%) or with coal tar and kerosene (1:2) during March –April and November- December.

2. Rubber (*Hevea brasiliensis*)

Rubber tree (*Hevea brasiliensis*) is the most commercial source of natural rubber as industrial raw material of strategic importance recorded from its latex. The major rubber –growing countries are Indonesia, Thailand, Malaysia, China and India. Indonesia has maximum area under rubber but maximum rubber production in Thailand.

In India rubber is grown over an area of 5.45 lakh ha with a production of 5.84 lakh tones. In India Kerala is the predominant rubber growing state. Tamil Nadu, Kerala account for 98% of the total production. Crop losses resulting from diseases and pest in rubber plantations are considerable. Timely plant protection measure ensures healthy growth and economic production.

Abnormal leaf fall (*Phytophthora palmivora*) prolong wet weather coupled with humid atmosphere favours this diseases. First the fruit rot followed by premature leaf fall and die back of terminal twings leading to considerable crop loss. Prophylactic spray on foliage prior to the onset of south west monsoon with Bordeaux mixture 1%.

Powdery mildew (*Oidium heveae*) symptoms are prominently noticed on newly formed tender flush during the refoliation period of January-March. Tender leaves with ashy coating curl and crinkle edges roll inwards. In later stages white patches appear on older leaves. Infected flowers and fruits are shed. Dusting with sulphur alone or mixed with talc (70:30), spray bavistin 0.05% at fortnight intervals. Eradicate collateral host like Jatropha, Loranthus *etc.*

Pink disease (*Corticium salmonicolor*) rubbers plants are suffering about age group of 3-12 year are highly susceptible. The seat of infection in fork region of branches, where white or pink colored cobweb of mycelial growth are seen bark surface with latex oozing out from the lesions. This is followed rotting drying up and cracking of affected bark. In early stage of infection apply Bordeaux past up to 30 cm above and below the affected region.

Scale Insect (*Saissetia nigra*) and mealy bug (*Ferrisiana virgata*) these pests suck sap from the tender portions causing drying anddie-back. If infection is severe spray malathion 0.05% . Fish oil rosin soap is effective against mealy bug.

3. Tea (*Camellia sinensis*)

Tea (*Camellia sinensis*) is the most popular and the cheapest beverage consumed by two-thirds of world population. The estimulative effect of tea is due to the presence of caffeine (1.25-4.5%). The per cup value of caffeine in tea is three and a half time lesser than coffee and is less harmful than coffee. In addition, it contains useful flavonoids, tannins, volatile oil and vitamin b which are useful for health. The major tea growing states in India are Assam, West Bengal, Tamil Nadu and Kerala. If plant suffer diseases the quality will detoriate so need management of diseases. Various organisms causing disease to tea plant, on root, leaf and stem.

Root diseases can not be detected early and by this time symptoms are visible in the above ground parts of the plants. Root diseases are fungal diseases, they are brown rot (*Fomes noxious*), Charcoal stump rot (*Ustulinia zonata*), Black root rot (*Rosselinia arcuata*). The infected plant centering the diseased one must immediately be uprooted with the entire root system. One metre deep and 30 cm wide isolation trench should be dug followed by soil fumigation. Before filling in the area should be under rehabilitation crop for at least 2 years.

Leaf diseases blister blight (*Exobasidium vexans*) is most wide spread leaf diseases. It spreads very fast through its air-born spore under cool misty conditions. Infected leaf should be collected and burnt. Copper fungicide should be spray at weekly interval.

Black rot (*Corticium invisum*) attacks tea areas having poor ventilation. The sanitation and shade status of such sections should be rectified to control the diseases. Spraying of copper fungicide should be done twice at 15 day interval.

Stem diseases branch canker (*Poria hypobrunnea*) a wood parasite enter the plant through most widely prevalent stem disease. Careful removal of all dead woods prevents its spread . The cut surface should immediately be spared with copper fungicide.

Red rust (*Cephaleuros parasiticus*) secondary disease of tea which attack tea plants. Red rust is only algal disease often causing severe damage to tea. The diseases is found in most tea plantation in the country but it intensity may vary considerably depending upon climatic factors. The algal appear in the form of minute rusty spots on the leaves. The spots are characteristically reddish and hairy occurring often on upper leaf surface. It is usually epiphytic on the leaves and may also become parasitic. In the farmer case it grow on the leaf surface only , where as in the later penetrates the cuticle and epidermal cells extending up to the parenchyma and reducing the effective area of photosynthesis. It can control spraying of copper fungicide at 15 day interval.

4. Coconut (*Cocos nucifera*)

Coconut (*Cocos nucifera*) provide edible and industrial oils , Protein rich milk, invigorating water, timber , fiber , roofing and matting materials. Coconut plants suffering different pest and diseases some are important as given below.

Stem bleeding (*Ceratocystis paradoxa*) diseases is caused by fungus exudation of a reddish brown liquid through the growth cracks at about 5-10 cm above the ground level and bleeding patches higher up in the trunk are characteristic symptoms. Remove affected portion and apply hot coal tar or Bordeaux past, application of Calixin @1ml/lit of water as soil drenching one in 4 months is also effective.

Bud rot (*Phytophthora palmivora*) disease is severe during monsoon when humidity is high. Young palms are more susceptible. The early symptoms of the diseases are yellowing of one or two younger leaves surrounding the spindle. The tender leave base and soft tissue of crown rot into a slimy mass emitting a foul smell. In early stage of disease remove all affected tissue of the crown and apply Bordeaux past and protect it from rain till shoot emerges, spray Bordeaux mixture 1% on the crown of affected as well as neighboring palms as a prophylactic measure.

Grey blight (*Pestalotia palmarum*) diseases is characterized by the appearance of minute yellow spot which a grey brown margin on outer whorl of leaves. The spot coalesce to form large irregular necrotic patches and presents a blighted appearance. Remove older diseases affected leaves and spray Bordeaux mixture 15.

Burrowing nematode diseases caused by *Radopholus similes* is the most important coconut palms show stunting, yellowing reduction in leaves number and size, delayed flowering and reduced yield. Lesions are caused on root. Apply phorate @ 10 gm a.i./palm and 3 gm a.i./ seedling June –July and Sept.- Oct. . Avoid banana as shade crop and inter crop. Grow and apply green manure crop like *Crotolaria jumcea*.

Red ring nematode (*Bursaphelenchus cocophilus*) disease is most common and develops rapidly in young coconut tree 3 to 10 year old; it kills them within about three month from infection. The symptoms appear on older leaves nearest the pint on infection as yellowing from the tips inward and them as browning of the leaves. Several drying or dead leaves often break close to the stem and remain hanging. Yellowing and browning then spread to progressively younger leaves and finally the whole treetop dies. In bearing tree, inflorescences wither and nuts of all stages drop prematurely. In cross sections of infected stem an orange-red to brownish ring about 3-5 cm wide is present about 5 cm inside the stem periphery over the length of the stem. Nematode transmitted from palm to palm by palm weevil (*Rhynchophorus* sp.) and sugarcane weevil (*Metamasius marum*). Infected tree should be treated with systemic insecticide and nematicides to reduce the vector and nematode within the trees.

Palm weevil (*Rhynchophorus ferrugineus*) is most destructive pest associated with coconut palm. The stout grub feed on soft tissue of the stem and bud causing death of the palm. Presence of holes on stem oozing out of a viscous brown fluid and extrusion of chewed up fibers through the hole, longitudinal splitting of leaf base and wilting and central shoot indicate attack of the pest. Field sanitation should be given prime importance and avoid making injury on the tree trunks to reduce infection. Pheromones trap or trap with molasses can be set in the plantation to attack and trap adult weevil. Inject attacked palm with carbaryl 50% WP as an alternative Aluminum phosphide at 1-2 tab/tree.

5. Arecanut (*Areca catechu*)

Areca nut (*Areca catechu*) popularly known as betel nut or supari. India is the largest producer and consumer of areca nut. Areca nut palm is prone to a number of diseases during its different stages of development.

Kaleroga/ Mahali / Fruit rot caused by *Phytophthora arecae* is one of the majer diseases of areca nut . This occurs as an epidemic in the heavy rainfall area of Karnataka and Kerala. The disease fruit appearance after the first monsoon shower and extends through the monsoon period. The first symptom is the appearance of water soaked lesion on the nut surface near the calyx. The patches enlarge and nuts darken and they shed in large numbers. The fallen nut soon develop whitish mycelial mass all over surface. Some times the infected nuts may not be shed and remain mummified in the bunches, such type of infection if known as dry mahali. This can be checked by spraying Bordeaux mixture 1% twice a year one just before the onset of south-west monsoon and other 40 days later.

Anaberoga or Foot rot (*Ganoderma lucidum*) disease initial visible symptoms are yellowing of outer whorl of leaves which gradually extends to inner whorls followed by wilting and drooping. The fungus invasion interrupts uptake of water and nutrients by the palms, leading to yellowing and wilting. Strict phyto-sanitary measures are to be adapted by removing and destroying the stumps along with roots, drench the soil with capton0.3% and Bordeaux mix 0.1% before planting.

Inflorescence die-back (*Colletotrichum gloeosporiodes*) diseases symptoms appear button shedding followed by die-back of inflorescence is a severe problem in areca nut plantation.

Removal and burning of inflorescence help reduce the load on inoculums in the field. Spray Zineb (4gm/lit water) twice one just after female flowers are set and again 15-28 days later. Aureofungin 50 ppm concentration is also effective in controlling the diseases.

Conclusion

This is for researcher, breeder and scientist those are working on these areas on focus to manage diseases sustainable and ecofriendly pest and disease on the basis of resistant verities, cultural practices, bio-control and plant based origin as locally to effective and economically management of plant disease. Now days pathogens changing their race due to continue using resistant verities and pesticide in huge amount in field. Therefore required integrated management practices in the field. Precise and correct diagnosis of the diseases and its cause are must for effective management of diseases. Effective and economically management of plant diseases in the ultimate goal of the science of Plant Pathology.

References

Bugbeen, W.M. 1974: A selective media for the enumeration and isolation of Phoma betae from soil and seed. Phytopathology. 64:706-708

Bos, L. 1970: Symptoms of virus Diseases in Plants. Oxford & IBH, Publishing Co., New Delhi.

Ball, E.M. 1974: Serological test for Identification of plant viruses. Am. Phytopathological Soc. ST. Paul, Minnesota, 31.

Backmnan, P.A. and Rodriguez-Kabana, R. 1976: Development of a medium for the selective isolation of *Slerotium rolfsii*. Phytopthology.66: 234-236.

Briere, S.C. and France , G.D. 1998: Use of digital images to aid in diseases diagnosis at the University of Wyoming , Extension Plant Pathology Lab. Phytopathology, 88:11.

Commonwealth Mycological Institute, 1970: Description of Plant viruses. CMT, Kew, Surrey England.

Holmes, G.J., Brown E.A., and G. Ruhl 2000: What's a picture worth? The use of modern telecommunication in diagnosis of plant disease. Plant Disease. 84:1256-1265

Ko, W.H. and F. K. Hora 1971: A selective media for the quantitative determination of Rhizoctonia solani in soil. Phytopathology. 61: 707-710

Klausner, A. 1987: Immunoassays flourish in new markets. Biothechnology. 5:551- 556.

Lockerman, R.H., Putman, A.R., Rice, R.P. and W.F. Meggitt 1975: Dignosis and prevention of herbicide injury. Cooperative extension service, Michigan State University Extension Bulletin E 809.

Miller, S.A. and Martin, R.R. 1988: Molecular diagnosis of Plant disease, Annu. Rev. Phytopathol. 26: 409- 432.

Miller, S.A. 1990: Developmentof modern dignosistic test and benefits to farmers. In: Shots, A (Ed) Monoclonal antibodies in Agriculture, Pudoc. Wagenigen. Netherlands, 15

Miller, S.A. 1992: From the research bench to the market place, development of commercial diagnostic kits. In: Deencan, J.M., and Torrance, L. (Eds) . Techniques from the Rapid Diagnosis of Plant Pathogens, Blackwell Oxford.

Mai, W.F. and Lyou , H.H. 1975: Pictorial key to genera of plant parasitic nematodes 4th ed., Cornell Univ. Press, Ithaca , New York.

Mathews, R,E,F, 1991: Plant Virology, 3rd ed., Academic Press, San Diego, California.
Papavizas, G.C. 1967: Evaluation of various media and antimicrobial agents for isolation of Fusarium from soil. Phytopathology. 57:848-852.
Papavizas, G.C. and N.G. Klay 1975: Isolationand quantitative determination of Macrophomina phaseolina from soil. Phytopathology. 65:182-187.
Smith, K.M. 1972: A textbook of Plant virus Diseases, 3rd ed., Longmans, New York. 652.
Tsao, P.H. 1970: Selective media for isolation of pathogenic fungi, Anu. Rev. Phytopathol. 8:157-186.
Tuite, J.F. 1969: Plant pathological methods fungi and bacteria, Minneapolis, Minnesota 128.

❑❑❑

10

Climate Change and Effectiveness of Pesticides in Vegetable Crops

D. Sharmah, A.K. Singh and M. Datta

Abstract

Tripura is expected to be highly prone to the consequences to climate change because of its geo-ecological fragility, strategic location, its trans-boundary river basins and its inherent socio-economic instabilities. Environmental security and sustainability of the state is and will be greatly challenged by the impacts of climate change in coming years.

Climate and weather can substantially influence the development and distribution of insects. Current best estimates of changes in climate indicate an increase in global mean annual temperatures of 1°C by 2025 and 3°C by the end of the next century. Such increases in temperature have a number of implications for temperature-dependent insect pests. An increase of 3°C in mean daily temperature would cause the carrot fly, Delia radicum (L.) *to become active a month earlier than at present and temperature increases of 5 to 10°C would result in completion of four generations each year, necessitating adoption of new pest control strategies. An increase of 2°C will reduce the generation turnover of the aphid, Rhopalosiphum padi* (L.) *by varying levels, depending on the changes in mean temperature. Climate change may trigger major changes in diversity and abundance of insect pests, geographical distribution of insect pests, population dynamics, insect biotypes, herbivore plant interactions and efficacy of crop protection technologies in vegetables. Major insect pests such as the pod borers, aphids and white flies may cause greater damage in grain legumes and vegetables.*

Climate change will also reduce the effectiveness of bio-pesticides and synthetic chemicals for pest management in vegetables. Therefore, there is a need to generate information on the likely effects of climate change on insect pests to develop robust technologies that will be effective in future under climate change.

Keywords: Climate change, Insect pests, Vegetable crops, Pesticides.

Introduction

Plant Protection Measures require an understanding of the ecology of the cropping system including that of the pests and their environment. For example, weather conditions can influence insect pest populations, wind-driven rain is a major source of mortality to nymphs of the potato leafhopper on potatoes and thrips on onions.

Climate change will trigger major changes in diversity and abundance of insect pests and diseases, geographical distribution of insect pests, population dynamics, insect biotypes, herbivore plant interactions, activity and abundance of natural enemies, species extinction, and efficacy of crop protection technologies. Changes in geographical range and insect abundance will increase the extent of crop losses, and thus, will have a major bearing on crop production and food security. Distribution of insect pests will also be influenced by changes in the cropping patterns triggered by climate change.

Climatic change will also reduce the effectiveness of host plant resistance, transgenic plants, natural enemies, bio-pesticides and synthetic chemicals for pest management. Therefore, there is a need to generate information on the effects of climate change on insect pests of vegetables and impact of climate change on effectiveness of bio-pesticides and chemical insecticides to develop robust technologies that will be effective in future under climate change.

1) Impact of Weather on Major Insect Pest Problem of Vegetables in Tripura:

a) *Impact of Temperature change on population dynamics of insect pests*

Climate is the long term effect of the factors and determines what insect pests are present and how many generations are possible in a single active season, while weather is the short term or day to day effect and plays primary role in influencing insect pests abundance and damage. In Tripura, maximum temperature recorded is 35° C to 38° C and minimum is 7° C to 10° C which is congenial for growth and multiplication (Optimum temperature 21° C to 35° C) of insect pests. Theses insect -pests causes colossal damage to crops resulting in to low production in Tripura.

Low temperatures are often more important than high temperatures in determining geographical distribution of insect pests, extending their geographical range and sudden outbreaks of insect pests can wipe out certain crop species and also encourage the invasion by exotic species. For all the insect species, higher temperatures, below the species upper threshold limit, will result in faster development and rapid increase in pest populations as the time to reproductive maturity will be reduced considerably. In addition to the direct effects of temperature changes on development rates, improvement in food quality due to abiotic stress may result in dramatic increases in growth of some insect species, while the growth of certain insect pests may be adversely affected. Pest outbreaks are more likely to occur with stressed plants as a result of weakening of plants' defensive system, and thus, increasing the level of susceptibility to insect pests.

Table 1: Following table shows the Major Insect Pest Problem of Vegetables in Tripura.

Major Vegetable Crops	Major insect pests problems
Brinjal, Potato, Tomato, Cole crops, Okra, Beans, Cow pea,Cucurbits, Chilli, etc.	Insect pests fruit and shoot borer (Brinjal), aphid (cole crops, cowpea,), fruit fly (cucurbits), fruit borer (Brinjal, Tomato, Okra), cabbage caterpillar (Cole crops, tomato), cut worm (Cole crops, tomato, potato), termite, red ant (Potato), epilechna beetle (Cowpea), DBM (Cole crops)DiseasesDamping off (chilli, tomato), wilt (brinjal, tomato, potato), late blight (Potato), leaf spot, fruit rot (Chilli), Leaf curl (brinjal, tomato, potato) etc.

Climate change will lead to earlier infestation by Helicoverpa armigera (Hub.) (Sharma 2010), resulting in increased crop loss. Rising temperatures are likely to result in availability of new iches for insect pests. Temperature has a strong influence on the viability and incubation period of H. armigera eggs (Dhillon and Sharma 2007) which decreases with an increase in temperature from 10 to 27 °C, and egg age from 0 to 3 days (Dhillon and Sharma 2007). An increase of 3°C in mean daily temperature would cause the carrot fly, Delia radicum (L.) to become active a month earlier than at present and temperature increases of 5 to 10°C would result in completion of four generations each year, necessitating adoption of new pest control strategies. An increase of 2°C will reduce the generation turnover of the aphid, Rhopalosiphum padi (L.) by varying levels, depending on the changes in mean temperature. Increase in this temperature may cause overwintering of Diamond back moth (Plutella xylostella L.). If this problem becomes common, the status of diamond backmoth will increase dramatically in cabbage crop. It has been estimated that with a 20 °C temperature increase insects might experience one to five additional life cycles per season (Yamamura & Kiritani 1998).

Table 2: Expected responses of vegetable bugs under two scenarios of temperature change

Categories of responses	Slight temperature increase (<2°C)	Substantial temperature increase (>2°C)
Abundance	Likely to increase in multivoltine species with flexible life cycles	Likely to change, depending on the community response
Distribution range	Likely to shift in some species, especially those capable of long distance flights and associated with ornamental plants and/or urban habitats	Likely to shift in many species
Physiology and behaviour	Slight/ undetectable changes	Evident/detectable changes (e.g. in parameters of photoperiodic responses)
Community structure	Similar to currently observed substantial changes in structure	Increased species richness;

b) *Impact of Relative Humidity on population dynamics of insect pests*

On the other hand, the relative humidity of Tripura naturally varies from 60% to 90% which greatly influences the development of insect pests attack. Relative air humidity from 50% to 80% is probably the best range for most external plant pests. The plant also helps to maintain favourable humidity for most plant pests attacking them internally.

However, an insect pest would move with a crop into a new habitat will depend on other environmental conditions such as the presence of overwintering sites, soil type, and moisture. Morning relative humidity had significant positive correlation with leafhopper and aphids and significant negative correlation with population of whiteflies, mites and *Chrysoperla* in case of Okra crop. Leafhopper and aphid populations showed significant positive correlations with afternoon relative humidity. A significant negative correlation is also observed between population of whiteflies and afternoon relative humidity and also between afternoon relative humidity and population of *Chrysoperla*.

c) *Impact of Rainfall on population dynamics of insect pests*

Changes in Rainfall are of greater importance for agriculture than temperature changes, especially in Tripura where most of the cultivable area are under upland condition rain fall may be a limiting factor for crop production (Parry 1990). The average annual rainfall of Tripura is 1800-2200mm may increase as a result of intensification of the hydrological cycle, which will cause disruption of agriculture as the cropping systems and the composition of fauna and flora will undergo a gradual change. High mobility and rapid population growth will increase the extent of losses due to insect pests. Pest outbreaks might occur more frequently, particularly during extended periods of drought (e.g. pea aphids are not tolerant of drought) followed by heavy rainfall. Some insects are sensitive to precipitation and are killed or removed from crops by heavy rains There is a significant negative correlation is observed between rainfall and mites and also with Chrysoperla in case of okra crop.

2) Effect of Weather on Insect Pests Management in Vegetables

a) *Change of species diversity, its influence on pest management*

Biodiversity plays an important role in abundance of insect pests and their natural enemies (Sharma and Waliyar 2003). There is a need to increase functional diversity in agro-ecosystems vulnerable to climate change to improve system resilience and decrease the extent of losses due to insect pests (Newton *et al.* 2009). However, changes in cropping patterns as a result of climate change will drastically affect the balance between insect pests and their natural enemies. Since climate change will lead to a shift in cultivation of crops in non-traditional areas and crop rotations, this may influence the prevalence and importance of specific pests.

b) *Change on activity and abundance of natural enemies*

As a result of climate change relationships between insect pests and their natural enemies may change, resulting in both increases and decreases in the status of individual pest species. Changes in temperature will also alter the timing of diurnal activity patterns of different groups of insects, and changes in inter-specific interactions could also alter the effectiveness of natural enemies for pest management. Quantifying the effect of climate change on the activity and effectiveness of natural enemies for pest management will be a major concern in future pest management programs. The majority of insects are benign to agro-ecosystems, and there is considerable evidence to suggest that this is due to population control through inter-specific interactions among insect pests and their natural enemies. Oriental armyworm, Mythimna separata (Walk.) populations increase during extended periods of drought (which is detrimental to the natural enemies), followed by heavy rainfall because of the adverse effects of drought on the activity and abundance of the natural enemies of this pest (Sharma *et al.* 2002). Aphid abundance increases with an increase in CO_2 and temperature, however, the parasitism rates remain unchanged in elevated CO_2. Temperatures up to 25°C will enhance the control of aphids by coccinellids (Freier and Triltsch 1996). Temperature not only affects the rate of insect development, but also has a profound effect on fecundity and sex ratio of parasitoids (Dhillon and Sharma, 2008, 2009).

c) *Change on the effectiveness of bio-pesticides and synthetic insecticides*

Higher temperatures will make dry seasons drier and conversely, may increase the amount and intensity of rainfall, making wet seasons wetter than at present.

Natural plant products and synthetic pesticides are highly sensitive to the environment. Increase in temperatures and UV radiation and a decrease in relative humidity may render many of these control tactics to be less effective and such an effect will be more pronounced on natural plant products and the biopesticides (Isman 1997). Rapid dissipation of insecticide residues due to increases in temperature and precipitation will require more frequent application of insecticides. Therefore, there is a need to develop appropriate strategies for pest management that will be effective under situations of global warming in future.

Pesticide use effectiveness is influenced by many factors of which presence of pests and their natural enemies is one of the main factors. The same is again influenced by the weather and in the midterm by climatic changes. Climate change might therefore have an influence on pesticide use. Tilman *et al.* (2001) foresee a 2.4 to 2.7-fold increase in pesticide use by 2050 related to population growth and conversion of natural ecosystems to agriculture. It has been revealed by investigating the relationship of temperature, precipitation and pesticide costs for several vegetable crops. A simulation by applying different climate change scenarios showed uniform increases in average pesticide costs for soybeans and potatoes.

It is likely that farmers will experience extensive impacts on insect management strategies with changes in climate. Entomologists expect that insects will expand their geographic ranges and increase reproduction rates and overwintering success. This means that farmers in the Tripura will have more types and higher numbers of insects to manage. Hence, type and frequency of use of insecticide and expense are likely to increase with the change of climate. Insecticides and their applications have significant economic costs for growers and environmental costs for society.

Table 3: Information on use of Plant Protection Chemicals (Kg or L) in South Tripura District

Chemical	2001/02	2002/03	2003/04	2004/05
Mancozeb	1350.87	2020.87	2244.20	1746.69
Carbendazim	377.50	1027.70	1595.27	1632.51
Copper Oxy chloride	282.59	883.08	1651.34	1136.14
Quinalphose	38.195	107.35	614.50	898.58
Deltamethrin	17.345	282.42	101.77	34.31
Captan	172.19	140.146	255.585	11.18
Plantomycin	99.638	6.185	2.165	1.687
Chlorpyriphos	2394.20	4676.33	5131.28	5718.06
Endusulfan	4581.41	14743.63	9728.79	11417.15
Malathion	1229.65	448.45	534.20	244.15
Cypermethrin	3206.615	10891.40	7077.80	4174.85
Phosphamidon	1385.20	1933.45	968.40	502.10
Carbofuran	7294.82	17819.45	7774.75	7086.00
Hinosan/Kitazin	215.30	444.10	646.50	655.00
Neem based product	369.60	75.572	20.51	22.092
Phozalone	139.96	10.008	1.29	00.00

Additionally, some classes of pesticides (pyrethroids and spinosad) have been shown to be less effective in controlling insects at higher temperatures (Musser & Shelton 2005). It can be predicted that additional generations of important pest insects of vegetables may occur in Tripura due to change in climate conditions, probably necessitating more insecticide applications to maintain populations below economic damage thresholds. A basic rule of thumb for avoiding the development of insecticide resistance is to apply insecticides with a particular mode of action less frequently (Shelton *et al.* 2001). With more insecticide applications required, the probability of applying a given mode of action insecticide more times in a season will increase, thus increasing the probability of insects developing resistance to insecticides.

The diamondback moth, *Plutella xylostella* (Linnaeus) has consistently remained the most destructive insect pest of crucifer vegetables in Tripura. The major outbreaks of *P. xylostella* are more likely in the fields that are sprayed frequently and heavily with insecticides. The absence of natural enemies and fast development of insecticide resistance are believed to be the major causes of increasing pest status of *P. xylostella* in most parts of the India. Currently, this insect has become resistant to almost all classes of insecticides used against it in south-east Asian countries (Dhaliwal and Koul, 2010). An outbreak of diamondback moth, *Plutella xylostella* (Linnaeus) on cauliflower was reported in Aligarh, western part of Uttar Pradesh, during September to first fortnight of October during 2006. The infestation increased gradually from first fortnight of August and leads to total loss (100 % yield loss) of the crop (Ahmed *et al.*, 2009). Moreover, the current climatic change may lead to increase in severity of this pest in many regions of the India. For instance, *P. xylostella* may have two additional generations per year in coming year.

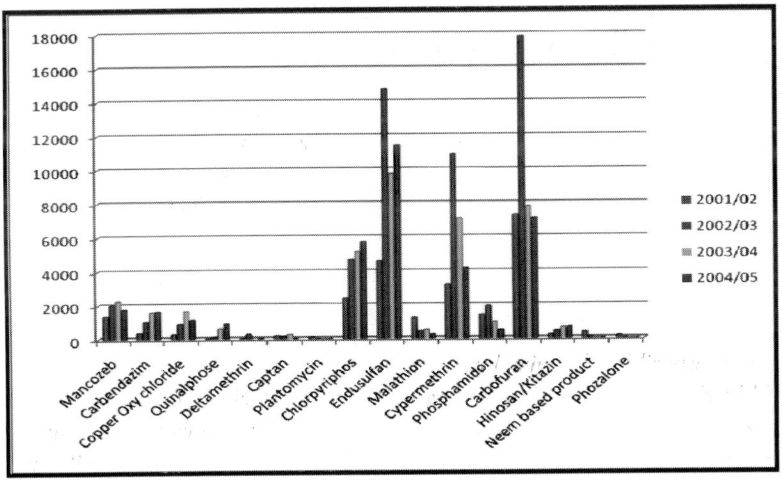

Fig 1: Bar diagram of pesticides used in South Tripura

Table 4: Some insect pests and their control measure by changing sowing time

Crop	Pests problems	Sowing time
Vegetable Brinjal, Potato, Tomato, cole crops, okra, Beans, Cucurbits	Shoot borer	Timely sowing
	Cucumber beetle	Adjustment of sowing time, clean cultivation reduces the pests' population.
	Thrips	Damage in some early storage varieties (100-115 days) may be reduced when plants are planted in late June or early July.
	Cutworm	Flooding of field

d) *Change on effectiveness of transgenic crops for pest management*

Factors like temperature have strong influence on the expression of *Bacillus thuringiensis* (*Bt*) toxin proteins deployed in transgenic plants. Cotton bollworm, *Heliothis virescens* (F.) destroyed *Bt*-transgenic cottons due to high temperatures in Texas, USA (Kaiser, 1996). Similarly, *H. armigera* and *H. punctigera* (Wallen.) destroyed the *Bt*-transgenic cotton in the second half of the growing season in Australia because of reduced production of *Bt* toxins (Hilder and Boulter 1999). Cry1Ac levels in transgenic plants decrease with the plant age, resulting in greater susceptibility of the crop to insect pests during the later stages of crop growth. Possible causes for the failure of insect control in transgenic crops may be due to inadequate production of the toxin protein, effect of environment on transgene expression, *Bt*-resistant insect populations and development of resistance due to inadequate management (Sharma and Ortiz 2000).

Conclusions

The relationship between crop protection costs and the resulting benefits will change as a result of climate change. This will have a major bearing on economic thresholds, as greater variability in climate will result in variable impact of pest damage on crop yields. Increased temperatures and UV radiation and low relative humidity may render many of these control tactics to be less effective, and therefore, there is a need to:

- Predict and map trends of potential changes in geographical distribution, and study how climatic changes will affect development, incidence, and population dynamics of insect pests.
- Understand the influence of climate change on species diversity and cropping patterns, and their influence on the abundance of insect pests and their natural enemies.
- Understand the changes in expression of resistance to insect pests and

identify stable sources of resistance and pyramid the resistance genes in commercial cultivars.

- Study the effect of weather on the efficacy of transgenic crops in pest management.
- Assess the efficacy of various pest management technologies under diverse environmental conditions and develop appropriate strategies for pest management to mitigate the effects of climate change.
- System diversity can be exploited to enhance the resilience of agro-eco-systems, improve resource utilization, and stabilize yields to cope with the effects of climate change on food security (Sharma and Waliyar 2003).
- The interactions between insect pests and their natural enemies need to be studied carefully to devise appropriate methods for using natural enemies in pest management.

References

Adamczyk Jr., J.J., Adams, L.C., and Hardee, D.D. 2001. Field efficacy and seasonal expression profiles for terminal leaves of single and double Bacillus thuringiensis toxin cotton genotypes. Journal of Economic Entomology 94: 1589-1593.

Coley, P.D. and Markham, A. 1998. Possible effects of climate change on plant/herbivore interactions in moist tropical forests. Climatic Change 39:455-472.

Dhaliwal, G.S., Arora, R., and Dhawan, A.K. 2004. Crop losses due to insect pests in Indian agriculture: An update. Indian Journal of Ecology 31: 1-7.

Dhaliwal, G.S., Jindal, V., and Dhawan, A.K. 2010. Insect pest problems and crop losses: Changing trends. Indian Journal of Ecology 37: 1-7.

Dhillon, M.K. and Sharma, H.C. 2007. Effect of storage temperature and duration on viability of eggs of Helicoverpa armigera (*Lepidoptera: Noctuidae*). Bulletin of Entomological Research 97: 55-59.

Dhillon, M.K. and Sharma, H.C. 2008. Temperature and Helicoverpa armigera food influence survival and development of the ichneumonid parasitoid, Campoletis chlorideae. Indian Journal of Plant Protection 36: 240-244.

Dhillon, M.K. and Sharma, H.C. 2009. Temperature influences the performance and effectiveness of field and laboratory strains of the ichneumonid parasitoid, *Campoletis chlorideae*. BioControl 54: 743-750.

Isman, M.B. 1997. Neem and other botanical pesticides: Barriers to commercialization. Phytoparasitica 25: 339-344.

S.K. 1998. Expression and segregation of genes encoding Cry1A insecticidal proteins in cotton. Crop Science 38: 1-11.

Sharma, H.C. 2002. Host plant resistance to insects: Principles and practices. Pages 37-63 in Resources Management in Plant Protection (Sarath Babu, B., Varaprasad, K.S., Anitha, K., Prasada Rao, R.D.V.J., and Chandurkar, P.S., eds.). Volume 1. Rajendarnagar, Hyderabad, Andhra Pradesh, India: Plant Protection Association of India.

Sharma, H.C. (ed.). 2005. Heliothis/Helicoverpa Management: Emerging Trends and Strategies for Future Research. New Delhi, India: Oxford & IBH, and Science Publishers, USA. 469 pp.

Sharma, H.C., Dhillon, M.K., Kibuka, J., and Mukuru, S.Z. 2005. Plant defense responses to sorghum spotted stem borer, Chilo partellus under irrigated and drought conditions. International Sorghum and Millets Newsletter 46: 49-52.

Sharma, H.C., Srivastava, C.P., Durairaj, C., and Gowda, C.L.L. 2010. Pest management in grain legumes and climate change. In: Climate Change and Management of Cool Season Grain Legume Crops (Yadav, S.S., McNeil, D.L., Redden, R. and Patil, S.A., eds.). Dordrecht, The Netherlands: Springer Science + Business Media. pp.115-140.

Sharma, H.C., Sullivan, D.J., and Bhatnagar, V.S. 2002. Population dynamics of the Oriental armyworm, Mythimna separata (Walker) (*Lepidoptera: Noctuidae*) in South-Central India. Crop Protection 21:721-732.

❑❑❑

11

Efficacy of *Beauveria bassian* (Balsamo) Vuillemin - A Potential Entomopathogen against Rice Hispa (*Dicladispa armigera* Olivier)

K.C. Puzari, Pranab Dutta, L.K. Hazarika and P. Das

Abstract

As such in search of alternative to chemical insecticides against insect pest of rice resulted in identifying white muscardine fungus, Beauveria bassiana an entomopathogen out of several entomogenous fungi prevalent in rice ecosystem as the potential mycoinsecticides. The fungus has 78-87% virulence against rice hispa with LC_{50} *value of 90.16. Under field situations it is required that the effectiveness of the bioagents are enhanced by combining them with compatible pesticides. A sublethal dose of a pesticide would make the insect physiologically weak which makes it much more susceptible to the attack of the entomogenous organisms. In view of this, compatibility test was conducted with 14 different commonly applied insecticides at two different doses i.e., recommended and half of the recommended dose. In small scale field trial chloropyriphos at half of the recommended dose along with B. bassiana was the best treatment in controlling the pest along with increased yield of the crop consecutively for two years. Similarly, in multilocational field demonstration trial conducted in hispa endemic areas of Assam viz., Sibsagar and Jorhat covering six different localities B. bassiana (@ 10 million spore/ ml dilution) and half of the recommended dose of chlorpyriphos proved superiority then the commonly applied insecticides in decreasing the pest population with increased yield of crop.*

***Keywords**: Beauveria bassiana*, Compatibility, Field efficacy, Rice hispa.

Introduction

Rice hispa, *Dicladispa armigera* (Olivier) (Coleoptera : Chrysomelidae) is one of the major pests of rice in Assam causing 35-65% loss in yield by attacking the vegetative stage of rice (*Oryza sativa* L.) (Puzari and Hazarika, 1992). Under post flood situation, the late transplanted rice crop suffers 100% loss because of its attack (Hazarika, 1999). The farmers heavily relied on use of insecticides for its control. Due to heavy rain and its occurrence in large areas concurrently in an epizootic condition, insecticides failed to suppress the pest, even after large scale and indiscriminate application of chemical insecticides. This invited several unwanted social and environmental problems (Hazarika and Puzari, 1997; Hazarika, 1999). The pest management system in present day context demand careful and compatible integration of chemical, biological and cultural techniques in order not only to reduce environmental and ecological hazards of pesticides but to increase economical benefits. Methods and technologies have been generated for the mass production and field application of *B. bassiana* (Samsinkava, *et al.*, 1981; Rombach *et al.*, 1988a; Mazumdar *et al.,* 1994; Puzari *et al.,* 1997) .*It* plays a significant role in rice pest management with special reference to Rice hispa became of its capacity to infect the developmental stages of hosts with varying intensity (Ferron, 1978; Fuxa, 1987; Agarwal, 1990; Hazarika and Puzari, 1995).

Biocontrol agents have been studied inadequately as a component of integrated pest managemenmt system. Our earlier work (Puzari *et al.*, 2005) confirmed the compatibility of *B. bassiana* with half of the commercial concentration of some chemical *viz*., Alphamethrin, Cypermethrin, Deltamethrin, Quinalphos, Endosulfan, Chloropyriphos. Combined application of entomogenous fungus with compatible chemical result in synergism between the two, an approach considered promising but little studied. With these in view, small scale and pilot field trials were carried out to study the efficacy of compatible insecticide against Rice hispa which could in future be recommended as a possible combination in IPM programmes for insect pests.

Materials and Methods

Fungus: *Beauveria bassiana* was originally isolated from the infected dead cadavers collected from rice hispa endemic rice field on potato dextrose agar medium (Puzari *et. al*., 1994). The infectivity in terms of corrected mortality percentage of the isolate calculated by Abbotts formula (Abbott, 1925), 1994 was recorded 90.16% in laboratory inoculation and is superior in performance in the field to that of the recommended pesticides (Puzari and Hazarika).

Mass culture of *B. bassiana*: To harvest large quantity of propagules , the isolate was mass cultured in polypropylene bags containing sterilized (autoclaved at 121°C fro 20 min. repeated for 2 successive days) rice hull : saw dust: rice bran medium (25 : 25 : 100) supplemented with 2 % dextrose (Mazumdar *et al*., 1995; Puzari *et* *al*., 1997). Polypropylene bags each having capacity of 150 g containing 50 g medium was inoculated with 1ml of 1 X 10^7 conidia /ml suspension and incubated at 24 ± 1^0 C for 20 days. To obtain uniform growth of the fungus, the bags were mechanically shaked thoroughly after 5 days. After 21 days of incubation the bags were kept in racks at room temperature and ready for application.

Preparation of spore/ conidial suspension of *B. bassiana*: The conidial density as obtained from the mass cultured medium (RH:SD:RB) was 39.33 x 10^7 conidia/ ml of water. A conidial concentration 1 x 10^7 conidia/ml of water was used spraying by suspending 1000g of mass cultured substrate with homogenous fungal growth in 5000 ml of water. Prior to spray the mixture was filtered through single layered muslin cloth to prevent the nozzle block.

Pot tests: To determine the effective dose of compatible insecticides with *B. bassiana* six chemicals *viz.* Deltamethrin (Decis 28EC) @ 0.001%, Alphamethrin (Farsa 10 EC) @ 0.004%, Cypermethrin (Kripcord 10 EC) @ 0.003%, Endosulfan (Thiodane 35 EC) @ 0.004% Chloropyriphos (Dursban 20 EC) @ 0.125%, and Quinalphos (Ekalux 25 EC) @ 0.025% were selected on the basis of the earlier experiments (Puzari *et al*., 2006).

The experiment was conducted in laboratory condition by growing four seedlings of 30 days old (cv. Ranjit) per plastic pots (500 ml Beaker) containing soil mixed with manure (2.5g farm yard manure) and fertilizer(0.11, 0.15, 0.04g Urea, Single Super Phosphate and Muriate of Potash respectively). Twenty laboratory reared adults 48 hr. after eclosion (starved for 6 hrs.) were released into each pot and caged with paired lantern chimney, top of which was covered with muslin cloth. The pots were inoculated with fungal propagules (1 x 10^7 conidia/ml) suspended over the surfactant Tween 80 (@ 0.023%) and mixed with the respective chemicals at the required concentration. Twenty ml of the suspension was sprayed on each treatment covering three replications (6.5 ml/hill with atomizer over leaf surface). Equal amount of inoculum mixed with Tween 80 was sprayed to test the efficacy of *B. bassiana* alone. Chloropyriphos @0.25% mixed with water and Tween 80 was sprayed for comparison in the respective pots. Controls were sprayed with water and Tween 80 solution

The experiment was conducted in the laboratory at room temperature (24±1°C). Mortality records were taken daily. Dead insects were allowed to remain in cages during the time of replacement of new seedlings. Mortality (%) were adjusted with Abbott's formula and subjected to analysis of variance. For convenience in comparison the cumulative mortality data obtained on days 10 of inoculation were analyzed.

Small-scale field test: The experiment on small-scale field trial was conducted for two years in two seasons *viz.,sali* and *ahu* season in Instruction Cum Research (ICR) farm of AAU,Jorhat, Assam in an area of 300 and 330 sq m plot respectively. The plots were subdivided into 24 small plots of 3 × 3 m^2.

Preparation of seedbed and main field for *Sali rice*: Well germinated and fungicide treated (Captan @ 0.2%) seeds @ 850 g/bed were sown in puddled seed bed (10 m length x 1.25 m breadth). For each bed Cowdung 30 kg, urea 80 g, Single super phosphate 80 g and Muriate of potash 40 g were applied during the time of puddling. Water level upto saturation condition (100% pore space was filled) was maintained through irrigation.

Well rotten farmyard manure @ 10 tonnes/ha was applied during the field preparation. In addition Urea @ 88 kg/ha, single super phosphate @ 125 kg/ha and Muriate of potash @ 33 kg/ha were also applied. Half of Urea and the entire quantity of Super phosphate and Muriate of potash were applied at the time of final puddling. The remaining part of Urea was applied at the panicle initiation stage.

One month old seedlings were transplanted in the main field (@ 2 to 3 seedlings per hill) at a distance of 4 to 5 cm. Standing water 5 ± 2 cm in the field was maintained from 2 – 3 days after transplanting up to 8 – 10 days before harvest. Two weeding one at 20 and another at 40 days after transplanting was done with paddy weeder.Treatment combination were same as mentioned else where except Endosulfan because of its toxic affect on aquatic animal.

Field Release of insect: In the experiment, considering the importance of rice hispa in this region of India (Hazarika and Dutta, 1991) special emphasis was given on this insect in addition to other insect pest of rice ecosystem. Since, during the year of experimentation the rice hispa was not observed in the experimental plot probably because of its endemic nature in certain major rice growing areas of Assam (Hazarika and Dutta, 1991) the insects were released on the plots (treatment wise) under mosquito prevent nylon net at economic threshold level (1 to 2 adult/hill) (Dhaliwal and Arora, 1998) after one month of transplanting @ 300 adults/plot. The population of rice hispa at both embryonic and post embryonic stages and the population of other insects at pre- and post-treatments were presented separately.

Observation on the average count of eggs, larvae, pupa and adults of rice hispa were recorded on different treatments initially at 7 days after application up to 1 month at an interval of every 48 hrs. in considering the period of life cycle of rice hispa for 22 to 25 days (Mishra, 1997). In case of larvae and adults, 15 hills were selected at random and their total populations were counted early in the morning when adults did not readily fly away. For eggs, five hills were selected and total populations were counted from their individual leaves. Average population was expressed as adults/hill, larvae/hill, pupae/hill and eggs/leaf. Data from each of the treatments were subjected to analysis of variance and compared between treatments based on critical difference at 0.05 probability level.

Field Application of *Beauveria bassiana* alone and in combination with other chemicals: The prepared conidial suspension of *B. bassiana* sprayed alone and in combination with sublethal doses of chemicals *viz.* Alphamethrin, Cypermethrin, Deltamethrin, Quinolphos and Chloropyriphos. In all the treatments Tween 80 (polyethylene sorbitan monolourate) was added @ 0.23 ml/litre (Puzari and Hazarika, 1994). Water mixed with Tween 80 at the same rate served as control. Chloropyriphos alone at lethal dose was applied for comparison with treatments.

The treatments were applied after 24 hrs. field release of rice hispa through high-volume knapsack sprayer @ 600 litres/ha and drifting were prevented by erecting aluminium sheets during spray operation.

Preparation of field for *Ahu rice* and application of manure and fertilizer: Five ploughing followed by laddering was done for soil preparation. Laddering was done properly to retain water uniformly in the field. Well rotten FYM 10 t/ha or 15 q/bigha was applied during the initial field preparation. Fertilizer like Urea @ 88 kg/ha, Single Super Phosphate @ 125 kg/ha and Muriate of Potash @32 kg/ha were also applied.

The full dose of phosphate fertilizer was applied at the time of final ploughing. Half of nitrogenous fertilizer and half dose of potassic fertilizer were applied at 20 days after seed germination or after first weeding. The second application with the remaining nitrogenous and Potassic fertilizer was given 45 days after seed germination or after the second weeding.

Rice seed (var. Luit) having a germination of >80% was selected and treated with Dithane M-45 (@ 2.5 g/kg of seed) for sowing. Seeds were sown in line @ 75 kg/ha. Treatment combinations were same as the study in *in vitro conditions.*

Pilot field trial: The best compatible treatment combination (sub lethal dose of Chloropyriphos @ 0.125% in combination with *B. bassiana* @ 1 x 10^7 conidia/ml of water) in terms of reduction of insect populations in embryonic and post embryonic stages of rice hispa and yield were selected for pilot field trials in two season viz., *Sali* and *Ahu* season.In both the seasons rice hispa endemic area were selected. For *Sali* season Konwarpur and Dikhumukh of Sibsagar district and Badulipar of Jorhat District were selected similarly, Banmukh, Lepaipather and Deesangmukh of Sibsagar district of Assam were selected for *Ahu* season. This treatments were selected on the basis of the findings obtained from the previous experiments under small-scale field trials, conducted in two winter crops (*sali*) and two summer crops (*ahu*) in the Instruction cum research farm, Assam Agricultural University, Jorhat-13, Assam. Comparison of the treatment was made with untreated control.

***Sali rice* / winter rice: Field preparation and method of application:** The methods of field preparation, seed sowing, seedling transplantation in the main field was same as followed for *sali* crop as mentioned earlier. The methods of media preparation, preparation of spore suspension field application of the suspension of *B. bassiana* and spore suspension of *B. bassiana* in combination with Chloropyriphos were also same as mentioned elsewhere. Suspension of *B. bassiana* obtained from mass cultured medium(RH: SD: RB) was prepared in presence of farmers of the localities (20 farmers in each of the three batches) with full explanation regarding its method of preparation, efficacy, eco-friendliness etc. They were also shown the also shown the method of mixing the chemical with the biopesticides before its field application.

***Ahu rice* / summer rice:** The method of field preparation, seed sowing, seedling transplanting in the main field was same as mentioned earlier. The method of media preparation, preparation of spore suspension, field application of the suspension of *B. bassiana* in combination with chloropyriphos is also same as mentioned elsewhere. The methodology for the preparation of the suspension was demonstrated to the farmers before its application.

Observation recorded: Observation of pretreatment and post treatment count of egg/leaf, larva/leaf, pupa/leaf and adult/hill was recorded. Pretreatment count was made one day before application of the treatment and post treatment count was made fifteen days after the treatments for two times at an interval of 15 days considering the life cycle of rice hispa for 22-25 days (Mishra, 1997) as well as for other insect pests present in the experimental field. Post treatment count on embryonic and post-embryonic stages was recorded as per the method described else where. The average of two post treatment count of different embryonic and post embryonic stages of rice hispa was used to calculate the percent decrease over the pretreatment count.

Results and Discussion

Pot tests: Table 1 shows the efficacy of compatible combination of six insecticides together with *B. bassiana* against rice hispa (in terms of mortality percentage). Chloropyriphos at lethal dose cause highest mortality (89.60%) of rice hispa followed by Chloropyriphos at sublethal dose + *B. bassiana* (86.27%) without having any significant difference among them. Ribban *et al.,* (1983) advocated that half of the commercial dose (i.e., sublethal dose of chloropyriphos in combination with *B. bassiana* to *Ostriana rubilais*(Hbn) .The mortality percentage recorded in Quinalphos at sublethal dose + *B. bassiana* (80.38%) was next to the above two treatments. This was followed by sublethal dose of Endosulphan in combination with *B. bassiana* (74.51%) and *B. bassiana* alone (70.58%) without any significant difference. Similarly earlier study on combine effect of chloronocotinyl insecticide imidaclorpid and *B.* bassiana exert significant synergestic effects on mortality and mycosis of Colorado potato beetle (*Leptonitarsa decemlineata)* (Eldert *et al.,* 1991; Boucias *et al.*, 1996). Comparatively a lower percent of mortality 52.94%, 54.90% and 56.86% was recorded respectively in Alphamethrin, Deltamethrin and Cypermethrin treated pot without showing any significant difference between them. The insecticides used in the present study may induce the changes in insect behaviour which may be a potential mechanism for enhanced infection.

Small scale field trail: When mean values of two years experiments (both for *sali* and a*hu* rice) were compared a significant difference among all then treatments with almost a similar treatment of efficacy was observed in reducing the embryonic and postembryonic stages of rice hispa (Table 2 & 3). Among all the treatments, although chloropyriphos at lethal dose (0.025%) performed best, sublethal dose of chloropyriphos (0.125%)along with *B. bassiana* (1x 10^{7} conidia/ml of water) was found to next to it in terms of decreasing the post embryonic stages of rice hispa without any significant difference between them. This was followed by *B. bassiana* alone and application of sublethal dose of Alphamethrin (0.004%) along with *B. bassiana* (Table 2 & 3). Effect of *B. bassiana* alone and in combination with chemical observed in the present study is in agreement with the earlier report of Hazarika and Puzari(1995) who showed *B. bassina* to be pathogenic to all the stages of rice hispa, including egg. Furthermore, Storch and Dill (1987) observed that larval population of *Leptinotarsa decemlineata*Say was at par in the potato(*Solanum tuberosum*L.*)*plots treated with insecticides and *B. bassiana*. Highest population count of egg/leaf, larva/hill and pupa/hill and infestation was recorded in the control plot where rice hispa were released alone without any biological and /or chemical treatment. The observations recorded in Cypermethrin and Deltamethrin at sublethal dose together with *B. bassiana* treated plot in terms of reducing the developmental stages of rice hispa was statistically *at par* with each other in both the crop seasons of the crop. The plot where Quinalphos was applied at sublethal dose along with *B. bassiana* was found least effective in the management of the pest.

The yield data of both the seasons for two years were compared and presented in Table 1 & Table 2. Highest mean yield of 56.62 q/ha and 22.26 q/ha in *sali* and *ahu* season respectively was recorded in Chloropyriphos (at lethal dose) treated plot. This was followed by Chloropyriphos (sublethal dose) + *B. bassiana* in *sali* (55.09 q/ha) and *ahu* (20.04% q/ha) seasons. A significant increase in yield was also observed in the treatment Quinalphos (0.025%) + *B. bassiana* (53.44 q/ha) followed by sublethal application of Deltamethrin + *B. bassiana* (51.30 q/ha) and *B. bassiana* alone (51.20 q/ha) in *sali* seasons. While in *ahu* seasons this trend of increase in yield was observed in case of the sublethal application of Alphamethrin (16.98 q/ha), followed by Deltamethrin (14.27 q/ha), Endosulfan (13.78 q/ha) and *B. bassiana* alone (13.28 q/ha). Lowest yield 45.73 q/ha and 8.98 q/ha were recorded in *sali* and *ahu* seasons respectively in the control plot where rice hispa was released alone without any chemical and/or biological treatment. Higher yield record observed in *B. bassiana* treated plot was similar with the earlier report of Hazarika and Puzari (2004) who reported higher yield of rice at various rice hispa endemic localities of Assam. They also reported that yield of rice was significantly higher in plots treated with *B. bassiana* than those treated with either Monocrotophos or neem seed oil. Similarly, Maniania(1993) also observed an increase in yield of maize (*Zea mays* L.) in the field treated with *B. bassiana*. *B. bassiana* could establish in the field and prevent the pest population to build up.

Bartlett and Lefevre (1934) and Hoi *et al.*,(1994)showed such a capability of *B. bassiana* in controlling the European corn borer (*Ostrinia nubilis*Hbn) and the red imported fire ant (*Solenopsis invicta* Buren) in maize field and Pasteur respectively.

Pilot field trial: Combine application of Chloropyriphos (at sub lethal dose) + *B. bassiana* showed a sharp decrease of population of all the embryonic and post embryonic stages of rice hispa in both *Sali* and *Ahu* season (Table 4 and 5) in all the different localities. A significant increase in the population of egg/leaf, larva/leaf, pupa/leaf and adult/hill was observed in the control plot of all the localities. In *Sali* season. Highest percentage of the decrease in the population of egg/leaf, larva/leaf and adult/hill was observed in Konwarpur and highest reduction of the pupa/leaf was recorded in Badulipar whereas in *Ahu* season Highest percentage of the decrease in population of larva/leaf and adult/hill was observed in Banmukh while egg/leaf and pupa/leaf were recorded in Deesangmukh. Best result found combine application of chloropyriphos(at sublethal dose) + *B. bassiana* might be due to the fact that sub lethal dose of a pesticides would make the insect physiologicallyweak up to adesired degree which made it much more susceptible to the attack of entomgens as reported by Fedorinehik,(1974). Earlier Puzari and Hazarika, 1991, observed *B. bassiana* (1X10^{7}conidia /ml) in combination with 0.023% aqueous solution of Sandovit, Triton-AE, Teepol 80 and Hamam toilet soap caused 81%, 74%, 81%, 96% and 92% mortality to the adult rice hispa respectively.

Table 1: Efficacy of compatible combination of insecticides and *Beauveria bassiana* against rice hispa (*in vitro*)

Treatment	Mortality	(%)	Corrected mortality (%)
T_1	Control	15.00	-
T_2	*B. bassiana* (@ x 10^7 conidia/ml)	75.00	70.58 (57.15)
T_3	Chloropyriphos (L: 0.25%)	91.16	89.60 (71.19)
T_4	Alphamethrin (SL: 0.004%) + T_2	60.00	52.94 (46.69)
T_5	Cypermethrin (SL: 0.003%) + T_2	63.33	56.86 (48.94)
T_6	Deltamethrin (SL: 0.001%) + T_2	61.66	54.90 (47.81)
T_7	Quinalphos (SL: 0.025%) + T_2	83.33	80.38 (63.71)
T_8	Endosulfan (SL: 0.175%) + T_2	78.33	74.51 (59.68)
T_9	Chloropyriphos (SL: 0.125%) + T_2	88.33	86.27 (68.25)
	S.Ed (±)		2.094
	CD (0.05)		4.400

Data are mean of three (3) replications
Data in the parentheses are angular transformed value

Yield data (q/ha) of rice recorded at the various localities shows that the Chloropyriphos at sub lethal dose + *B. bassiana* was significantly higher than with those of untreated control (Table 4). Highest yield (58.12 q/ha and 59.25 q/

Table2: Efficasy of *Beauveria bassiana* alone and in combination with different insecticides against rice hispa(*Sali* rice)

Treatment	Egg/leaf			Larva/hill			Pupa/hill			Adult/hill			Yield(q/ha)		
	I	II	Mean*	I	II	Mean*	I	II	Mean*	I	II	Mean*	I	II	Mean*
T1 Control	3.45	5.00	4.22	3.54	2.79	3.16	1.48	1.19	1.33	8.74	4.83	6.78	46.67	44.79	45.73
T2 B.bassiana	3.02	2.47	2.74	2.39	2.89	2.64	1.78	1.29	1.53	0.20	2.07	1.13	52.08	50.33	51.20
T3 Chloropyriphos(L)	1.28	1.33	1.30	0.83	0.91	0.87	0.09	1.10	0.53	1.00	0.67	0.83	58.33	54.92	56.62
T4 Alphamethrin(SL)+T2	2.89	3.67	3.28	1.90	2.44	2.17	0.37	2.20	1.28	1.00	1.93	1.46	54.17	52.77	53.47
T5 Cypermethrin(SL)+T2	4.08	5.47	4.77	4.30	3.10	3.70	3.17	1.27	2.22	13.80	5.93	9.86	49.75	46.57	48.16
T6 Deltramethrin(SL)+T2	1.83	6.53	4.18	2.73	3.61	3.17	1.28	1.33	1.30	1.30	2.80	2.05	52.50	50.05	51.30
T7 Quinalphos(SL)+T2	3.56	4.40	3.98	2.11	2.53	2.32	2.11	1.06	1.58	0.48	4.40	2.44	54.17	52.71	53.44
T8Chloropyriphos(SL)+T2	2.16	2.00	2.08	2.08	3.21	2.64	0.23	0.93	0.58	1.00	1.53	1.26	57.08	53.01	55.09
S.Ed.($\pm$)	0.083	0.348	0.568	0.102	0.166	0.134	0.915	0.265	0.296	0.134	0.126	0.140	0.471	0.531	0.488
CD (0.05)	0.165	0.688	1.163	0.205	0.33	0.274	0.183	0.515	0.607	0.268	0.252	0.286	1.760	1.984	0.998

*I = mean of 1st year II = mean of 2nd year * Data are mean of two years experiment L: Lethal dose, SL: Sub lethal dose*

Table 3: Insect pest populations other than rice hispa prevalent in the experimental plot of rice (Sali rice)

Treatment	Percent decrease over pretreatment count														
	GH			LF			BPH			YSB			Coccinellid		
	I	II	Mean*	I	II	Mean*	I	II	Mean*	I	II	Mean*	I	II	Mean*
T1 Control	0.00	0.00	0.00	0.00	0.0	0.00	0.00	0.00	0.00	0.00	0.00	0.00	0.00	0.00	0.00
T2 B.bassiana	66.67	55.00	60.84	75.19	50.38	62.79	67.00	60.48	63.74	50	50.75	50.38	0.00	49.62	24.81
T3 Chloropyriphos(L)	0.00	22.85	11.43	100	100	100	100	100	100	100	100	100	60.24	55.79	58.01
T4 Alphamethrin(SL)+T2	33.55	30.03	31.79	39.76	40.12	39.94	50	50	50	50	50	50	0.00	22.33	11.16
T5 Cypermethrin(SL)+T2	54.77	54.50	54.64	24.81	48.50	36.66	100	75	87.5	100	42.42	21.21	60.24	75.29	67.76
T6 Deltramethrin(SL)+T2	33.33	33.33	33.33	0.00	20.36	10.18	24.81	19.88	22.35	24.81	24.81	24.81	0.00	40.00	20.00
T7 Quinalphos(SL)+T2	0.00	0.00	0.00	34.00	40.12	37.06	34.00	11.00	22.50	0.00	0.00	0.00	0.00	0.00	0.00
T8 Chloropyriphos(SL)+T2	80.18	44.33	62.26	100	100	100	100	100	100	100	100	100	100	66.67	83.38
S.Ed.(+)			1.246			3.103			2.064			4.529			2.970
CD (0.05)			2.342			6.356			4.227			9.276			6.082

\I= mean of 1st year II= mean of 2nd year Data are mean of two years experiment*

Gh: Grass hopper; LF: Leaf folder; BPH: brown plant hopper; YSB: Yellow stem borer

Table 4: Efficasy of *Beauveria bassiana* alone and in combination with different insecticides against rice hispa (ahu Rice)

Treatment	Egg/leaf			Larva/hill			Pupa/hill			Adult/hill			Yield(q/ha)		
	I	II	Mean*	I	II	Mean*	I	II	Mean*	I	II	Mean*	I	II	Mean*
T1 Control	38.00	36.25	37.13	3.75	3.00	3.38	1.93	1.34	1.64	2.07	1.99	2.03	9.68	8.28	8.98
T2 B.bassiana	13.00	11.11	12.06	2.56	2.16	2.36	1.47	1.27	1.36	1.62	1.76	1.69	13.64	12.92	13.28
T3 Chloropyriphos (L)	8.25	6.22	7.24	1.69	1.68	1.69	0.36	0.38	0.37	0.94	0.98	0.96	21.86	22.66	22.26
T4 Alphamethrin(SL)+T2	9.67	8.25	8.96	2.40	2.34	2.37	1.38	1.18	1.28	1.47	1.58	1.53	16.91	17.05	16.98
T5 Cypermethrin(SL)+T2	11.42	9.22	10.32	2.86	2.44	2.65	1.62	1.26	1.44	1.75	1.78	1.77	11.48	12.00	11.74
T6 Deltramethrin(SL)+T2	10.66	10.28	10.47	2.16	2.16	2.16	1.42	1.32	1.37	1.56	1.41	1.49	14.68	13.86	14.27
T7 Quinalphos(SL)+T2	21.25	19.00	20.13	3.08	3.66	3.37	1.69	1.66	1.68	1.89	1.59	1.74	10.55	10.60	10.58
T8 Endosulfan(SL)+T2	11.17	10.66	10.92	2.48	2.08	2.28	1.44	1.39	1.42	1.60	1.58	3.18	13.55	14.00	13.78
T9 Chloropyriphos(SL)+T2	9.83	8.24	9.04	1.80	1.90	1.85	0.49	0.55	0.52	1.13	1.10	1.12	19.86	20.22	20.04
S.Ed.(+)	1.721	1.100	1.084	0.396	0.19	0.905	0.337	0.312	0.683	0.378	0.315	0.466	1.57	1.820	1.125
CD(o.o5)	3.532	2.250	2.220	0.832	0.390	1.853	0.708	0.655	1.399	0.776	0.646	0.954	4.39	4.550	2.303

I = mean of 1st yea, II = mean of 2nd year, L: Lethal dose, SL: Sub lethal dose

*Data are mean of two years experiment

Table 5: Insect pest populations other than rice hispa prevalent in the experimental plot of rice (*ahu* rice)
TreatmentPercent decrease over pretreatment count

	GH			LF			BPH			YSB		
	I	II	Mean*	I	II	Mean*	I	II	Mean*	I	II	Mean*
T1 Control	0.00	0.00	0.00	0.00	0.00	0.00	0.00	0.00	0.00	0.00	0.00	0.00
T2 B.bassiana	28.33	28.33	28.33	36.97	33.50	35.23	37.50	23.37	30.44	37.50	37.50	37.50
T3 Chloropyriphos(L)	39.93	37.43	38.68	98.12	97.00	97.56	95.80	61.30	78.55	59.88	96.11	77.99
T4 Alphamethrin(SL)+T2	12.73	9.33	11.03	16.67	22.18	19.43	16.50	17.49	16.99	14.16	16.50	15.33
T5 Cypermethrin(SL)+T2	22.63	18.60	20.61	52.57	22.82	37.70	23.04	20.85	21.94	14.16	31.33	22.74
T6 Deltramethrin(SL)+T2	12.50	0.00	6.25	36.51	22.82	29.66	10.38	19.09	14.73	16.50	13.63	15.06
T7 Quinalphos(SL)+T2	0.00	0.00	0.00	16.88	11.33	14.11	13.05	16.88	14.96	16.88	21.67	19.27
T8 Endosulfan(SL)+T2	22.00	24.56	23.28	52.17	49.76	50.97	22.00	25.00	23.50	13.16	52.61	32.88
T9 Chloropyriphos(SL)+T2	37.83	29.33	33.58	62.55	58.05	60.30	71.24	61.10	66.17	66.50	85.84	76.17
S.Ed.(+)			1.443			1.886			2.636			2.252
CD(o.o5)			2.955			3.863			5.398			4.612

*I= mean of 1st year II= mean of 2nd year * Data are mean of two years experiment*
Gh: Grass hopper; LF: Leaf folder; BPH: brown plant hopper; YSB:Yellow stem borer

ha) was recorded in the treated plot of Badulipar, Jorhat district and Banmukh, Sibsagar district respectively in *Sali* and *Ahu* season. Yield recorded in *Sali* season in the treated plot of Dikhowmukh and Konwarpur was 56.66 q/ha and 54.83 q/ha respectively. Whereas in *Ahu* sasson it was 56.00 q/ha and 50.05 q/ha in Deesangmukh and Lepaipathar, respectively. Higher yield recorded in chloropyriphos (sub lethal dose) + *B. bassiana* treated field might bre enhanced action of *B. basiiana* in presence chemical thereby causing a quick death of the host resulting in reduced damage of the plant surface. Ribban *et al.* earlier reported such observation, (1983) also observed enhanced susceptibility of *B. bassiana* to *Ostrinia nubilalis*(Hbn) by sublethal dose of chloropyriphos.

References

Abbot, W.S. 1925. A method of computing the effectiveness of an insecticide. Journal of Economic Entomology 18: 265-267

Agarwal, G.P. 1990. Entomogenous fungi in India and management of insect pests. Indian Phytopathology 43:131-142.

Bartlett, K. A. and Lefebvre, C. L. 1934. Field experiments with *Beauveria bassiana* (Bals.) Vuill., a fungus attacking the European corn borer. Journal of Economic Entomology, 27:147-157.

Boucias, D.G., Stokes, C., Stroey,G and Pandland, J.C. 1996. The effect of imidacloprid on the termite *Reticulitermes flavipes* and its interaction with the mycopathogen *Beauveria bassiana. Pflanzenschutz-Nachr.* Bayer 49:103-144.

Dhaliwal, G.S. and Arora, R. 1998. Principles of Insect Pest Management (Ed.) Dhaliwal, G.S. and Arora, R. 1998. Kalyani Publ. New Delhi, pp. 297.

Elbert, A., Becker, B., Hartwig,J and Erdelen, C. 1991. Imidacloprid- a new systemic insecticide. Pflanzenschutz-Nachr. Bayer 44:113-136.

Fedorinehik, N.S. 1974. In: *Biological control, method for plant protection* (ed) E.H. Chunsakova, G.V. Gusev, N.S. Fedorinehik. Pp. 163-307, Moscow; Kolos (in Russian).

Ferron, P. 1978. Biological control of insect pests by entomogenous fungi. Annual Review of Entomology 32:409-442.

Fuxa, J. R. 1987. Ecological considerations for the use of entomopathogens in IPM. Annual Review of Entomology 32:225-251.

Hazarika, L. K. 1999. Rice hispa and Strategies for its management in Assam. A paper presented in the Interface between the SDA, Assam and AAU, Jorhat at Guwahati, Assam on 14th October.

Hazarika, L.K. and Dutta, B.C. 1991. Reaction of rice cultivars to rice hispa. International Rice Research Newsletter. 16:14-15.

Hazarika, L.K. and Puzari, K.C. 1995. White muscardine fungus (*Beauveria bassiana*) pathogenic to different stages of rice hispa (*Dicladispa armigera*). Indian Journal of Agricultural Science 65:63-66.

Hazarika, L. K. and Puzari, K. C. 1997. Field efficacy of white muscardine fungus (*Beauveria bassiana*), Indian Journal of Agricultural Science 67:463-465.

Hazarika, L.K. and Puzari, K.C. 2004. Entomopathogenic fungi in rice pest management. *In*: R. J. Rabindra., Hussain, S.S. and Ramanujam, B. (Eds). *Proceedings of ICAR-CABI workshop on Biopesticide formulation and application.* 9-13 December, 2002, PDBC, Bangalore, pp. 139-149.

Hoi, H. D., Pereira, R. M., Stimac, J. L. and Wood, L. A. 1994. Field application of *Beauveria bassiana* for control of the red imported fire ant (Hymenoptera : Formocidae), Journal of Economic Entomology 87: 623-630.

Majumder D, Puzari K C and Hazarika L K 1995. Mass culture of *Beauveria bassiana* on different substrates, Indian Phytopathology 48:275-278.

Majumder, D., Puzari, K. C. and Hazarika, L. K. 1994. Mass culture of *Beauveria bassiana* on rice hull. International Rice Research Newsletter 19:18-19.

Maniania, N. K. 1993. Evaluation of three formulation of *Beauveria bassiana* (Bals.) Vuill. for control of the stem borer *Chilo partellus* (*Swinhoe*) (*Lepidoptera, Pyralidae*), Journal of Applied Entomology 115:266-272.

Mishra, M.S. 1997. Insects pest of crops – Principles and Practices. In: Handbook of Agricultural Science (ed.) S.S. Singh, P. Gupta and A. K. Gupta, 1997. Kalyani Publisher. New Delhi, pp. 325-401.

Puzari, K.C. and Hazarika, L.K. 1991. Efficacy of *Beauveria bassiana* combined with various stickers or spreaders against rice hispa. International Rice Research Newsletter 16, 21.

Puzari, K.C. and Hazarika, L. K.1992. Entomogenous fungi from North-East India, Indian Phytopathology, 45: 35-38.

Puzari, K.C., Hazarika, L. K. and Deka, N.1994. Pathogenicity of *Beauveria bassiana* on rice hispa, Indain Journal of Agricultural Science 64, 47-49.

Puzari, K.C., Hazarika, L.K. Dutta, Pranab and Das, Purnima (2006) In vitro inhibition of *Beauveria bassiana* (Bals.) Vuill. Growth by different commonly used insecticides in rice. Journal of Biological Control 20:51-55.

Puzari, K. ., Sarmah, D. and Hazarika, L. K. 1997. Medium for mass production of *Beauveria bassiana* (Balsam) Vuillemin., Journal of Biological Control 11: 97-100.

Puzari, K. C. and Hazarika, L.K. 1991. Efficacy of *Beauveria bassiana* combined with stickers or spreaders against rice hispa (RH). International Rice Research Newsletter 16: 21.

Puzari, KC, Hazarika, L.K. 1992. Entomogenous fungi from North East India. Indian Phytopathology 45:35-38.

Ribban, G., Marcandier, S., Richard, G and Larget, I. 1983. Susceptibility of the maize pyralid (*Ostriana rubilalis*) (*Lepidoptera : Pyealidae*) to entomopathogenic hyphomycetes. Entomophaga, 28: 55-64.

Rombach, M.C., Aguda, R. M. and Roberts, D.W. 1988. Production of *Beauveria bassiana* (*Deuteromycotina : Hyphomycetes*) in different media and subsequent condition of dry mycelium. Entomophaga 93: 315-324.

Samsinokova, A., Kalakova, S., Vlcek, V. and Kybal, J. 1981. Mass production of *Beauveria bassiana* for regulation of *Leptinotarsa decemlineata* populations, Journal of Invertebrate Pathology 38: 169-174.

Storch, R. H. and Dill, J. F. 1987. *Beauveria bassiana* for control of potato beetle (*Coleoptera : Chrysomelidae*) in Maine, Maine Agricultural Experimental Station Technical Bulletin 128, 8.

❑❑❑

12

Mulberry Diseases and Pest Control Measures

T.K. Chakma, B.C. Dabbarma, D. Saha and S. Chakraborty

Abstract

Mulberry is a fast growing deciduous woody perennial plant. It has a deep root system. the leaves are simple, alternative, stipulate, petiolate, entire or lobed. The number of lobes varies from one to five. Plants are generally dioecious. Inflorescence is catkin with pendent or drooping peduncle bearing unisexual flowers. Inflorescence is always auxiliary. Male catkins are usually longer than the female catkins. Male flowers are loosely arranged and after shedding the pollen, the inflorescence dries and falls off. These are four presistent parianth lobes and four stamens implexed in bud. Female Inflorescence is usually short and the flowers are very compactly arranged. There are four presistent parianth lobes. The ovary is one-celled and the stigma is bilid. The chief pollinating agent in mulberry is wind. Mulberry fruit is a sorosis, mainly violet black in colour. Most of the species of the genus Morus and cultivated varieties are diploid, with 28 chromosomes. However, triploids (2n=(3x)=42) are also extensively cultivated for thier for thier adaptability, vigorous growth and quality of leaves. The key pests of mulberry are Maconellicoccus hirsutus (mealy bug)- causing tukra in mulberry Diaphania pulverulentalis- Leaf roller Spilarcita obliqua-Bihar hairy caterpiller (sporadic pest) adaptability, vigorous growth and quality of leaves. Minor pests of mulberry are Thrips, jassids, scale insects, shorthorned grasshopper. Main disease of mulberry are Foliar. Leaf spot, leaf rust, powdery mildew, leaf blight and bacterial blight Soilborne. Root rot and root knot Nursery diseases. Stem canker, cutting rot, coller rot and dieback.

Keywords: Mulberry disease, Cultivative organism, Remedial measures.

Introduction

Morus, a genus of flowering plants in the family Moraceae, comprises 10-16 species of deciduous trees commmonly known as mulberreies growing wild and under cultivation in many temperate world regions.

The closely related *Broussonetia* is also commonly known as mulberry, notably the *Paper Mulberry, Broussonetia papyriera*. Mulberries are swift-growing when young, but soon become slow-growing and rarely exceed 10-15 m (33-49 ft) tall. The leaves are alternately arranged, simple, often lobed, more often lobed on juvenile shoots than on mature trees, and serrated on the margin. Depending on the species, they can be monoecious or dioecious. The mulberry fruit is a multiple fruit, 2-3 cm (0.79-1.2 in) long. Immature fruit are white, green, or pale yellow. In most species, the fruits turn pink then red while ripening, then dark purple or black and have a sweet flavor whenfully ripe. The fruits of the white-fruited cultivar are white when ripe; the fruit in this cultive. The ripe fruit is edible and is widely used in pies, tarts, wines, cordials and tea. The fruit of the black mulberry, native to southwest Asia, and the red mulberry, native to eastern North America, have the strongest flavor. The fruit of the white mulberry, an east Asian species which is extensively naturalized in urban regions of eastern North America, has a differnt flavor, sometimes characterized as insipid. The mature plant contains significant amounts of resveratrol, particularly in stem bark. The fruit and leaves are sold in various forms as nutritional suppliements. Unripe fruit and green parts of the plant havbe a white sap that may be toxic, stimulating, or mildly hallucinogenic. Black, red, and white mulberry are widespread in southern Europe, the Middle East, northen Aferica and Indian Subcontinent where the tree and the fruit have names under regional dialects. Jams and sherbets are often made from the fruit in this region. Balck mulberry was imported to Britain in the 17th century in the hope that it would be useful in the cultivation of silkworms. It was much used in folk medicine, especially in the treatment of ringworm. Mulberries are also widespread in Greece, particularly in the Peloponnese, which in the Middle Ages was known as Morea (Greek: Movpia, *Morias*), deriving from the Greek word for the tree (Greek: Movpia, *Mouria*).

Mulberries can be grown from seed, and this is opten advised as seedling-grown trees are generally of better shape ans health. But they are most often planted from large cuttings which root readily. The mulberry plants which are allowed to grow tall with a crown height of 5-6 feet from the ground level havinbg stem girth of 4-5 inches or more is called tree mulberry. They are specially raised with the hepl of well grown saplings of 8-10 months old with any of the varieties recommended for rain-fed areas like S-13 (for red loamy soil) or S-34 (black cotton soil) which are tolerant to drought or soil-moisture stress conditions. Usually the plantation is raised as block plantation with a spacing of 6 feet x 6 feet or 8 feet as plant to plant and row to row distance. The plants are usually pruned once a year during monsoon (July-August) at a height of 5-6 feet from the ground level and allowed to grow with maximum of 8-10 shoots at crown. The leaf is harvested 3-4 times in a year by leaf picking method under rain-fed or semi-arid conditions depending on the monsoon. The tree brances pruned during the fall season (after the levels have fallen) are cut and used to make durable baskets supporting agriculture and animal husbandry.

Mulberry leaves, particularly those of the white mulberry, are ecologically important as the sole food source of the silkworm (*Bombyx mori*, named after the mulberry genus *Morus*), the pupa/cocoon of which is used to make silk. Other Lepidoptera larvae also sometimes feed on the plant including common emerald, lime hawk-moth, and sycamore moth.

India is the second largest country in the world after China producing mulberry silk. Indian mulberry silk production is about 16,000 metric tones compared to China's 80,000 metric tones. All the silk produced in China is of bivoltine and in our country more than 95% of silk is of multivoltine silk. Our weavers prefer this multivoltine silk for making our traditional dress material on account of its luster, smoothness, colour absorption, get up, draping quality etc., which lacks in imported Chinese bivoltine silk. India is only is only country in the world producing all varieties of silk i.e, mulberry, tasar, muga and eri. Our silk fabric exporters need large volumes of biovoltine silk and this being imported from China. Presently, country is importing about 10,000 metric tones of silk from China and it is learnt that a substantial quality milk of silk is smuggled in to the country through Bangla and Nepal borders. It is evident from the above that there is a great demand for this fiber in spite of high cost compared to synthetic/coton fabric. It is estimated that one heactare of mulberry cultivation generates employment to 12 to 13 persons throghout the year. Realising this, the National Commission on Agriculture has rightly that large scale expansion of this industry would substantially benefit the vast number of small and marginal farmers, besides providing emplyment opportunities for millons in the rural semi-urban areas, thus restricting the rural folk migrating to urban areas for a bowl of food. Also expansion of mulberry aear will definitely improve our productivity of both multi and bivoltine cocoon production in achieving self in the years to come. In the interest of the country and keeping in mind our rural economy, it is not wise to import silk from China, as they are dumping silk to our country below the cost of prodcution and ultimately they will dominate the silk market, if corrective measures are not taken forthwith.

To-day, the area under mulberry cultivation is about 177943 hectares and silk production is about 16,000 metric tones i.e., less than 90 kgs. per annum (average). New technologies developed by the Research Institutes have proved beyond doubt cost effective and improved quality and productivity. New mulberry varieties introduced are giving higher yield so that, in the same unit area under mulberry the leaf production has doubled. Package of practice, disease control methods in silkworm reading and usage of new CSR hybrids (Bivoltine) for rearing have resulted in increase of productivity and quality. Introduction of new nulti-end reeling units have proved taht 2A and above grade silk is possible with this innovation. With this new technology now it is possible to produce more than 120kgs. of silk per annum. Though these new CSR hybrid races are suitable for temperate climate, silkworm rearers have been successfully rearing these in tropical climate with adoption of necessary technology during favourable season from August to March. These races are gaining momentum and popularity among the sericulturists of Karnataka, Tamilnadu, Andhra Pradesh and Maharashtra. In the traditional silk worm rearing areas of Karnataka, it is possible to harvest about 85 kgs. and above of multivoltine (cross breed) and about 65 to 75 kgs. of Bivoltine hybrid varieties per 100 Dfl. Due to man power shortage, silk worm rearers are practicing mechanization in cultivation and silk worm rearing. Now they follow shoot rearign system. With this annual crop will be 10 as against 11 to 12 when they were following leaf plucking and feeding system. Silk and Milk is the proverb of the area. Every silk worm rearer is also engaged in milk production. After Gujarat, Karnataka may be the largest milk producer in the country. Direct income from milk will provide some cushion to the farmer and also motivate him to use farm yard manure to the field. Dung is also used effectively for production of bio-gas. Enrich slurry from the gas plant is very goog for the soil health and many farmers have taken up vermi compost production on large scale. Slurry derived from bio-digester is very nutritious to the plant and this will definitely bring down the amount of inorganic fertilizers thus bringing down the cost of cultivation. Regular use of farm yard manure/vermi compost etc. is useful in maintaining the soil health. Also cultivation of horse gram, sunhemp, dhaincha in between the rows of mulberry garden which are leguminous in anture will provide sufficient nitrogen. Mulberry pests and diseases are powdery mildew, leaf rust, leaf spot, sooty mold, bactrial blight and root knot diseases.

Diseases

The leaf of mulberry (Morus spp.) is the basic food of silkworm (Bombyx mori L). The role of left quality in the success of silkworm rearing and cocoons yield is well established.

Mulberry is a fast growing plant and leaf contents are rich in protein, sugar and moisture. The nutritional value of left, however, varies greatly owing to a number of factors among which the diseases play an important role. Being a perennial plant, mulberry is exposed to the environmental vagaries and pathogens prevailling in the region throughout the year. As a result, fungi, bacteria and nematodes cause number of diseases.

Feeding the leaves of infected plant to the silkworms adversely affects their growth and health resulting in partial or sometimes complete failure of cocoons production. Hence, a close watch, timely control measures and regular monitoring are very essential to manage the problems posed by diseases.

Cercospora Leaf Spor (Causative organism: Fungus, *Cercospora moricola)*

Symptoms: Initial symptoms of the disease are small light brown specks on leaves. These spots later enlarge into nearly circular, dark brown lesions with chlorotic margin. Whitish conidial masses appear on the lesions solitarily or in a circular fashion. Severely infected leaves turn yellow and fall- off prematurely. The loss due to the diseases ranges from 10-20 per cent depending on the age of the garden, temperature and frequency of rain.

Seasonal incidence: The disease occurs during south- West monsoon season and severity is seen during August to October. The disease may also prevail on plant during November to December depending on the receipt of North-East monsoon. Young leaves within the age of 18 days are susceptible to infection.

Control: All the infected leaves should be collected and burnt prior to receipt of rain to reduce availability of the garden for further spread.

Spray 0.2% Carbendazim (2 gram of chemical/ litre of water) twice at 15 days interval is used as chemical control. Safety period of Carbendazim is 15 days.

Powdery Mildew (Causative organism: *Fungus, Phyllactinia corylea)*

Symptoms: White powdery patches appears on the under surface of leaves. Corresponding to these patches on the upper surface, the leaves turn chlorotic.Under favourable condition; the symptoms may progress to form an almost total white superficial covering on the under-surface of leaves. As the lesions become older, they turn black due to the growth of a hyperparasite on the mildew fungus. Infected leaves turn yellow, lose moisture and become easily friable leading to defoliation. The loss incited by the disease ranges from 10-15 per cent.

Seasonal incidence: The disease generally occurs during August to January. Maximum severity is noticed during October- December. Close spacing of plants and excessive nitrogen application favours disease severity.

Control: Adopting wider spacing to facilitate adequate aeration and penetration of sunlight helps to reduce disease severity. Removal of weak branches and weeds checks the humidity build-up under the plant canopy thereby reducing the severity of the disease.

Chemical control of mildew becomes necessary in hilly areas and regions receiving heavy rainfall. Fungicidal spray of 0.2 per cent Carbendazim/ Dinocap (2 gram/ ml of chemical in one litre of water) should be applied as soon as the symptoms appear on plants which may be repeated after 15 days, if necessary. Leaves can be fed to worms 15 days if Carbendazim is sprayed and 20 days if Dinocap is sprayed.

Leaf Rust (Causative organism: *Fungus, Peridiospora mori)*

Symptoms: Numerous, small, light brown specks appear on the under surface of leaves which later turn into dark brown rusty spots.

Seasonal incidence: The disease is most common during winter months of November- January. Medium and coarse leaves are susceptible to infection. The loss due to this disease is about 5 per cent.

Control: Timely utilization of leaves during winter months is most essential to prevent inoculum build up in the garden.

Chemical control using foliar spray of 0.2 per cent Carbendazim / Chlorothalonil (2 gram/ml of chemical / litre of water) can reduce the disease severity. Leaves can be fed to silkworms after 15 days.

Bacterial Leaf Spot(Causative organism*: Bacterium Pseudomonas mori)*

Symptoms: Lesions initially consist of small, irregular, water soaked areas on the under surface of leaves. Later they enlarge and become dark brown to black necrotic lesions. In wet weather, sticky bacterial ooze can be seen on the lesions. If young leaves are infected, they curl and drop off, while medium aged leaves turn yellow prior to defoliation. The pathogen can also cause lesions on tender stem portions.

Yield loss as a result of disease is generally not high. Occasionally it may cause 1-2 per cent loss in yield.

Seasonal incidence: The disease occurs in most mulberry growing areas during June- December.

Control: Infected leaves should be removed and burnt to reduce the inoculum load in the garden. Chemical control using 100 ppm Stteptocycline solution (1 gram of chemical in 10 litres of water) as foliar spray twice at 4 days interval has been found to be effective. Residual toxicity to silkworm is absent.

Stem Canker (Causative organism: Fungus, *Lasiodiplodia theobromae)*

Mulberry is propagated mainly through stem cutting. The cut ends of the cuttings provide the main portal for the entry of many pathogens. One such important pathogen is Lasiodiplodia theobromae. (Otryodiplodia theobromae), a would parasite causing stem canker in mulberry. Canker is a disease causing death and/or decay of the bark. The disease occurs throughout South India assuming epidemic proportions in nurseries killing the cuttings. Lately, occurrence of stem canker is observed in established gardens also causing death of plant.

Disease symptoms in nurseries

- Failure of cuttings to sprout.
- Sudden withering and death of sprouts.
- Discolouration and drying of stems and buds above the soil.
- Rotting and peeling off bark on stem below the soil surface.
- Black mycelial strands seen below infected bark.
- Black eruptions on the bark of the infected stem portion.

Disease symptoms on established plants

- Greyish brown discolouration of the bark at the cut ends of the stem.
- Delayed sprouting.
- Death of buds and sprouts.
- Black eruptions on the bark in the infected region.
- Death of plants.

The above symptoms can be observed few days after the plants are pruned.

Disease management: The pathogen is soil-borne and spreads through irrigation, farm activities, insects and wild. Effective management of stem canker can be achieved through cultural and chemical methods.

Cultural methods

a) **Roguing:** - this method involves removal and burning of the infected cuttings and saplings in nurseries. In established gardens, the affected stem parts should be cut and burnt.

b) **Field sanitation:** - field sanitation is essential especially in the control of L. theobromae which survives in the soil and on host debris. Destruction of crop debris by burning reduces inoculum build-up. Stocking of pruned stems and twigs by the side of mulberry garden should be avoided since it aids in the growth, multiplication and survival of the pathogen which serves as an important source of inoculum.

c) **Land use:** - Raising saplings repeatedly in the same side should be avoided as it helps in inoculum build-up in the soil, making it sick. Such sick soil supports a very high incidence of disease and mortality of cutting.

d) **Planting season:** - Planting during winter months should be avoided since cold weather delays normal sprouting thereby predisposing the cutting to infection.

e) **Proper pruning:** - Damaging the bark or buds while pruning should be avoided to prevent entry of pathogen. Pruning the stem close to the buds helps normal sprouting and prevents fast colonization by the fungus. Cutting close to the buds should also be followed while preparing cuttings to avoid delayed sprouting.

Chemical control

a) **Pre-treatment of cutting**: the cuttings should be dipped in 0.2 per cent Carbendazim solution (2 grams of chemical in one litre of water) for 30 minutes prior to planting, followed by a spray after 15 days of planting.

b) **Protecting established plant**: soon after purning, the cut surfaces of the stems should be dressed with a spray/smear of 0.2 per cent Captan or Carbendazim solution.

   This is a very important step to be followed in gardens where shoot harvesting is practiced for rearing silkworm. The treatment forms a toxic barrier on the wound surface preventing the entry of pathogenic inoculum.

c) **Soil disinfection**: this method is suitable for the treatment of sick soils in nursery beds. Carbendazim or Captan at the rate of 8kg/acre should be broadcasted after loosening the soil followed by irrigation. After 10-12 days, well decomposed farmyard manure should be applied and thoroughly mixed with soil for planting the cuttings. Post treatment application of farmyard manure is a pre- requisite prior to planting in treated soils.

Biological control: Nursery Guard, a bioformulation of Trichoderma pseudokoningii can be used for long term protection, better survivability and growth of saplings.

Mix 1 kg of Nursery Guard with 60 kg of finely powdered well decomposed farmyard manure/neem cake. Moisten the mixture to a maximum of 30% moisture, store it under shade in the form of a heap and cover with a wet gunny cloth. This enhances the multiplication of biocontrol agent in the mixture.

Broadcast the mixture @ 2 gk/m^2 over the well prepared nursery beds and thoroughly mix with soil by light digging. Apply the mixture only to the planting area in nursery beds.

Root Rot (Causative organism: Fungus, *Fusarium solani* and *F. oxysporum*)

Fusarium solani is more virulent and causes maximum damage. The disease appears in all types of soil and climate throughout the year. The disease incidence with plant mortality ranges form 2-10%.

Symptoms and nature of damage: Sudden wilting/ withering and defoliation of leaves followed by drying/death of affected plants. The underground symptoms include decaying of root bark and rotting of primary and secondary roots. Rotten roots turn black and contain black powdery debris below the bark. The severely affected plants loose the hold in the soil.

The disease is soil-borne and appears in isolated patches in the gardens and quickly spreads to surrounding areas through irrigation and farm operation.

Chemical control: Chemical control involves dipping the root of saplings in 0.2% Carbendazim solution for half an hour and planting in pits dusted with 5g Mancozeb.

For existing plantation (disease in initial stage): As soon as the initial symptoms like witting of leaves and drying of branches are noticed, dust 5g of Mancozeb/ plant after removing the soil around plants to a depth of 15-20 cm. And also apply the same dose of Mancozeb to the plants in 2-3 surrounding rows around the infected plants.

For re-plantation

- Uproot the severely affected plants and destroy. Prepare one foot deep pit and remove the soil including rotten stump of roots and allow the pit for sun drying for a month before planting. Prior to planting, dip the roots of the saplings in 0.2% Carbendazim solution (2 grams of chemical in one litre of water).
- Dust 5g Mancozeb/pit and plant the treated saplings in the pit and cover with fresh soil.
- Also apply the Mancozeb alone @ 5g/plant for 2-3 rows of plants surrounding the infected plants.
- Apply second dose of Mancozeb @ 5g/ plant a month after replantation.

Biocontrol: A biofungicide 'Raksha', containing Trichoderma harzianum has proved to be effective and is under large-scale filed trial. The biofungicide can be used for the management of the disease.

Method of Application: reparation of mixture of 'Raksha'

- Mix one kg 'Raksha' with 50kg FYM/neem cake (Sufficient for 1000 plants) and keep under shade for a week covered with a wet gunny cloth to enhance the multiplication of Trichoderma.

For existing plantation (Disease in initial stage)

- As soon as the initial disease symptoms appear, apply the mixture of ' Raksha' @ 500g/pit plant at a depth of 15-20 cm below the soil including surrounding 2-3 rows of plants.

For new plantation, apply' Raksha' @ 500g/pit and plant the Mancozeb (0.2%) treated sapling in the pits (Mancozeb is not toxic to biocontrol agent). Apply second dose of 'Raksha' after 3 months of replantation.

Lasiodipladia Root Rot Causative organism *Lasiodiplodia theobromae (Botryodilodia theobromae).*

Symptoms

- Sudden wilting and withering of leaves.
- Drying and death of branches and plant.
- Appearance of back eruption (Pycnidial clusters) on the bark near the soil surface.
- Rotting and pealing of bark from roots.
- Grayish black mycelial strands appear on the under surface of bark and on wood portion of root.
- In presence of excess soil moisture, rotting and pealing of bark extends even to the basal portion of the stump.

Seasonal occurrence: The disease occurs throughout the year.

Disease Management

- Pruned stem portion should not be stacked near the garden to prevent multiplication of the pathogen and further spread.
- Uproot and burn the diseased plant along with roots.
- Soil from where the infected plant was uprooted should be thoroughly drenched with 0.2% Carbendazim or Captan solution. After 15 days, apply well decomposed farmyard manure and plant a new sapling with its root pre-treated with 0.2% Carbendazim solution.
- Immediately after pruning, the cut surfaces and the stumps should be sprayed with 0.2% Carbendazim.

Root Knot

Root knot is one of the major diseases limiting crop production throughout the world. It can occur any time of year mainly in sandy soils low in organic matter. The severity of the disease increases with increased age of the garden. The estimated yield loss due to the disease is 15-30 per cent. Infected plants become weak and predisposed to other diseases while severely infected plants ultimately die.

Causative organism: The organism causing root knot disease, is a nematode (Meloidogyne incognita), an endoparasite inhabiting mulberry roots. The disease is more prevent in irrigated gardens especially in sandy loam soils.

Symptoms

- Stunted growth.
- Poor and delayed sprouting.
- Reduced leaf size and yield.
- Chlorosis and marginal necrosis of leaves, yellowing and wilting of leaves inspire of adequate soil moisture availability.
- Death of plants in severe cases.

These symptoms first appear in isolated patches, slowly spreading over the entire garden.

Symptoms on the underground parts

- Formation of galls/knots on roots.
- Reduced root system.
- Retarded root growth.
- Necrotic lesions on the root surfaces and death of active rootlets.

Disease Management

Cultural methods

- Saplings free of nematode infestation should be used for planting to prevent the introduction of nematodes into a disease free area.
- Heavily infested soils should be deep ploughed to expose the nematodes and their eggs to solar radiation. Keeping the land free of vegetation during summer kills the nematodes due to desiccation by solar radiation.
- Application of neem cake @ 2 MT/ha/year in 4 split doses reduces the disease incidence.

Chemical methods

Chemical methods of nematode management become necessary in sick soils and heavily infested gardens as they give quick results. Furadon 3G (40 Gk/ha) can be applied either in furrows or broadcast to the soil after light harrowing followed by irrigation. The leaves from treated plots can be fed to silkworms after 45 days.

Biological control: A bionematicide Verticillium chlamydosporium can be applied with neem cake. One kg bionematicide mixed with 200 kg FYM and 24 kg neem cake is sufficient for 1000 plants. Prior to application, the mixture is kept in shade for a week covered by wet gunny cloth. Apply the mixture @ 200/g plant around the root zone and repeat the treatment every 4 months for 2 years.

Note: Do not use any fungicide/insecticide to the soil when bionematicide is applied.

Pests

Mulberry Leaf-Roller, Diaphania *pulverulentalis*

The leaf-roller Diaphania pulverulentalis (Hampson) (Lepidoptera: Pyralidae) is a pest of mulberry since 1995 in Karnataka, Andhra Pradesh and Tamil Nadu. It has also spread to Kerala. Even through the infestation was predominant in Mysore, Mandya, Bangalore and Kolar districts earlier, it has spread to other sericultural area in subsequent years. The leaf-roller infestation causes considerable reduction in leaf yield, resulting in economic loss to sericulturists. The pest infests all the commercial mulberry varieties namely, Local, M5, MR2, S36, S54, V1, etc. the infestation is observed in mulberry plantations from 15 days through 70 days after pruning or leaf harvest. Up to 100% infestation is recorded in several areas.

Period of occurrences: Through, leaf-roller infestation is observed round the year but for the month of May, infestation is severe from July to November.

Symptoms and nature of damage: The young caterpillar within the unopened young leaves secrete delicate, white, silky filaments, which bind the leaf blades together and feed on the soft green tissues on the leaf surface. Sometimes a single leaf is roller up and the caterpillar is found within this roller up leaf formed by the filaments and hence the name leaf-roller. Grown up caterpillars feed voraciously on tender leaves and their faecal matter can be seen on the leaves below the affected portion.

The leaf-roller not only feeds on the leaves at the apical portion of the affected plants, but also damages the young shoot. As a result, the plants become stunted. In such plants, tender leaves required for chawki rearing become scarce. In addition, the leaf-roller is highly susceptible to pebrine disease of silkworm. The pest larva infected with pebrine spores in the field can contaminate mulberry leaves. When such contaminated leaves are fed to silkworm, it can cause pebrine disease in silkworms. Thus, this pest causes dual damage.

Life – cycle: Moths are greyish brown in colour and measure about 1 cm in length. Wings have brownish wavy stripes. Moth lays 1 or 2 eggs on the apical portion of each shoot. Each female lays about 100 eggs. Incubation period is 3-4 days. Grown up caterpillars are greenish brown in colour with black markings on the lateral and dorsal regions of the body segments. The head is black in colour.

Larval duration lasts for 10-15 days. Pupae are dark brown in colour. Pupation occurs in dry leaves and pupal duration is 8-10 days. The pest completes its life-cycle in 21-29 days.

Management practices: Adopt an integrated Pest Management (IPM) strategy comprising cultural/ mechanical, chemical and biological control methods.

- Clip the affected apical portions, collect them in polythene containers and dip in 0.5% soap solution (5g of soap in 1 litre of water) to destroy the larvae.
- Press and kill the caterpillar, when the apical portion of the plant is affected.
- Collect and burn dry leaves and weeds found on the surface of soil in infested mulberry gardens to destroy pupae.
- Dress mulberry shoots before transportation in order to check the pest from spreading.
- Promote the activity of bird predators by them with water in small containers in and around mulberry plantations.

Chemical control: Spray 0.076% DDVP [1 ml of DDVP (76%EC) in 1 litre of water] on the apical portion of plants when infestation is noticed. A second spray is necessary 10 days after the first spray, if infestation exceeds 20%. Use mulberry leaves for silkworm rearing 10 days after spray. About 150 litres of DDVP solution is required / sprays/ acre. Freshly prepared solution should be used.

The aforesaid management practices should be adopted simultaneously by sericulturists in pest infested areas.

Precautions

- Undertake the insecticides spraying during the cooler hours of the days.
- Do not use insecticides after expiry date.
- Keep the container of the insecticide airtight after use.
- Keep the insecticides always from reach of children and pet animals.

Turka *(Maconellicoccus hirsutus)*

"Tukra" is caused due to infestation of mulberry plants by the mealy bug, Maconellicoccus hirsutus (Green) (Hemiptera: Pseudococcidae). Through the pest infests majority of mulberry plantations in Karnataka, Tamil Nadu and Andra Pradesh, the infestation is severe in Tamil Nadu. The pest infests most of the commercial mulberry varieties namely, M5, Local, Viswa, MR2, S36, S54, V1, etc. the infestation is about 34% in Karnataka.

Period of occurrence: Through "Tukra" infestation is observed throughout the year, it is severe in summer.

Symptoms and nature of damage: Malformation of the apical shoot due to flattening and/or thickening of the affected portion of soot, reduction in intenodal distance and wrinkling and curling up of apical leaves is observed. Affected leaves become dark green in colour and turn pale yellow, in the case of severe attack. Affected portion of plants become brittle. Presences of white mealy substance on infested portion of plants confirm mealy bug infestation.

The mealy bugs suck the sap, resulting in stunted growth of the plant and reduction in leaf yield.

Life-cycle: Each gravid female deposits about 250 eggs. The eggs hatch in 5-8 days. The crawlers are orange in colour and this stage lasts for a day. They settle on the plant and secrete a white mealy substance on their body. Females have three nymphal instars, while males have four. Life- cycle is completed in 24-26 days. Adult males are slender and possess a pair of wings and caudal filaments, each. They predominantly reproduce pathenogenetically.

Management practices: Adopt an integrated Pest Management (IPM) approach comprising culture/mechanical, chemical and biological methods.

Cultural/mechanical practices

- Clip the affected portions of the plants, collect them in polythene containers and dip in 0.5 % soap solution (5g of soap in 1 litre of water) for 3 to 4 minutes to destroy the pest. This practice should be employed about a week before commencement of leaf harvest.
- Do not grow altemate host plants such as croton, hibiscus, guava, grapes, lady's finger, cotton, etc. in the vicinity of mulberry garden.

Chemical control: Spray 0.15% DDVP [2 ml of DDVP (EC-76%) in 1 litre of water] in 0.5% soap solution on pruning or leaf harvested plants, twice at a an interval of 10 days. Leaves can be fed to silkworm, 15 days after spraying the insecticide. This practice should be adopted during summer (February- May).

Biological control: *Release the predatory* coccinellid Ladybird Beetle, Crytolaemus montrouzieri (Muls) (Coleoptera: Coccinellidae) @ 125 adults per acre twice, during August- September and October- November.

Precautions

- Before purning the infested mulberry plantation, clip the affected portions, collect in a polythene bag and dip in 0.5% soap solution.
- When cutting from "Tukra" affected plantations are procured for planting, dip them in 0.15% DDVP solution for 2-3 minutes before planting.

- Ants play an important role in spreading mealy bugs in mulberry plantations and treat them with 5 % Malathion dust soon after pruning or leaf harvest.

Bihar Hair Caterpillar, Spilosoma oblique

Bihar hair caterpillar, Spilosoma oblique. Walker (Lepidoptera: Arctidae) is a poiyphagous pest infesting several crop including mulberry, sunflower, green gram, Soya, jute. Etc. It occurs sporadically in most areas in the sericulture belt os India. Occasionally, it is a severe pest of mulberry in Mysore, Bellary and Chamarajanagar district of Karnataka.

Period of occurrence: It occurs during the monsoon and post-monsoon periods and the insect aestivates during summer. Infestation is severe from August to January.

Symptoms and nature of damage: Young caterpillars are gregarious and they feed on the green, soft tissues of tender leaves, leaving behind only the veins. Grown up caterpillars are solitary and feed voraciously on entire mulberry leaves, thus defoliating the plants.

Life-cycle: Moths have light brown coloured wings, brick red abdomen, peppered with dark row of spots laterally and dorsally. Gravid females lay 1,000 to 2,000 eggs in a group on the under surface of mulberry leaves. Eggs are pale green in colour and hatch in 5-7 days. Caterpillars moult 6 times. Grown up caterpillars measure 4-5 cm in length and are covered with bristles. Anterior and posterior ends of the caterpillar are black in colour, while the rest of the body is reddish brown. There are seven instars and the larval duration is 27-32 days. They pupate in crevices found between stones on bounds, tree trunks and corners of nearby buildings. Pupae are dark brow in colour and measure about 2 cm in length. Pupal duration is 12-14 days. The life-cycle is completed in 44 to 53 days.

Management practices: Adopt an Integrated Pest Management (IPM) strategy comprising culture/ mechanical, chemical and biological control methods.

Culture/ Mechanical measures

- Collect and destroy eggs masses and gregarious young caterpillars by crushing them.
- Use light traps to attract moths by placing a tray with 0.5 % soap solution below the light source.

Chemical control: Spray 0.15% DDVP [2 ml of DDVP (EC-76%) in 1 litre of water], 15 days after releases of T. chilonis. Mulberry leaves can be fed to silkworms 10 days after spraying.

Biological control: Trichogramma chilonis is an egg parasitoid of many lepidopteran pests. It is widely used a biocontrol agent of several crop pests.

Release twice T. chilonis at the rate of 5 tricho-cards (20,000 parasitised eggs in each tricho-card) per acre, at an interval of 3 days. Parasitoid releases have to be undertaken 20 days after purning or harvesting.

Economics (Per acre, per year)
Expenditure

Item	Quantity Reqd.	Man days Required	Rate (Rs)	Amount (Rs)
Culture/ Mechanical measures	-	8 (2 man days x 4 crops)	60	480
Biological	30 eggs cards (15 cardsx 2 crops)	-	35	1050
Chemical (DDVP)	1.2 litres (400ml x3 crops)	-	450	540
			Total	**2,070**

Receipts

a) Average recovery in mulberry left = 2,386 kg.
b) Value of mulberry left (@ Rs.2/- per per kg) = Rs.4,772
c) Net gain = Rs. 2,702
d) Cost- benefit ratio = 1:2.

Wingless Grasshopper, *Neorthacris acuticeps nilgirenisis.*

The wingless grasshopper, Neorthacris acuticeps nilgirenisis Uvarov (Orthoptera: Acrididae) is a serious pest in rained mulberry plantations of black cotton soils in Mysore and Chamarajanagar districts of Kanataka. The rate of infestation ranges from 8 to 95%. Besides causing exensive damage to mulberry, this pest infests sunflower, ragi, groundnut, beans, potato, jowar, etc.

Period of occurrence: The pest occurs during monsoon and post monsoon periods. The peak infestation is observed during October and then declines. Infestation dose not occur from January onwards till the onset of monsoon.

Symptoms and nature of damage: Nymphs and adults feed voraciously on sprouting buds and leaves of mulberry. Sometimes, they also feed on green bark. When the infestation is severe, mulberry plants are completely devoid of leaves.

Life-cycle: Adults of the grasshopper are light green in colour. Gravid females lay 6-8 eggs in loose soil at a depth of 2-3 cm. each egg pod contains 8-18eggs. Eggs hatch in 28-31 days. The I, II and III instar nymphs are straw coloured, whereas the remaining instars are light green in colour. Nymphs pass through 6 instars in 90-95 days. Females are larger in size than males and live for 45-60 days. The pest completes its life-cycle in 163-186 days.

Management practices

Cultural/mechanical measures

- Plough the infested mulberry plantations immediately after the onset of monsoon, to expose the egg masses to sunlight and predators.
- Maintain field sanitation by keeping the mulberry garden free from altemate host plants.

Chemical control: Spray 0.076% DDVP [1 ml of DDVP (EC-76%) in 1 litre of water], on mulberry foliage to kill nymphs and adults. If infestation is severe, a second application is necessary, 10 days after the first spray. Harvest mulberry left for silkworm rearing 10 days after the last spray.

Cutworm (*Spodoptera litura*)

The cutworm *Spodoptera litura* F (Lepidoptera: Noctuidae), is a polyphagous pest infesting several crops, such as mulberry, tobacco, castor, several vegetable crops, etc. it occurs sporadically on mulberry, where vegetable crops are commonly growth..

Period of occurrence: It occurs from August to February months.

Symptoms and nature of damage: The cutworm caterpillars attack shoots of young mulberry plants and cut them, hence the name cutworm. The cut portion of the shoot dries up and falls off. They also feed on mulberry leaves, voraciously.

Life-cycle: Moths are stout; forewings are dark with wavy white markings. Hindwings are white with brown margins. Eggs are laid in clusters of 200-300 underneath, the leaves and are covered with brown bristles. Incubation period is 4-5 days. Full grown caterpillars are stout, cylindrical and pale greenish brown in colour, with dark markings. They have transverse and longitudinal grey and yellow bands. The caterpillar is nocturnal in habit, and lies hidden in crevices of loose soil at the base of the plant during day time. It measures 3.5-4 cm in length and pupates in soil. Larval period is 2-3 weeks. Pupae are dark brown in colour and measure 1 to 1.5 cm in length. Moth emerges from pupa in about 2 weeks. Life-cycle is completed in 36-40 days.

Management practices: Adopt Integrated Pest Management approach comprising of culture/mechanical, chemical, and biological control methods.

Cultural mechanical practices

- Collect and destroy egg masses and young caterpillars.
- Plough the pest-infested garden and dig near the base of mulberry plant to expose the pest to sunlight and predators.

- Use light traps to attract months by placing a tray with 0.5% soap solution below the light source and kill them.

Chemical control

- Spray 0.15% DDVP [2 ml of DDVP (EC-76%) in 1 litre of water], in the evening, 20 days after purning or leaf harvest. Feed silkworms 10 days after pesticide application.
- Dust 5% Malathion near the base of the plants immediately after purning or leaf harvest.

Biological control: Use 'Spodolure' a pheromone trap @ 2 lures/acre twice at an interval of 15 days staring from 25 days after purning or leaf harvest to attract and kill male moths.

Preparation of Inseticides

(*Source*: Diseases and Pest of Mulberry and their control, CSR&TI, Mysore, 1991).

Calculation of dilutions: The active toxicant in a commercial formulation suitable for direct application is rarely available because small quantity of active ingredient has to be distributed on a large area. Dilution, therefore, becomes anther important aspect for consideration.

To obtain the weight or volume of soil or liquid formulation of commercial insecticide available in the market for the preparation of a soil or liquid containing a desired (or recommended) percentage of toxicant, the following formula can be applied.

$d = (a+b)/c$

Where,

a = of toxicant desired (recommended)

b = Wt (in g or kg) of dust or volume (in ml or lt) of liquid required for application.

c = % of toxicant available in the commercial insecticide formulation.

d = wt (in g / kg) or volume (ml or lt) of commercial formulation required.

After obtaining the value of d, the following calculation should be carried out to obtain the weight or volume of diluent (e),

Value of (b) – value of (d) = e.

Mix well the diluent (value of e) by weight (in case of soil formulation or by volume (in case of liquid formulation) to commercial formulation (value of d) to obtain the required quantity of desired strength of particular insecticide.

Therefore, to obtain 150 litre of 0.2% dimethoate from dimethoate 30% EC (Commercial formulation), 1 litre of dimethoate 30% EC should be added to 149 litre of water. In case of dust formulation, levigated china can be used as inert material to reach commercial formulation for application.

To determine the weight of soil of formulation soil in market for the preparation of a liquid for spray containing a desired (recommended) percentage of toxicant, the step mentioned above under (B) need not be carried out. After obtaining the volume of d (in g/kg) the determined weight of soil formulation should be added to the volume of liquid diluent considered fro b.

Example:- Prepare 200 litres of 0.5% BHC from commercial formulation BHC 50% WP available in the market. Follow step mentioned above under (A).

$$\frac{0.5\text{x } 200(\text{lts})}{50} = 2.00 \text{ kg or } \frac{0.5 \text{ x } 200000 \text{ (ml)}}{50} = 2000\text{g.}$$

Mix 2 kg or 2000g of BHC 50% WP in 200 litres of water.

Waiting period for various insecticides available commercially

Insecticides are widely used for the control of insect pests of several field crops. Improper use and their application results in undesirable drift residues to adjacent filed crops. Accordingly, mulberry gardens have also been found contaminated due to use of insecticides in the nearby crops.

Since silkworms are highly sensitive to insecticides, harvest of mulberry leaves during the waiting period should be completely avoided. In case, this is not followed scrupulously and contaminated leaves are harvested from the garden either located adjacent to field crops applied with insecticides or directly treated with insecticides, silkworm larvae develop toxic symptoms followed by the loss of the crop. These symptoms include.

- Vomiting of the digestive juice.
- Swinging of the anterior half of the body.
- Shortening of the body due to loss of the body fluid.
- Muscle contraction and.
- Paralysis followed by death of the silkworm larva.

To avoid loss of silkworm crops (due to feeding of contaminated leaves), mulberry leaves should be harvested on or after the number of days shown against each insecticide given below.

Safe periods of various insecticides against silkworm

Name of insecticide	Concentration (% active ingredient)	Waiting period **(in days)
Demeton	0.01	11
(Metasystox)	0.05	13
Demeton	0.01	07
(Metasystox)	0.05	11
Aldrin	0.01	11
(Aldrex)	0.05	13
Phosphomidon	0.01	11
(Demeton)	0.05	13
DDVP	0.01	07
(Dichlorovos, Nuvan)	0.05	11
Methyl parathion	0.01	07
(Metacid)	0.05	13
Carbary	0.05	09
(Sevimol)	0.10	11
Quinalphos	0.05	07
(Ekalux)	0.10	13
Dimethoate	0.05	07
(Rogor)	0.10	11
Endosulfan	0.05	09
(Thiodan)	0.10	17
Phosalone	0.05	09
(Zolone)	0.10	11
BHC	0.05	07
(BHC)	0.10	11
Chlordane	0.05	11
(Termex)	0.10	13
Malathion	0.50	13
(Cythion)	1.00	17

** Names in parentheses are the trade names of insecticides.*

** Waiting periods have also been indicated with each insecticide recommended for mulberry pest control.*

Meaning of the commonly used terminology on the containers of insecticides

Acute toxicity: How poisonous it is after a single exposure.

Chronic toxicity: How poisonous it is when a man or animal is exposed to small repeated doses.

Dermal toxicity: How poisonous it is when absorbed through skin.

Hazard: The chance that it may harm the beneficial organism is including man from the use of toxicant formulation. Hazard should not be confused with toxicity.

Inhalation toxicity: How poisonous it is when inhaled.

LC_{50}: Acute inhalation toxicity (Lethal concentration) necessary to kill 50% of the test organism. LC_{50} values are measured in mg/lt.

LD_{50}: Amount (Lethal dose) of toxicants of active ingredient that can kill 50% of the tested population through oral ingestion or dermal absorbtion. LD_{50} is given as mg/kg of body weight.

Lethal: Deadly.

Orol toxicity: How poisonous it is to man or animal when it is swallowed.

Residue: Toxicant deposit is remaining on the treated crops for sometime after the application.

Tolerance: The maximum amount of residue which may safely remain on a harvested crop. It is measured in parts per million (ppm). Expanded forms of various abbreviations used on the labels of containers are given while describing different formulations of insecticides.

Toxicity: How poisonous.

References

Chinnaswamy, K.P. and Hari Prasad, K.B. 1995. Fuel energy potentiality of mulberry. Indian Silk 34 (4): 15-18.

Choudhury, P.C., 1972. *Handbook of Silkworm Rearing. Agriculture Technique* Manual 1, CSRTI, Mysore, Tokyo, Japan, Fuji Publishing Co. Ltd.

Choudhary, P.C. 1976. Chemical composition and quality of mulberry leaves, p. 154-165. Text Book of Tropical Sericulture.

Jalaja S. Kumar, Sarkar, A . and Datta, R.K. A breakthrough in mulberry breeding in sustainable cocoon production. In Global silk scenario – 2001. Proceedings of the International Conference on sericulture, p. 242-247. Oxford and IBH Publishing Co. Pvt. Ltd.

Rangaswami, S., Narasimhanna, M.N., Kasi Viswanathan K. and Sastry, C.R. 1976. Manual on Sericulture Vol. 1, Rome, FAO.

Ravindran, S., Ananda Rao, A. Girish Naik, V., Tikader, A., Mukherjee, P. and Thangavelu, K. 1997. Distribution and Variation in Mulberry Germplasm. Indian J. Plant Genet. Res., 10(2):233-242.

Ravindran, S., Tikader, a. Girish Naik, V. Tikader, A., Mukherjee, P. 1988. Distribution on Mulberry Species in Indian and its Utilisation. Poster Paper Presented in National Dialogue.

Sarkar, A. and Jalaja S. Kumar and Datta, R.K. 1999. Potentially of Victory-1 under irrigated conditions in South India and potential programme for varietal selection. International Symposium on Sericulture Science, Hangzhou, China.

Sarkar, A., Jalaja S. Kumar & Datta, R.K. 1999. Potentiality of Victory-1 under irrigated conditions of South India. Indian Silk, May 1999.

Susheelamma, B.N., Jolly, M.S., Sengupta, K., Gridhar, K., Baksh, S., Mogili, T., Mallikarjuna, R.s. & Pavan Kumar, T. 1992. Statistical analysis of adaptability of drought resistant mulberry genotypes. Sericologia, 32(4): 619-628.

Sastry, C.R., 1984. Mulberry varieties, exploitation and pathology, Sericologia, 24(3):333-359.

Ullal, S.R. and Narasimhanna, M.N. 1987. Handbook of Practical Sericulture, Bangalore Central Silk Board.

Watt, G. 1981. Dictionary of economic products of India Vol. 5: 279-284, Delhi Periodical Experts.

❑❑❑

13

Role of Fungal Pathogens as Biological Agents of Rice Pests in Changing Climatic Conditions

Pranab Dutta, S. Kalita, K.C. Puzari and L.K. Hazarika

Abstract

Rice being the major food crop in the world requires a serious attention to protect the crop from variable pest attack under change in climatic scenario. More then 100 species of insects have been recored as rice pest affecting the productivity as well as maintaining sustainability in rice production. Bio-control agents for the control of rice pests are to be systematically evaluted. More than 1500 species of pathogens are formulated as microbiological insecticides. So our current awareness in pest- pathogen-abiotic interactions should be made use in designing the crop protection strategy.

Keywords: Pest outbreak in rice; Biological control of insect pests; Microbial formation of pesticides.

Introduction

Half of the world population lives in Asia, majority of which are dependent on rice. In general, rice is cultivated by small and marginal farmers of the under developed and developing countries of the continent. As such rice cultivation is not associated with prosperity, since majority of Asian countries are under developed and developing. However, introduction of dwarf gene has slightly improved this situation during the last 20 years. At the present rate of population growth,by 2020, 7% additional mouths are to be fed requiring 758 million tonne of rice every year, i.e., 350 tonne more than what was needed in 1992 from the same cultivated land. In order to increase rice production, high yielding cultivars have been introduced widely along with adoption of intensive crop management practices. This although resulted in substantial increase in yields, but they have invited newer problems of pests. They attack rice in different stages of growth resulting in yield losses ranging from 30-40 %. Thus farmers are faced with the problems of protecting their produces, for which they are mostly relied on the chemical control. Indiscriminate use of pesticides led to resurgence of pests, such as, brown plant hopper, secondary pest outbreak (eg. leaf folder, hispa), toxic hazards and residues and environmental pollution. In order to overcome these problems, major emphasis has been given on integrated pest management (IPM), which has become popular and gained momentum in several countries. IPM is a "management system" consisted of programmes to promote favourable ecological, economical and sociological outcomes. The use of appropriate scouting tactics, proper diagnosis of pest, the use of action thresholds and conservation of naturally occurring biocontrol agents (defenders/farmer's friend) are fundamental components of a sound IPM system.

Insect Pests of Rice

More than 100 species of insects have been recorded as pests of rice (Beevi *et al.*, 2001), which are the major constrains not only for increasing the productivity but also for sustaining it because of their repeated outbreaks in different parts of the country (Table 1). However, the major insect pests of rice are brown plant hopper, *Nilaparvata lugens*, gallmidge, *Orseolia oryzae*, yellow stem borer, *Scirpophaga incertulus*, leaf folder, *Cnaphalocrosis medinalis*, white backed plant, *Sogatella furcifera*, Gandhi bug, *Leptocorisa varicornis*, mealy bug, *Rioersia oryzae*, swarming catterpiller, *Spodoptera mauritia*, climbing cut worm, *Mythimna separata*, rice hispa, *Dicladispa armigera*, green leaf hopper, *Nephotettix virescens*, thirps, *Baliothrips biformis*, termites, *Odontotermes* spp., case worm, *Nymphulla depunctalis*, root weevil, *Echinocnemus oryzae*. Their occurrence in various states of India is shown in Table 2. It may be noted that these pests attack different parts of the plants having different damage potentials (Table 3).

Table 1: Serious pest outbreaks in India

Sl No	Pest	State	Year
1.	Hispa	Assam, Andhra Pradesh, Punjab	1918-2000 (in a cyclic manner since 1918 till 2001) 1981-1981
2.	Brown plant hopper	Kerala, Tamilnadu	1973-1984
3.	Yellow stem borer	Andhra Pradesh	1987-1990
4.	Gall midge	Andhra Pradesh	1987-1990
5.	Leaf folder	Punjab	1987-1990
6.	Ufra nematode	Assam	1987-1990
7.	RTV	Bihar, Uttar Pradesh, West Bengal	1969
8.	Armyworm	Uttar Pradesh	1981

Source: Beevi, *et al*., 2001

In Asia rice is mostly grown as rain fed. Due to heavy rain and occurrence of certain pests in endemic form, insecticides failed to suppress the pest; instead, large scale and indiscriminate application of chemical insecticides invited several unwanted social and environmental problems as discussed earlier (Hazarika and Puzari, 1997). As alternatives to insecticides, microbial especially, fungal pathogens may be excellent biocontrol agents in the rice ecosystem, which is endowed with high humidity and temperature. Epizootics may easily occur under such conditions.

Table 2: Prevalence of insect pests of rice in different states

States	Stem borer	Gall midge	Brown plant hopper	Green leaf hopper	Leaf folder	White backed plant hopper	Cut worm	Rice hispa
Andhra Pradesh	***	**	***	***	***	*	**	
Assam	**			*				***
Bihar	*	**		*		*		**
Gujarat	*			*	*	*		
Haryana					*	*		
Himachal Pradesh				*	*	*		
Jammu & Kashmir				*	*			
Karnataka	**	**	*		**			
Kerala	**	**	***	*	**		*	
Madhya Pradesh	**	***	*	***	*	**	*	
Maharastra	***	*	*		**			
Manipur		***		*	*			
Meghalaya	*			*				
Nagaland	*			*		*		
Orissa	***	***	**	***	*	*	*	**
Punjab	*			*	*		*	
Rajasthan				*				
Tamil Nadu	**	*	**	**	***	*	*	*
Tripura		*		*	*			
Uttar Pradesh	*	*		*	*	**	**	**
West Bengal	**	*	**	*	*			**

*Low, **Moderate, ***Severe *Source:* Manual on Integrated Pest Management in rice, Publ. By Directorate of Plant Protection, Quarantine and Storage, GOI, Faridabad.

Table 3: The stages of occurrence of various pests and their economic threshold level (ETL)

Crop stages	Pest	ETL
A. Nursery	Green leafhopper	1-2 insects/sq. meter
	Gall midge	1 silver shoot (gall)/ sq. meter
	Stem borer	1 moth/ 1 egg mass/sq. meter
B. Planting to pre planting	Leaf folder	2 freshly damaged leaves/hill
	Yellow stem borer	5% dead hearts or one egg mass or one moth/sq. meter.
	Gall midge	1 gall/sq. meter in epidemic areas or 5% affected tillers in non endemic areas.
	Brown plant hopper	5-10 insect/hill
	Green leaf hopper	10 insects/hill (in Rice Tungro Virus endemic areas 21 insects/hill)
	White backed plant hopper	10 insect/hill
	Rice hispa	2 adults or 2 damaged leaves/hill
C. Mid tillering	Leaf folder	2 freshly damaged leaves/hill
	Stem borer	10% dead heart or 1 egg mass or one egg mass/ sq. meter.
	Gall midge	5% silver shoots
	Brown plant hopper	10 insects/hill
	Green leaf hopper	10-20 insects/hill
	Rice hispa	2 adults or 2 damaged Leaves/hill
D. Panicle initiation to booting	Stem borer	1 egg mass or moth /sq. meter.,
	Leaf folder	2 freshly damaged leaves/hill
	Green leaf hopper	20 insects/hill
	Brown plant hopper	15-20 insect/hill
	White backed plant hopper	15-20 insects/hill
E. Flowering and after	Brown plant hopper	25-30 insects/hill
	Climbing cutworm	4-5 larvae/sq. meter
	Gandhi bug	1 or 2 bug/hill

Source: Manual on Integrated Pest Management in rice, Publ. By Directorate of Plant Protection, Quarantine and Storage, GOI, Faridabad.

Biological Control Approaches of Rice Insect Pest

There are three major categories of biological control; these are-natural biological control, augmentation and introduction or classical biological control. A large number of biological control agents are active in the field and are naturally performing biological control functions. This phenomenon is known as naturally occurring biological control or natural biological control. Natural biological control can be the most important component of pest management in crop fields and in its absence crop production would be extremely difficult.

Natural biological control is often limited by factors such as low plant diversity, pesticide use or highly seasonal planting of field crops that reduce natural enemy populations. If pests enter crops early in the season, and start to grow rapidly on abundant food, it is difficult for natural enemies to keep up with them. Therefore, in order to create favorable situations for natural enemies, interventions are often necessary. There are three major interventions: 1) conservation, 2) augmentation and 3) introduction or classical biological control. An important goal in pest management is to maintain or restore the normal role of natural biological agents in regulating pest population even though they alone may not cause adequate control. Several methods may increase the number of biocontrol agents in an ecosystem more than what would occured normally. These methods are a) environmental manipulation b) addition of host or food resources and c) routine release of mass-produced biocontrol agents.

Conservation is the modification of the environment or habitat in combination with judicious use of pesticides in order to conserve natural enemies and enhance biological control. Evaluating the success of biological control is, however, a daunting problem far different from the "spray-and-count" evaluation methods of pesticide-based approaches. Despite this, environment management to enhance biological control has proven valuable in many settings and appears to be very important for rice insect pest management. Although a wealth of native natural enemies occurs in the rice ecosystem, their utilization in pest control has not reached the desired level due to various reasons. Except for advocating the conservation of these natural enemies, no attempt has so far been made to mass multiply and make innundative releases. Considering the potential of naturally occurring bio-agents in rice pest management, it is evident that conservation of promising indigenous biocontrol agents deserves special attention. There is a definite need to systematically evaluate the use of potential entomopathogens either alone or in combinations, for the management of key pests of rice.

Entomopathogenic Fungi

More than 1500 species of pathogens are known to attack arthropods. These include bacteria, viruses, fungi, protozoa and nematodes. Some pathogens have been formulated as microbiological insecticides. Others have been genetically engineered for commercial exploitation. Rice insect pests have had long, and close associations with their natural enemies, allowing stable relationships to develop. For example, predator and parasitoid guilds recorded in five irrigated rice ecosystems in the Philippines belong to 10 orders and 57 families of predators and three orders comprising 40 families of parasitoids, indicating the great diversity and richness of the natural enemy community in rice. The interactions of such predators, parasitoids and insect pathogens are the cornerstone of modern integrated pest management programs in rice. The collective importance of natural enemies to pest population regulation is clearly demonstrated by observing outbreaks of insect pests after the removal of natural enemy communities by a broad-spectrum insecticide. Recognition that fungi have potential for biological control dates back to Louis Pasteur. During 1880s *Metarhizium* species was used to control wheat chafer, *Anisoplia austriaca* and the sugar beet curculio, *Cleonis punctiventris*

Table 4: Field occurrence of entomopathogenic fungi on rice pests in India and abroad

Sl No	Pest	Bioagent/Formulation	Dose used	Result	Reference
1.	*Nilaparvata lugens*	*Beauveria bassiana*	-	60-90% infection 15 days after incubation	Li (1985)
2.	*N. Lugens*	*B. bassiana*	-	Field, laboratory	Ambethgar (1996) Landa *et al.*(1994)
3.	*N. Lugens*	*B. bassiana* andbuprofezin	-	Suppression	Rombach *et al.* (1988)
4.	*N. Lugens,Nephotettix virescens*	*Beauveria bassiana* dust	@11 x 10^8 conidia/g	Field cage in China	Li (1986)
5.	*N. lugens*	(1). Suspension of conidia of *M. anisopliae, M. flavolividae, B. bassiana, Hirsutella citriformis*	@ 4-5 x 10^{12} conidia/ha	3 weeks after application, mortality was 63-98%	Rombach *et al.*, (1986)
		(2). *Pendora delphacis*	@ 2.25 kg/haor 40 kg/ha but 100% to nymph	Mortality was 61% to adults	Narayanasamy (2001)
6.	*N. lugens*	*Metarhizium* sp.	-	-	Bandra and Ahangama, (1994)
7.	*Cnaphalocrosis medinalis*	*B. bassiana*	-	-	Joshi *et al.* (1987)
8.	*C. Medinalis*	*B. bassiana*	-	Field, laboratory	Padmanabhan, (1993); Ambethgar, (1997)
9.	*C. medinalis*	(i)*Zoophthora radicans*	-	84% mortalityHighly pathogenic	Narayanasamy (2001) Battu *et al.* (2000)
		(ii) Deuteromycotina (=near *Nomuraea* sp.)			
10.	Rice cut worm	*Nomuraea rileyi*	-	90-92% mortality	
11.	*Hierglyphus banian*	*Aspergillus flavipes*	-	Field occurrence: 0.12% and mortality in laboratory: 36.21%	Hazarika *et al.* (1994)Baruah *et al.* (2003)

(*Contd.*)

(*Contd.*)

Sl No	Pest	Bioagent/Formulation	Dose used	Result	Reference
12.	*Sitotroga cereallela*	*M. anisopliae, B. bassiana*	2.6 x 10^9 conidia $50g^{-1}$ and *N. rileyi*	-	Ekesi *et al.* (2000)
13.	Brown Plant Hopper	*Hirsutella citriformis* Speare	-	Not very effective	Van Vreden and Ahmadzbidi (1986)
14.	Black rice bug, *Scohinophora* sp.		*Paecilomyces farinosus* (Holnex Gray) Bown & Smith	-	Heavily infective Van Vreden and Ahmadzbidi (1986)
15.	*Spodoptera mauritia*	*Nomuraea rileyi*			Van Vreden and Ahmadzbidi (1986)
16.	*Tryporyza incertulas*	*Beauveria* sp.		Field	Padmanabhan (1993)
17	*Oebalus poecilus*	*Metarhizium anisopliae*	5 x 10^{13}/ha 5 x 10^{11}/ha 5 x 10^{10}/ha	5x10^{13}/ha was most effective concentration	Martins and Magalhaes (1987)
18.	*Sesamia inferens*	*B. bassiana*	-	-	Li (1993)
19.	*Scirpophagas incertulas* *Chilo auricilius*	*Beauveria bassiana* *Arthinium urticae*	-	*B. bassiana* was present in larvae of *S. incertulas* and *C. auricilius* Field occurrence 1.50% and mortality in laboratory: 7.00%	Israel and Padmanabhan (1978) Baruah *et al.* (2003)
20.	*Chilo suppressalis, Sesamia inference*	*Beauveria* sp	-	Field, laboratory	Padmanabhan (1993)
21.	*Nephotettix nigropictus* *N. Virescens* *Chilo auricilus*	*B. bassiana*	-	First record of the natural occurrence of *B. bassiana* on *N. nigropictus* and adult *C. auricilius*	Nayak and Srivastava (1979)
22.	*Nephotettix* sp.	*Erynia delphacis* (Hori) Humber	-	High mortality in cages	Van Vreden and Ahmadzbidi (1986)

(*Contd.*)

Sl No	Pest	Bioagent/Formulation	Dose used	Result	Reference
23.	*Marasmia patnalis*	*Beauveria* sp	-	Field	Ambethgar (1997)
24.	*Marasmia ruralis*	*Beauveria* sp	-	Field	Ambethgar (1997)
25.	*Melanitis ismene*	*Beauveria* sp	-	Field	Padmanaban (1993)
26.	*Nymphula depunctalis*	*Beauveria* sp	-	Field	Padmanaban (1993)
27	*Creatonotos gangias*	*B. bassiana*	-	Field occurrence 0.25% and mortality in laboratory: 74.66%	Baruah *et al.* (2003)
28.	*Dicladispa armigera*	*B. bassiana*	X 10^7 spore ml @ 700 l/ha	Field efficacy was at par with the insecticides to suppress the pest population, but superior to increase the yield	Hazarika and Puzari, (1994)
29.	*Leptocorisa acuta*	*Conidiobulus thromboides*	-	Field occurrence 3.60% and mortality in laboratory: 24.31%	Baruah *et al.* (2003)
30.	*Nezara viridula*	*Fusarium oxysporum*	-	Field occurrence 36.69% and mortality in laboratory: 88.43%	Baruah *et al.* (2003)

(Milner, 1987). The genera such as *Metarhizium*, *Beauveria, Verticilium, Nomurea, Entomopthora, Neozygites* etc are commonly encountered in nature. More than 400 species of entomopathogenic (EP) fungi have been recognized as entomogenous in nature. These fungi are found under five subdivisions: Mastigomycotina, Zygomycotina, Ascomycotina, Basidiomycotina and Deuteromycotina Fungal epidemics occur periodically and can cause high levels of mortality in affected arthropod populations. Only a few species of fungi have been reported from the rice ecosystems. However, though the system may serve as the paradise for fungi because of high (80-85%) relative humidity (RH) and other topographical factors. Most terrestrial species of fungi depend on high RH at several points of their life cycle (Ferron, 1978). Table 4 shows the efficacy of various entomopathogens in different parts of India and aboard. Table 5 shows the occurrence of EP fungi on rice pests either in the field or laboratory situation. The species diversity and total number of natural enemies in tropical rice are impressive, but usually a few species dominate the complex. It is difficult to generalize about the precise role and relative importance of individual species.

Table 5: Pathogenicity of *B. bassiana* to different pests of rice

	Pest species	Stages infected	% Infectivity	Reference
Lepidoptera				
1.	*Scripophaga incertulas* (Wlk.) (Pyralidae)	Larva		Israel and Padmanabhan (1980)
2.	*Chilo suppresalis* (Wlk.) (Pyralidae)	Larva		Fukaya (1951), Ndoye (1977), Itoh *et al.*, (1994)
3.	*Chilo auricilius* Dudgn (Pyralidae)	Larva		Nayak and Srivastava (1979)
4.	*Cnaphalocrosis medinalis* (Guen) (Pyralidae)	Larva		Ambethgar (1995)
5.	*Sesamia inferens* Wlk. (Noctuidae)	Larva		Li (1985)
Hemiptera				
6.	*Nilaparvata lugens* (Stal) (Delphacidae)	Nymph First & second instar	80-100	Li (1985)
7.	*Nephotettix virescens* (Distant) (Cicadellidae)	Nymph and adult 100		Nayak *et al.*, (1978)
8.	*N. nigropictus* (Stal.) (Cicadellidae)	Nymph and adult		
9.	*N. cincticeps* (Cicadellidae)		60-90	
10.	*Sogatella furcifera* (Horv.) (Delphacidae)		60-90	Li (1985)
11.	*Nigia atrovenosa*		60-90	

(*Contd.*)

Pest species		Stages infected	% Infectivity	Reference
12.	*Laodelphax striatella*		60-90	
13.	*Saccharosydue procerus*		60-90	
14.	*Empoasca subrafa*		60-90	
15.	*Deltocephalus dorsalis*		60-90	
16.	*Macrosfeles fascifrons*		60-90	
Coleoptera				
17.	*Dicladispa armigera* (Olivier) (chrysomelidae)	Egg, larva, pupa and adult	1-44 in the field and 90-100 in the laboratory	Hazarika and Puzari (1995)

Likewise, several fungal pathogens infect rice leafhoppers and planthoppers. The fungus grows into the body of the infected hopper; when the insect is dead, fungus grows out of the body through the joints. Common pathogens are *Metarhizium anisopliae* (Metchnikoff) Sorokin (Moniliales: Moniliaceae), *Metarhizium flavoviride* Gams and Roszypal (Moniliales: Moniliaceae), *Beauveria bassiana* (Balsamo) Vuillemin (Moniliales: Moniliaceae),*Hirsutella citriformis* Speare (Moniliales: Stilbaceae),*Erynia delphacis* (Hori.) (Entomophthoraceae), *Entomophthora coronata* (Constantin) Kevorkian (Entomophthoraceae), *Fusarium* sp. Narayanasamy (2001) utilized *P. delphacis* for the management of BPH in Tamil Nadu by utilizing two formulations: 70% WP and 10% dust. From 15 years of research work conducted at Assam Agricultural University, Jorhat, Assam, technology has been generated to mass produce formulate and field application of *B. bassiana,* for management of *D. armigera* in Assam (Hazarika *et al.*, 2005; Puzari *et al.,* 2006).

Beauveria bassiana has a wide host range and is reported from all over the world; however, it is found to be more predominant in the tropics and sub-tropics under moist and wet conditions. Nevertheless, it was also reported from the temperate regions. The rice ecosystem, being moist and wet, serves as a favourable environment for exploitation of mycoinsecticides including *B. bassiana* and *Pandora delphacis* Table 5 shows the pathogenicity of *B. bassiana* to pests from different rice growing countries: larvae of *Chilo* spp., *Scirpophaga incertulas*, *Sesamia inferens* and *Cnaphalocrosis medinalis*, nymphs and adults of leaf and plant hoppers were infected by this mycopathogen in India, China and Malaysia causing 60 to 90% mortality. In addition to, *Sesamia inferens, Tryporyza incertulas, Chilo auricilius, Panara spp, Heiroglyphus* spp., *Oryza* sp. were also infected by *B. bassiana* in the field (Rao, 1975). Other pathogen recorded rice pests in India is *Nomurea* sp. (Battu *et al*, 2000), *Zoopthora radicans* and *P. delphacis* (Narayanasamy, 2001). *B. bassiana* was successfully used for controlling *D. armigera* (Hazarika and Puzari, 1997), *Nephotettix virescens* and *N. lugens* (Gillespie *et al.*, 1990). *Cordyceps* is another white muscardine that infects stem borers. Field occurrence of *Arthinium urticae,* an entomogenous fungus of the rice stem borer was recorded in rice field of Assam by Boruah *et al.,* (2003). The fungus appears as white fluffy mycelial growth on 1.5% larvae in the field. Mycelia being visible inside the tunnels. However, laboratory inoculation resulted in 7 % infection of the hosts.

Metarhizium anisopliae, the green muscardine fungus, has been the subject of experiments at intervals extending over 80 years, but from the forgoing account it is evident that in no case has any result been demonstrated as sufficiently successful to warrant its adoptability as a means of destroying insects. In Brazil a number of semi commercial preparations are in vogue in the various regions against diverse pests. From the assessment made by Ferron (1978,1981), on this fungus, in particular entomopathogenic fungi in general, we now have much better understanding of arthropod dynamics, especially the factors that affect disease incidence.

Another pathogen, *Hirsutela* sp., is specific to rice mite, *Oligonychus indicus.* The cadavar is usually attached to leaves at the top canopy of rice. The whole body of the mite is engulfed in mycelial filaments radiating in all directions. The locomotory appendages are left uninfected. Early symptoms of the disease are difficult to diagnose due to small size of mites. The first sign of infection is the sluggish movement of mites and later, change their body colour from normal lemon yellow to deep yellow brown.

Baruah *et al.,* (2003) reported many opportunistic fungi attacking different rice pests. For example, diseased nymphs and adults of grasshopper, *Hierglyphus banian* infected by *Aspergillus flavipes* exhibited grayish brown fungal outgrowths on their lateroventral sides. Hyphae sometimes develop conidiophores and emerge out even from the live hosts. It is of significance that most records of premortem emergence or sporulation by Hyphomycetes concern species of *Aspergillus.* Another entomogenous fungus, *Conidobolus thromboides,* causes infection naturally on the rice bug, *Leptocorisa acuta* and appear as white cottony mycelial growth on meso and meta thoracic eusternites and abdominal pleurites. *Nomuraea rileyi,* infect larval stages of the rice cut worm Coppery green fungal spores are deposited all over the body except the head capsule. Thus severely infected larvae are mummified and the body becomes brittle after exposure to sunlight. The mycelia are pinkish white in colour, which upon sporulation, become typical malachite green.

The genus *Cordyceps* is a cosmopolitan Ascomycete confined almost entirely to insects (Mains, 1958). About 200 species are recognized attacking a great variety of immature and adult insects. *Ophiocordyceps* is a closely related genus. *Cordyceps* is characterized by the growth of long fruiting stems, sometimes branched, arising from a sclerotium within the body of the insect. Certain of the imperfect genera as *Isaria, Botrytis* and *Hirsutella* in part are considered as conidial stages of *Cordyceps*. When a *Cordyceps* spore under favourable conditions strikes an insect, the germ tube penetrates the integument. On reaching the haemocoele, they soon fill the cavity, absorb most of the blood and the host dies. The mycelium attacks and absorbs most of the tissue distending the insect integument to near normal size. The sclerotium thus formed may remain dormant for months or in favorable situations develops the perfect conspicuous fruiting stems.

Mode of action of entomopathogens

In most cases of entomopathogens, spreading infection in an insect population, pre-infection requirements are adhesion and germination of spores to form either germ tubes or appresoria. Invasion through the cuticle is facilitated with these structures and cuticle degrading enzymes. In order to understand the mode of entry by the fungus, it is necessary to analyse the chemical composition of the insect cuticle through which this group of fungi could affect the host. The insect cuticle is composed of wax, lipid, protein and chitin; chitin, constitute 25-50% of the dry weight of the cuticle. Fungal cell wall also contains chitin as the main structural component. Thus, insects and fungi share a key structural polymer chitin, a ß-1, 4-linked *N*-acetalglucosamine polymer and its production is essential to them (Despande, 1998). The proliferation of hyphal bodies or yeast like cells in haemolymph and the production of enzymic or non-enzymic components are involved in the insect killing process. Indeed, the fungus has to combat with the host defense responses, such as, melanization, cellular reactions like phagocytosis, nodulation, encapsulation and polyphenol oxidase activation leading to the production of reactive molecules (Charnley, 1989; Gillespie and Moorhouse, 1989; Leger, *et al*., 1987, Phukan *et al.,* 2006).

Present need for microbial pesticides

The increasing demand for food production at competitive rates has also accelerated intensive agriculture leading to the outbreak of new pests and diseases. The changing dynamics of biotic stress on one hand and non-sustainability of chemical pesticides on the other have projected microbial as potential alternatives. However the current level of production of microbial agents is not commensurate with the potential demand in India (Hazarika, *et al.,* 2005). Considering the area under cultivation of different crops in India and projecting target coverage of a 10 per cent area with biocontrol agents, the potential requirement of various microbial biopesticides is quite substantial In India presently 23 private companies are producing different microbial pesticides. It is estimated that out of the different entomopathogens, the country need 341, 6117, 9082 MT of *Verticilium, Metarhizium anisoplae* and *Nomurea releyi* per year respectively. The market value is estimated as 0.02 and 0.01 corers for *Verticilium* and *Metarhizium anisopleae* respectively (Rabindra, 2005). Quite often microbials amy nt brng about desired control of the targeted pests, under such situations, combination of microbials with chemical pesticides may prove to be useful for which compatibility of pesticides and entomopathens must be tested (Puzari *et al.*, 2006). In a similar study, Puzari and Hazarika (1991) showed that efficacy of *B. bassiana* was enhanced when it was mixed with Sandovit E and many other agrochemicals. Judicious use of pesticides is an important aspect for conservation of natural enemies including microbials.

The destruction of natural biocontrol agents by broad-spectrum pesticides leads to reduced natural biological control and the subsequent resurgence of some pests and diseases. At the early stages of rice crop growth, abundant populations of detritus-feeding and plankton-feeding insects supports large populations of generalist predators in rice fields. The abundance of alternative prey gives the predator populations a "head start" on later developing pest populations. Therefore, pesticides use is most damaging early in the season. Pesticides use can also create heavy selection pressure for strains of pests that can even overcome previously resistant cultivars. Such circumstances create outbreaks of minor pests such as brown plant hopper and impair biological control of some key primary pests such as stem borers.

Prophylactic use of insecticides in rice is common and widespread. A common tactic used by farmers to control stem borers is the use of granular insecticides as a preventive method, without any consideration of the pest population density or economic benefit. Farmers' rationale for insecticide use appears to be based on anticipated yield losses due to insect pest attack. Such attitudes are detrimental to biological control, pest management as a whole, the environment and human health. Judicious use of insecticides is essential for natural enemy conservation and for the enhancement of natural biological control.

In order to conserve indigenous and enhance natural biological control, some strategies for insecticide use should be followed. Prophylactic or calendar-based use of insecticides should be stopped rather curative use of insecticides should be followed as the last resort, early season spraying during the first 40 days after transplanting should be avoided, selective insecticides should be used only when pest population density reaches a damaging level, when using insecticides, use appropriate formulations in the right dose. Apply at the optimum time of intervention following proper application methods, broad-spectrum hazardous insecticides should be avoided, use host plant resistance rice varieties to major insect pests.

Introduction or Classical Biological Control

In only two cases did the introduction of exotic parasitoids result in successful biological control in rice production. In the first case, populations of immigrant rice leaf folder, *Marasmia exigua* (Butler) (Lepidoptera: Pyralidae), in the Fiji

Islands were apparently reduced by the introduced larval parasitoid, *Trathala flavorbitalis* (Careron) (Hymenoptera: Ichneumonidae) from Hawaii in 1928. and the second case involved the control of introduced rice striped stem borers *Chilo suppressalis* (Walker) (Lepidoptera: Pyralidae) in Hawaii through the introduction of an egg parasitoid – *Trichogramma japonicum* Ashmead (Hymenoptera: Trichogrammatidae) from Japan and two larval parasitoids – *Eriborus sinicus* (Holmgren)(Hymenoptera: Eulophidae) and *Bracon chinensis* Szepligeti (Hymenoptera: Braconidae) from China in 1929. All three parasitoids established and helped to reduce infestation of this stem borer.

Several other attempts have been made to introduce exotic parasitoids to control indigenous rice insect pests in South and Southeast Asia but none was successful. One such example involved the introduction of a larval parasitoid *Paratheresia claripalpis* (ven der Wulp) (Diptera: Tachinidae) from Trindad to Malaysia to control the dark-headed stem borer *Chilo polychrysus* (Meyrick) in 1951. A more recent example involved the introduction of two egg parasitoids *Psix lacunatus* (Johnson and Masner) (Hymenoptera: Scelionidae) and *Telenomus cyrus* (Nixon) (Hymenoptera: Scelionidae) to Palawan Island from the Island of Luzon, Philippines during 1985-87 to control Malayan black bug, *Scotinophara coarctata* (F.) (Hemiptera: Pentatomitidae). So far known, this project was not successful.

For making any biological control programme effective, selection of agents is very important for which some criteria are fixed. These are true for entomopathogens as well. These are a) Indigenous to a particular environment, b) Laboratory virulence should be high, c) Field performance should adequate and satisfactory, d) Genetic stability in wide environment, e) Easy to mass produce, f) Stability of conidia or spore in storage for a ling time in a wide range of environments, g) Field persistence, h) Target specific but with somewhat wider host range.

Progression

Scientist like Fawcett (1944) reflecting on the general lack of applied research with entomopathogenic fungi summarized the situation at the time: "One apparent cause for neglect of the field has been the hasty generalization that because of failure to get outstanding results with certain fungi in initial trials, the whole field has little promise of practical results. This "hasty generalization" or pessimism is still shared today by many involved in crop protection, or agriculture in general.

The main reason for considering myco-insecticides with skepticism is their poor field performance, particularly their instability and costs when compared. Moreover, most of the successful trials reported are not based on scientific data and do not take into account the impact of natural inoculum. The improvement of this situation, up until now too speculative, will consolidate the commercial exploitability of entomopathogenic fungi: without investment the science will not progress, with more failures the concept will be rejected. The future is delicately balanced. There are many problems concerning to the production, application and marketability of mycoinsecticides which remain to be resolved and none of the few products which have reached the crop protection stage are house hold names and have been unable to compete with the use of entomopathogenic fungi as biological control agents is impressive as these are, safe, in comparison with chemical control methods, environmentally sound, since they do not upset the ecological balance, generally persistent, possibly giving on-going control, compatible with other control agents, being readily incorporated into integrated control programmes, less possibility of pest resistance, compared with chemical insecticides, genetically manipulative.

On the debit side the production and commercialization costs are extremely high, while the safety testing of mycoinsecticides can be expansive. The difficulties involved with the patenting of organisms or their biotechnologies and the overall concept of biological control, with all its marketing difficulties and uncertain profits, are not compatible with the policies of most large companies, with short-term investors or shareholders to satisfy. The potential of such organisms is high input, westernized agriculture would appear to be limited by market forces. However as demonstrated in China and USSR, their use as biological control agents of crop pests seems both feasible and apparently successful, when there is a cooperative, non-commercially oriented production system operating at the factory or cottage-industry level. The latter could also be exploited in developing or Third world countries, as long as there is an organized extension service, and thus help to reduce their dependence on insecticides, often dumped because of environmental restriction sin the developed world. A surprising occurrence has been the acceptance of fungal biological control agents in the highly commercialized, competitive Brazilian market, which probably owes more to the entrepreneurial skills of a few individuals rather than to a state or company-controlled policy. However, claims of the effectiveness of many of these preparations should be treated with caution. As many workers in the field have appreciated, entomopathogenic fungi are not the solution to pest control, they form a part of the answer and should be used as such in integrated pest management programmes.The natural control of pests by entomopathogenic fungi is significant in many agricultural ecosystems and they provide examples of pest problems, which have been exacerbated following the excessive use of fungicides and the subsequent suppression of the beneficial mycoflora. Therefore, it is obvious that future crop protection programmes must make the best possible use of these organisms, through, the judicious use of fungicides, to maintain the natural equilibrium, and here it would be necessary initially to identify all the factors that are involved in the natural control of a potential pest, the manipulation of these natural populations to obtain maximum benefit from them, for example: the employment of irrigation or cover crops to increase humidity levels; the use of low doses of chemical insecticides; the use of mathematical models to forecast the appearance of natural epizootics, the induction of epizootics particularly for those fungus which are difficult to mass produce, the application of mycoinsecticides, only feasible for those fungi which can be readily and cheaply mass produce, the inclusion of low doses of chemical insecticides in mycoinsecticides spray mixes has shown promise in many countries and research on this aspect should be pursued, particularly the synergistic effects with other products or control organisms. A similar ploy along these lines, would be to apply chemical insecticides during the driest part of the crop season or during periods of low humidity and then apply the mycoinsecticides during the wetter months. Modeling of pest populations would be essential, of course, in this type of crop protection system for forecasting both the timing of application with a view to eliminating the epizootic lag phase , and the quantity of product to use. It is imperative, of course, that the interactions of climate, pathogen and host are completely understood, otherwise the application of fungi in biological control of insect pests must continue to be largely empirical (Madeline, 1966), and therefore, doomed to failure and rejection by those involved in crop protection. Hopefully, our current awareness of pest-pathogen-abiotic interactions is now sufficiently advanced to avoid such an empirical approach in the future.

It is worth considering that only a fraction of the entomopathogenic fungi that occur in nature has been tested till date. Certainly our stocks need to be replenished and the genetic base broadened. More surveys are needed particularly in tropical areas which have been relatively understudied in this respect, compared with temperate regions.

References

Ambethgar, V. 1996. Biological control of Brown plant hopper, Nilaparvata lugens with entomogenous fungi. Madras Agricul. J. 8: 203-204.

Ambethgar, V. 1997. Record of white muscardine fungus, *Beauveria bassiana* (Bals.) Vuill. on rice leaf folder complex from Karaikal, Pondicherry Union Territory (India). J. Entomol. Res. 21: 197-199.

Ambethgar, V. 1995. Natural suppression of populations of the leaf folder, Cnaphalcrocis medilalis Guenee in fields by an entomogenous fungus, J. Entomol. Res. 19: 87- 89.

Bandara, J.M.R.S. and Ahangama, D. 1994. *Metarhizium* sp.: a new biocontrol agent for brown plant hopper management in rice fields, Intl. Rice Res. Newsl. 19: 19

Baruah, G., Puzari, K.C., Dutta, Pranab, Hazarika, L.K., Das, P. 2003. Entomogenous fungi from North East India-II. Bioved.14: 101-108.

Battu, G.S., Singh, P. and Singh, J. 2000. *Deuteromycotina* (= near Nomuraea sp.) infecting *Cnaphalocrosis medinalis* (Guenee) on rice in Punjab. Indian J. Ent. 62: 442-444.

Beevi, S.P., Hazarika, L.K. and Katti, G.S. 2001. Augmentative biocontrol within paddy IPM-Indian scenerio, pp. 93-117. In: Singh, S.P., Murphy, S.T. and Ballal, C.R. (Eds.) Proceedings of the ICAR-CABI Workshop, CABI Bioscience, UK and PDBC (ICAR), India.

Charnley, A.K. 1989. Mechanisms of fungal pathogenesis in insects. pp. 85-117 In: Whips, J.M and Lumsden, B.(Eds.) Biotechnology of Fungi for improving Plant Growth, Cambridge University Press, Cambridge.

Despande, M.V. 1998. Novel Biopesticides.CSIR Golden Jubilee Series, CSIR, New Delhi.

Ekesi, S., Akpa, A.D, and Ogunlana, M.O. 2000. Entomopathogenicity of Beauveria bassiana and Metarhizium anisoplae to the cow pea aphid, *Aphis cracivora* (Homoptera : aphididae). Arach. Phytopath. Pl. Prot. 33: 171-180.

Fawcett, H.S. 1944. Fungus and bacterial diseases of insects as factors in biological control. Bot. Rev.10: 327-348.

Ferron, P. 1978. Biological control of insect pests by entomogenous fungi.Ann. Rev. Entomol. 23: 409-442.

Ferron, P. 1981. Pest control by the fungi Beauveria Metarhizium. pp.465-482. In: H.D. Burges, H.D (Eds.) Microbial control of pests and plant disease 1970-1980, Academic Press, London & New York.

Fukaya, M. 1957. On the theoretical base for predicting the occurrence of the rice stem borer in the 1st generation. Ber Ohara Inst., 9: 357-375.

Gillespie, A.T. and Moorhouse, E.R. 1989. The use of fungi to control pests of agricultural and horticultural importance. pp55-84. In: Whips, J.M and Lumsden, B (Eds.) Biotechnology of Fungi for improving Plant Growth, Cambridge University Press, Cambridge.

Hazaraika, L.K., Puzari, K.C. and Bhuyan, M. 2005. Biopesticide technology and entrepreneurship development, pp. 27-33. N Doloi, S.K. and Mahanta, C (Eds.) Science and Technology for Regional Development: Case for North East India. Tezpur University, IIT Guwahati and C-MMACS, Bangalore.

Hazarika, L.K. and Puzari, K.C. 1995. White muscardine fungus (*Beauveria bassiana*) pathogenic to different stages of rice hispa (*Dicladispa armigera*), Indian J. Agric. Sci. 65 : 63-67.

Hazarika, L.K., Puzari, K.C. and Dutta, Pranab. 2005. Beauveria bassiana in rice hispa management in Assam. Biocontrol News and Information: 106N – 108N.

Hazarika, L.K. and Puzari, K.C. 1997. Field efficacy of white muscardine fungus (*Beauveria bassiana*) on rice hispa (*Dicladispa armigera*). Indian J. of Agric. Sci. 67: 465-467.

Hazarika, L.K. and Puzari, K.C. 2004. Entomopathogenic fungi in rice pest management. In:R. J. Rabindra., Hussain, S.S. and Ramanujam, B. (Eds). Proceedings of ICAR-CABI workshop on Biopesticide formulation and application. 9-13 December, 2002, PDBC, Bangalore, pp. 139-149.

Hazarika, L.K., Puzari, K.C. and Baruah, G. 1994. Fungi as microbial insecticides on mole cricket. Two and A Bud 41: 22-25.

Israel, P. and Padmanabhan, S.Y. 1978. Biological control of stem borers of rice in India, Final Tech. Report (U.S.P.L. 480 Pub).155 pp., CRRI, Cuttack.

Itoh, K., Ichihawa, K. and Nahagome, T. 1994. Microbial control of insect pests on paddy field by entomopathogenic fungi. Pathogenicity of Beauveria bassiana to the larvae of the rice stem borer (*Chilo suppressalis Wlk.*), Res. Bull. Aichi Ken. Agril. Res. Centre, 26: 79- 82.

Joshi, R.C., Cadapan, E.P. and Heinrich, E.A. 1987. National enemies of rice leaf folder, Cnaphalocrosis medinalis Guene (Pyralidae: Lepidoptera)- A critical review (1913-1983). Agri. Rev. 8: 22-34.

Landa, Z., Osborne, L., Lopez, F and Egal, J 1994. A bioassay for determining pathogenicity of entomogenous fungi on white flies. Biol. Cont. 4: 341-350.

Leger, R.J., Cooper, R.M. and Charnely, A.K. 1987. Production of cuticle degrading enzymes by the entomopathogen Metarhizium anisopliae during infection of cuticles from Calliphora vomitoria and Manduca sexta. J. Gen. Micribiol. 133: 1371-1382.

Li, H. K. 1985. Entomopathogenic micro-organisms of rice hoppers and leaf hoppers in China Intl. Rice Res. Newsl. 10: 13-29.

Li, H.K. 1986. Beauveria bassiana for controlling brown plant hopper (BPH) and green leaf hopper (GLH), Intl. Rice Res. Newsl. 11: 22.

Madelin, M.F. (1966). Fungal parasites of insects. A. Rev. Ent. 11:423-448.

Mains, E.B. 1952. North American entomogenous species of Corydyceps. Mycologia 50: 169-222.

Martins, S.F.D. and Megegalhaes, B.P. 1987. effect of the fungus *Materhizium anisoplae* (Metsch) Sorak on the rice stink bug Oebalus paecillus (Dallas, 1851) (*Heteroptera: Pentatomidae*) Annals da Sociedada Entomologica do Brazil, 16 : 81-91.

Milner, R.J. 1987. From Chinese medicine to mycoinsecticides: Man's exploration of entomogenous fungi. Mycologist (Oct.) 147-151.

Narayanasamy, P. 2001. Indigenous pest suppression, pp. 87-111. In : Upadhyay, R.K., Mukerji, K.G. and Chamola, B.P. (Eds). Biocontrol Potential and its Exploitation in Sustainable Agriculture, Vol 2. Insect Pests. Kluwer Academic/ Plenum Publishers, New York.

Nayak, P., Srivastava, P.P. and Naik, G. 1978. Pathogenicity of Beauveria brongniartii Tenella on the rice skipper, Parnara mathii F., Indian. J. Ent., 40 : 418-419.

Ndoye, M. 1977. The susceptibility of larvae and pupae of Chilo suppresalis Walker to various entomogenous fungi and the factors that determine susceptibility to the mycosis caused by *Beauveria bassiana* (Bals.) Vuill., Bull. Institute Fund. Afrique Noire, 39:303-317.

Padmanabhan, B. 1993. Studies on the natural infection of the entomogenous fungus Beauveria bassiana Sans. & Evans on some pests of low land rice ecosystem, J. Biol. Cont. 7: 109-111.

Phookan, M., Hazarika, L.K., Baruah, M and Puzari, K.C. 2006. Interaction of Dicladespa armigera : Chrysomelidae) haemocytes with *Beauveria bassiana* (In Preparation).

Puzari, K.C. Hazarika, L.K and Dutta, Pranab 2006. Beauveria bassiana as biocontrol agent of rice pests with special reference to rice hispa (*Dicladispa armigera*) In: V. Ambethghar (Ed.) Advances in Pest Management.

Puzari, K.C., Hazarika, L.K., Dutta, Pranab and Das, P. 2006. Compatibility of Beauveria bassiana (bals.) Vuill. with different common insecticides in rice. J. Biol. Contrl.

Puzari, K.C.and Hazarika, L.K. 1991. Efficacy of Beauveria bassiana combined with stickers or spreaders against rice hispa (RH). Int. Rice Res. Newsl. 16: 21.

Puzari, K.C., Hazarika, L.K. 1992. Entomogenous fungi from North East India. Indian Phytopath. 45:35-38.

Rabindra, R. J. 2005. Current status of production and use of microbial pesticides in India and way forward. Proceedings of the ICAR_CABI workshop on Biopesticides formulation and application PDBC, 9-13 Dec, 2002, p. 1-12.

Rao, P. S. 1975. Widespread occurrence of *Beauveria bassiana* on rice pests. Curr. Sci. 44: 441-442.

Rombach, M.C., Aguda, R.M., Roberts, D.W. and Shepard, B.M. 1985. Suppression of brown plant hopper, Nilaparvata lugens (*Homoptera : Deiphacidae*) by *Beauveria bassiana* and buprofezin, Malay. Pl. Prot. Soc., 5: 109-113.

Rombach, M.C, Aguda, R.M and Roberts, D.W. 1988. Production of *Beauveria bassiana* (*Deuteromycotina: Hyphomycetes*) in different liquid media media and subsequent conditions of dry mycelium. Entomophaga 33:315-324.

Vreden, V., G. and Ahmadzabidi, A. L. 1986. Pests of rice and their natural enemies in Peninsular Malaysia. Pudoc. Wageningen.

❑❑❑

Section C: Climatic Influences Over Agricultural and Horticultural Crops

14

Climate Influences on Potato Cultivation

Pulak Chaudhuri

Abstract

Global warming is predicted to have significant effects in global potato production. Like many crops, potaoes are likely to be affected by changes in atmospheric carbon dioxide, temperature and precipitatiom, as well as interations between these factors. As well as affecting potatoes directly, climate change will also affect the distributions and populations of many potato diseases and pests. Potato is one of the world's most important food crops. Potato prodcution must be adapted to climate change to aviod reductions in crop yields. Potato plants and potato crop are predicted to benefit from increased carbon dioxide concentrations in the atmosphere. The major benefit of increased atmospheric carbon dioxide for potatoes (and other plants) is an increase in thier photosynthetic rates which can increase their growth rates. Potato crop yields are also predicted to benefit because potates partition more starch to the ediable tubers under elevated carbon dioxides levels. Higher levels of atmospheric carbon dioxide also results in potatoes having to open thier stomata less to take up equal amount of carbon dioxide for photosynthesis, which means loss through transpiration from stomata. As a result the water use efficiency (the amount of carbon assimilated per unit water lost) of potato plants is predicted to increase.

Potatoes grow best under temperate conditions. Tuber growth and yield can be severely reducted by temperature fluctuations outside 5-30^{0}C. The Intergovernmental Panel on Climate Change predicts global temperatures to increase between 1.1 and 6.4 C by 2100. The effect of increasted *temperatures on potato production in specific areas will very depending partly on the current temperature of that area. Temperatures above 30^0 C can have a range of negative effects on potato, including: Slowing tuber growth and initiation. Less partitioning of strach to the tubers. Physiological damage to tubers (e.g. brown spots)., Shortened/non-exiestent tuber dormancy, making tubers sprout too early.*

Climate change predictions by the IPCC include likely change to water resource availability acros much of the globe. In many areas water resources are expected to decrease, particulary in semi-arid areas. Increases in extreme weather events including flash flooding are predicted, even in areas where overrall average rainfall is predicted to decrease.

Potates are sensitive to soil water deficts compared to other crops such as wheat, and need frequent irrigation, especially while tubers are growing. Reduced rainfall in many areas is predicted to increase the need for irrigation of potato crops. As well as affecting potatoes directly, climate chnage is predicted to affect many potato pests and diseases.Late blight benefits from higher temperatures and wetter conditions. Adaptation of potato farming practices and potato varieties to changing conditions caused by climate change could help maintain crop yields and allow potato to be gworn in areas with predicted conditions unsuited to current commercial potato cultivars. Methods to adapt potatoes to climate change include shifting production areas, improving water use and breeding new tolerant potato varieties.

Keywords: Potato production, Weather parameters, Disease and Pest attack.

Introduction

The potato is starchy, tuberous crop from the perennial *Solanum tuberosum* of the Solanaceae family. It is the word's fourth-largest food crop, following rice, wheat and maize.

Wild potato species also occur. The potato was originally believed to have been domesticated independently in multiple locations-but later genetic testing of the wide variety of cultivars and wild species proved a single origin for potatoes in the area of present-day southern Peru and extreme northwestern Bolivia (from a species in the *Solanum brevicaule* complex), where they were domesticated 7,000-10,000 years ago. Following centuries of selective breeding, there are now over a thousand different types of potatoes. Of these subspecies, a variety that at one point grew in the Chiloe Archipelago (the potato's south-central Chilean sub-center of origin) left its germplasm on over 99% of the cultivated potatoes worldwide.

The annual diet of an average global citizen in the first decade of the 21st century included about 33kg of potato. However, the local importance of potato is extremely variable and rapidly changing. It remains an essential crop in Europe (especially eastern and central Europe), where per capita production in still the highest in the world, but the most rapid expansion over the past few decades has occurred in southern and eastern Asia. China is now the world's largest potato-producing country, and nearly a third of the world's potatoes are harvested in China and India.

After potato plants flower, some varieties produce small fruits that resemble green cherry tomatoes, each containing up to 300 true seeds. Potato fruit contains large amount of the toxic *alkaloid solanine* and is therefore unsuitable for consumption. All new potato varieties are grown from seeds, also called "true seed" to distinguish it from seed tubers. By finely chopping the fruit and soaking it in water, the seeds separate from the flesh by sinking to the bottom after about a day (the remnants of the fruit float). Any potato variety can also e propagated vegetatively by planting tubers, spices of tubers, cut to include at least one or two eyes, or also by cuttings, a practice used in greenhouses for the production of healthy seed tubers. Some commercial potato varieties greenhouses for the production of healthy seed tubers. Some commercial potato varieties do not produce seeds at all (they bear imperfect flowers) and are propagated only from tuber pie.

Potato Varieties

There are about five thousand potato varieties worldwide. There thousand of them are found in the Andes alone, mainly in Peru, Bolivia, Ecuador, Chile, and Colombia. They belong to eight or nine species, depending on the taxonomic school. A part from the five thousand cultivated varieties, there are about 200 wild species and subspecies, many of which can be cross-bred with cultivated varieties, which has been done repeatedly to transfer resistances to certain pests and diseases from the gene pool of wild species to the gene pool of cultivated potato species. Genetically modified varieties have met public resistance in the United State and in the European Union.

The major species grown worldwide is Solanum tuberosum (a tetraploid with 48 chromosomes), and modern varieties of this species are the most widely cultivated. There are also four diploid species (with 24 chromosomes): *S. stenotomum, S. phureja, S. goniocalyx*, and *S. ajanhuiri*. There are two triploid species (with 36 chromosomes): *S. chaucha* and *S. juzepczukii*. There is one pentaploid cultivated species (with 60 chromosomes): *S. curtilobum*. There are two major subspecies of Solanum tuberosum: andigena, or Andean; and tuberosum, or Chilean. The Andean potato is adapted to the short-day conditions prevalent in the mountainous equatorial and tropical regions where it originated. The Chilean potato, native to the Chiloe Archipelago, is adapted to the long-day conditions prevent in the higher latitude region of southern Chile.

The International Potato Centre, based in Lima, Peru, holds an ISO-accredited collection of potato germplasm. The international Potato Genome Sequencing Consortium announced in 2009 that they had achieved a draft sequence of the potato genome. The potato genome contains 12 choromosomes and 860 million base pairs making it a mediumsize plant genome. More than 99 percent of all current varieties of potatoes currently grown are direct descendants of a subspecies that once grew in the lowlands of south central Chile. Nonetheless, genetic testing of the wide variety of cultivars and wild species affirms that al potato subspecies derive from a single origin in the area of present-day southern Peru and extreme northwestern Bolivia (from a species in the *Solanum brevicaule* complex).

Most modern potatoes grown in North America arrived through European settlement and not independently from the South American sources. However, at least one wild potato species, *Solanum fendleri*, is found as far north as Texas and used in breeding for resistance to a nematode species that attack cultivated potatoes. A secondary centre of genetic variability of the potato is Mexico, where important wild species that have been demissum, as a source of resistance to the devastating late blight disease. Another relative native to this region, *Solanum bulbocastanum,* has been used to genetically engineer the potato to resist potato blight.

Potatoes yield abundantly with little effort, and adapt readily to diverse climates as long as the climate is cool and moist enough for the plants to gather sufficient water from the soil to form the starchy tubers. Potatoes do not keep very well in storage and are vulnerable to molds that feed on the stored tubers, quickly turning them rotten.

India produces potato crop annually in two harvest seasons, *Kharif* and *Rabi*. *Kharif* produce is available in the market from August to October, which accounts for 15% to 20%. The Rabi crop is harvested from mid-December to mid-April, which accounts for 80% to 85%. Only, the *rabi* potato is stored till November and is made available steadily for domestic as well as export market.

The various climatic factors that influence the potato are:

1. Temperature.
2. Relative Humidity.
3. CO_2 concentration
4. Photoperiod
5. Radiation (sunshine hours & intensity) etc. through their direct effect and the effect of interaction among themselves.

Temparature: Potato grows optimally between 15-30°C. Temperatures higher than the optimal constitute heat stress. But the optimal temperatures are different for foliage growth and tuber growth. Whether or not potato produces tuber depends largely on the minimum night temperature and not on daily mean temperature. Potato plants require cool nights for optimum tuberization and bulking. When potato plants are exposed to high temperature it reacts in the following ways i.e. (i) Photosynthesis is reduced, (ii) Respiration rate is increased, (iii) Haulm growth is increased, (iv) at high night temperature translocation of sugar from leaves to tuber is hampered. Thus the inability of the potato to tuberize under high temperature has often been cited as the major barrier to its production under high temperature. At night temperature of 25°C and above, there may not be any tuberization at all. The best night temperature has been observed being 10–18°C for tuberisation and bulking. As a result, potato is a cool climate crop which could not spread to tropical areas of the world where night temperature is relatively high.

A potato grower of Khowai, Tripura

Deviation in temp. in different months and the yield realised in Tripura

Months	2006-07		2007-08		2008-09		2009-10		2010-11	
	Max.	Min.	Max.	Min.	Max.	Min.	Max.	Min.	Max.	Min.
Nov	-0.37	1.21	-0.80	1.64	-0.34	0.07	0.09	1.04	2.18	1.97
Dec	-0.05	1.25	-1.42	1.21	-1.25	1.05	-1.43	1.43	-0.16	2.06
Jan	-1.97	0.02	-1.77	2.29	-0.70	1.78	-1.86	-0.03	-2.00	0.20
Feb	-2.20	1.02	-2.29	-0.70	0.37	0.66	0.03	0.13	-1.76	1.33
Yield (t/ha)	**15.03**		**14.58**		**17.43**		**16.64**		**18.00**	

No. days in which minimum temperature was recorded less than 10°C

Year	2006-07	2007-08	2008-09	2009-10	2010-11
No. of days	17	7	7	23	12
Yield (t/h)	15.03	14.58	17.43	16.64	18.00

However, most of the importance of potato as a food crop exists in colder regions of the world either highland tropics or northern and southern latitude temperate climate. But, population in these areas is low and transporting potatoes from these areas is costly. So, extending potato cultivation to the tropical areas of the world become imperative to overcome the food shortage. Potato is important for these areas because it is capable of producing twice as much dry matter and protein per unit area & time as compared to cereals. Moreover, potato provides high quality proteins.

The recommended agronomic practices for modifying heat stress are – Use of quality seed tubers of correct physiological age & planting at 7cm depth, closer spacing & mulching, intercropping with maize, planting in the cool facing side in hills, crop rotation to avoid wilt diseases, use of right genotype etc.

The cultivation of potato is restricted to sub-tropical and temperate zones of the world. But, for the temperate areas like Europe and Scandinavian countries, in areas of higher latitudes- the growing season will expand- may bring prosperity to yield. The most vulnerable areas will be the sub-tropical plains where further shorter growing season will lead to reduced yield.

Light intensity as effective in a certain photo-periodism, it is found that, a clear photoperiodic reaction occurred (Bodlacnder, 1959) at light intensity of 5-10 lux and a complete change from SD to LD characteristic occurred at 40-50 lux.

Moisture supply: - Hukkeri *et al* (1970) observed that moisture stress during the early stage of growth after planting was not as significant enough to cause a yield reduction, as in the stolon formation stage. Therefore, in hills of India where potato is grown during summer problem may make it worse due to early drought and excessive water in the later stage of growth.

It is expected that the change in temperature will be uniform, but not that of rainfall. There will be an increase in rainfall in mid- to high latitudes - areas with no water shortages - and in subtropical zones, where we will see rainfall on fewer days but with greater intensity. Rain-fed conditions are likely to be of strongly affected by water shortages, with yield reductions up to 50%. In arid regions, there will clearly be a drop in yield.

Carbon-Dioxide supply: Gross fluctuation of CO_2 concentration in the field condition does not occur generally. Chapman (1953) observed a significant effect of reduction in photosynthesis due to lesser CO_2 concentration during early hours of a day. Concentration of CO_2 to the level of 500ppm is the best for photosynthesis during the morning hours.

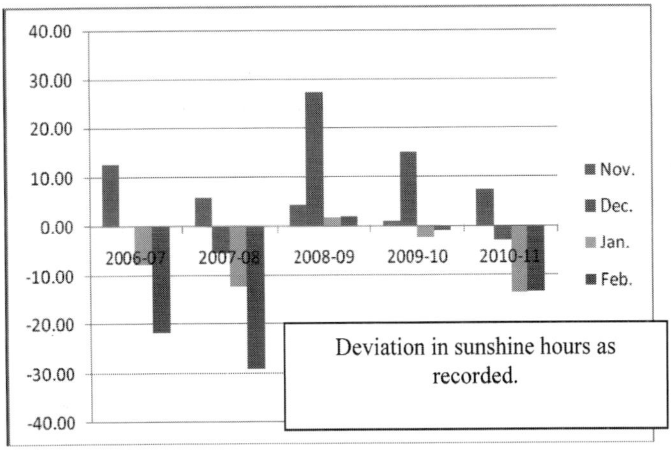

Deviation in sunshine hours as recorded.

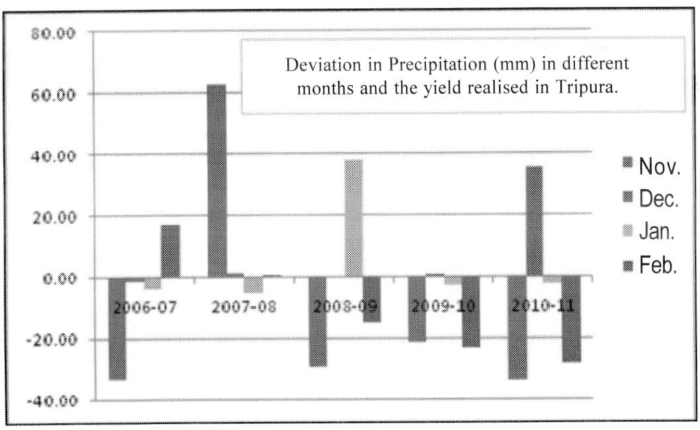

Deviation in Precipitation (mm) in different months and the yield realised in Tripura.

Effect of higher concentration of CO_2 as apprehended due to global warming are:-

Higher concentration will lead to more yield.

Increasing CO_2 concentrations - little effect on production of biomass above ground, but below ground biomass will be significantly enhanced.

Increasing levels of ozone - an overall reduction of photosynthetic efficiency and a significant decrease in tuber starch content, but an increase in the ascorbic acid concentration.

Presently, the level of CO_2 is about 385 ppm and the latest IPCC emission scenarios projected levels by the year 2100 ranging from 540 to 970 ppm. The yield increases by about 10 percent for every extra 100 ppm.

Now, due to global warming, the Scientists apprehend that a soaring temperature, changed pattern of precipitation, higher concentration of CO_2 and a change in solar radiation is going to influence the potato scenario in future. Such situation coupled with short days will lead to production of more foliage of pale colour due to lesser chlorophyll synthesis.

Effect on diseases: Given the thermal limit for late blight - 22°C - increases above that temperature threshold in Europe may prevent infections. At the present northern limit of potato cultivation in the USA, Canada and central Russia, late blight could increase significantly as temperature increases, but at the same time, warming further north may open up new zones for potato production, with minimal late blight risk.

Effect on pests: Increasing quantity and frequency of rainfall would also create conditions more favourable to viral disease vectors. Other researchers have predicted an increase in the Colorado potato beetle's area of diffusion in Europe, as well as in the area infested by potato cyst nematode. In conditions of long period having higher humidity favoured more insect attack in our state.

Experience: Since, there is no record of atmospheric parameters to study the effect of climate in the state on potato. However, when the data obtained from the automatic weather station established at HRC was analysed it was found that that the monthly mean max. temperature and monthly mean min. temperature for last five years, we found that only in four occasions the monthly mean maximum temperature were higher than the normal and only in two occasions the monthly (mean) minimum temperature was lower than the normal during the potato growing season i.e. from November to February. Since, the crop is mostly affected at less than 10°C, the number of days were maximum during the year 2009-10 followed by 2006-07.

As far as effective sunshine hours are concerned, the state is receiving lesser & lesser particularly in the month of January & February.

As far as relative humidity is concerned it was observed that it was never below the normal or at par with normal during the most effective period i.e. January & February either in the morning or evening, which has facilitated the outcome of late blight.

With regards to precipitation, establishment of any relationship with the yield was not possible for this short range of period under study, though deviation from the normal was observed.

Deviation in Relative Humidity in different months and the yield realised in Tripura

Months	Normal		2006-07		2007-08		2008-09		2010-11	
	Morn	Even	Morn	Even	Morn	Even	Morn	Even	Morn	Even
Nov	99.00	56.00	0.00	0.00	0.00	8.00	0.00	-3.00	0.00	1.00
Dec	91.00	60.00	7.00	-13.00	8.00	-6.00	8.00	-5.00	8.00	-8.00
Jan	84.00	44.00	14.00	3.00	14.00	12.00	15.00	13.00	14.00	8.00
Feb	80.00	37.00	4.00	26.00	17.00	8.00	15.00	1.00	18.00	2.00
Yield (t/ha)			**15.03**		**14.58**		**16.64**		**18.00**	

Early-maturing varieties allow the crop to escape stresses. They also ensure a harvest in the tropical highlands, for example where the high incidence of frosts causes big losses.

Adjustments of the date of planting, using different potato varieties and improving soil moisture status especially in dry regions might be of useful. But in practice, adaptation options could be a complex one. The planting time depends on factors like the preceding crop, water availability, occurrence of pests and diseases, and above all the prevailing market price. Such options may not be suitable to farmers in some regions. Another strategy may be shifting potato production towards areas where there is currently no potato production. In tropical high altitudes, cultivation could expand. In temperate areas, there could be considerable expansion of the potato area.

Conclusion

The erratic behavior of the climate has a significant effect on yield. In lower latitudes & lower altitudes the crop is bound to suffer more in coming days in areas like Tripura. Therefore, a keen eye is required to set for anything that seems out of the ordinary & a willingness to track down the cause. Moreover, actions to be made to find new genotypes as well as better agronomic practices which can withstand or escape the abiotic & biotic stresses.

Despite this, potato is likely to become more important as a crop against food security as temperatures rise, since it has higher water productivity, yielding more "crop per drop" - than rice wheat and maize.

References

Daccache, A., Keay, Jones, Weatherhead, Stalham, Knox 2012. "Climate change and land suitability for potato production in England and Wales: impact and adaption" Journal of Agricultural Science 150: 161-177.

Hijmans, Robert J. 2003. "The Effect of Climate Change on Global Potato Production" American Journal of Potato Research 80: 487-280.

Levy, David; Veilleux, R.E. 2007. "Adaptation of Potato to High Temperatures and Salinity A Review". American Journal of Potato Research 85: 487-506.

Liu, Zhenyu, Helterman (2006). "Identification and characterization of RB-orthologous genes from the late blight resistant wild potato species Solanum verrucosum". Physiological and Molecular Plant Pathology 69: 230-239.

Minhas, J., Rawat, Govindakrishnan, Kumar 14-17. "Possibilities of enhancing potato production in non-traditional areas". Potato Journal 38.

Zarka, K.A., Greyling, Gazendam, Olefse, Felcher, Bothma, Brink, Ouemada, Docuches 2010. "Insertion and characterization of the cry 1Ial Gene in the Potato Cultivar Spunta for Resistance to Potato Tuber Moth". Journal of the American Society for Horticultural Science 135: 317-324.

❑❑❑

15

Climatic Requirements for Horticultural Crops with an Approach to Forecast of Yield & Disease- Insect Occurrence

Biswajit Das

Abstract

India has diversified agro-climatic conditions with significant variability in climatic factors viz, rainfall pattern, temperature, humidity, sunshine hours and wind speed etc. A no of fruit crops and vegetable crops are grown in India with significant differences in climatic requirements for sustainable growth and production of these crops. Attempt has also been made to have off season production of horticultural crops. Weather variability could also play a pivotal role in disease and pest incidence. So, forecasting of yield along with disease pest incidence should be given top priority to save the crop from weather vagaries and increase the crop productivity.

Keywords: Weather variability, Fruit crops, Vegetable crops, Spices, Plantation crops; Disease- Insect incidence, Yield forecast.

Introduction

Horticulture has occupied prime position in Indian agriculture considering its higher profit and rich nutritive values. Horticulture as a whole comprises mainly fruits, vegetables, flower and Landscape plants, plantations crops, medicinal and aromatic plants and spices. Main research priorities are agro-climate specific selection and improvement of suitable superior varieties, development of sustainable and crop specific agro-techniques including disease and insect-pest management on integrated modules to increase productivity and overall production of quality horticultural commodities, further, technology generation for proper harvesting, post harvest handling and product development, as well as proper marketing of fresh produce and processed products are also prioritized under horticultural policies and planning. After achieving self sufficiency in food grain production through 'Green Revolution' in 60s, the main focus of Indian agriculture was directed towards anoverall development of'Horticulture' in mid Eighties as a means of diversification for making agriculture more profitable along with nutritional and livelihood security through more financial allocation and private investment, efficient land use, optimum utilization of natural resources including micro irrigation, generation of new and improvised production technologies. Over and above, several mega projects such as National Horticulture Mission, Horticulture Mission for North East and Himalayan States, All India Coordinated Research Project on:Sub-tropical fruits, Tropical Fruits, Arid Zone Fruits,Vegetable Crops, Floriculture, Potato, Tuber crops, Cashew, Mushroom, Spices, Palms, Betel vines and Medicinal and aromatic plants have been implemented with the prime objective to accelerate the research and development on different aspects of horticulture. Thus, the changing scenario encourages private investment to go for 'Hi-Techhorticulture' with micro-propagation, protected cultivation, drip irrigation, fertigation, and integrated nutrient and pest management, besides making use of latest post harvest technologies particularly in the case of perishable commodities. As a result, horticulture crop production has moved from rural confines to commercial ventures and has attracted youth, since it has proved to be intellectually satisfying and economically rewarding. Recently, specialized fields such as precision farming and Nano technologies have been also developed to increase efficiency of inputs for high value quality production.

Agro-climatic Zones in India and Classification of Fruit and Vegetable Crops

Diversified agro-climatic conditions are found in India which have been defined on the basis of the uniformity in major climatic factors such as rainfall patterns, temperature regimes, evapo-transpiration rate, soil types etc in relation to suitability for crop cultivation. Indian Council of Agriculture has recognized eight zone:

1. Humid Western Himalayan Region: Jammu & Kashmir, Himachal Pradesh, Uttarakhand.
2. Humid Bengal Assam Region: West Bengal and Assam.
3. Humid Eastern Himalayan Region: North Eastern states, Andaman Nicobar Islands.

Table 1: Area, Production and Productivity of Horticultural Crops in India

Horticultural Crops	Area (000 Ha.)	Production (000 MT)	Productivity (MT/Ha.)	Horticultural Crops	Area (000 Ha.)	Production (000 MT)	Productivity (MT/Ha.)
Banana	830	29780	35.9	Potato	1863	42339	22.7
Mango	2297	15188	6.6	Tomato	865	16526	19.1
Citrus	846	7464	8.8	Onion	1064	15118	14.2
Papaya	106	4196	39.6	Brinjal	680	11896	17.5
Guava	205	2462	12.0	Tapioca	221	8076	36.5
Apple	289	2891	10.0	Cabbage	369	7949	21.5
Pineapple	89	1415	15.9	Cauliflower	369	6745	18.3
Sapota	160	1424	8.9	Okra	498	5784	11.6
Grapes	111	1235	11.1	Peas	370	3517	9.5
Pomegranate	107	743	6.9	Sweet Potato	113	1047	9.3
Litchi	78	497	6.4	Chillies	792.1	1223.4	1.5
Coconut	1895.9	10840.0	5.7	Garlic	200.6	1057.8	5.3
Cashewnut	953.2	674.6	0.7	Turmeric	195.1	992.9	5.1
Arecanut	400.1	478.0	1.2	Ginger	149.1	702.0	4.7
Others	1265	7583	6.0	Others	2083	27557	13.2
Fruit and plantation Total					96322	86870.6	19.3
Vegetable and spices Total					9831.9	150530.1	33.9
Horticulture Total including aromatic, plantation, spices, nuts, flower and mushroom etc.					21825	240531	11.0

(NHB, 2012)

4. Sub-Humid Sutlez:Ganga Alluvial Plains comprising North India States namely Punjab, Uttar Pradesh, Bihar.
5. Sub-Humid to Humid Eastern: Eastern Madhya Pradesh, Odisha, Andhra Pradesh.
6. Arid Western Plains: Haryana, Rajasthan, Gujarat, Dadra and Nagar Haveli, Daman & Diu.
7. Semi-Arid Lava Plateaues: Maharastra, Western Central Madhya Pradesh and Central Island Goa.
8. Humid to semi-arid Western Ghats: Karnataka, Tamil Nadu, Kerala, Pondicherry and Lakshadweep.

Sub-humid climate is found in the temperate to sub-tropical conditions. Index or potential evapo-transpiration range is +20 to -20. Western & Central Madhya Pradesh, Vidarbha, Uttar Pradesh, Bihar, Tarai and Doon areas of Uttarakhand, Sub-mountain tracts of Himachal Pradesh and Jammu, West Bengal and Nilgiri hill. Hot humid conditions represents tropical climate and index or potential evapo-transpiration range is +20 to 100 and annual rainfall is even above 750 mm mostly during May to November. Areas are Konkan, Kerala, Coastal Tamil Nadu, Assam, Tripura, Manipur, Odisha, West Bengal, Eastern Madhya Pradesh and Andhra Pradesh. It is heavy rainfall zone with more than 200 cm annually. Semi-arid zone consists of dry farming areas with -20 to -40 index of potential evapo-transpiration. Annual rainfall is around 300-650 mm during June-September with hot and dry summer and cool winter in some areas, while, summer is hot and wet with dry winter.Areas are Northern and Eastern parts of Rajasthan, Gujarat, Marathwada of Maharastra, Malwa region of Madhya Pradesh, Mysore, Rayalaseema of Andhra Pradesh, western parts of Punjab, some parts of Delhi and Western Uttar Pradesh. Hot Arid zone comprises -40 to -60 index of potential evapo-transporation with the rate of 7-9 mm/day, 5.3-6.4 mm/day and 1.8-2.9 mm/day in summer, monsoon and winter, respectively.Annual rainfall of around 100-450 mm occurs mainly during June to September and temperature ranges from 40°C in summer to 14^{oC} in winter. Areas are Rann of Kutch to major western parts of Rajasthan. Some areas in Punjab, Haryana, Karnataka and Andhra Pradesh are also included in arid zone (Anonymous, 2000). Dry temperate areas are in Kinnaur and Chamba in Himachal Pardesh and Gurejand Dada areas in Jammu& Kashmir,whereas, cold arid areas are Lahul Spiti in Hamachal Pradesh and Ladakh in Jammu and Kashmir.

Fruit Crops

Temperate fruit crops are mainly deciduous in nature grown in cooler climate which experiences extreme to severe winter with snow and rainfall as precipitation and mild summer. These fruit crops require chilling temperature for breaking bud dormancy. Fruit crops are apple, pear, cherry, peach, apricot, plum, walnut, almond, chestnut, pistachionut, pecannut etc. North Western Himalayan states of India namely Jammu and Kashmir, Himachal Pradesh and Uttarakhand are major producer of these fruit crops. Areas under temperate fruit crops are also expanding in hilly tracts of North Eastern Himalayan states like Arunachal Pradesh, Sikkim and Nagaland, and some selected hilly parts of Western Ghat such as Otty and Kudaikanal. Subtropical fruits are adapted to mild to severe winter temperature with winter rainfall and veryhot summer. Moreover, these areas experience cool winter with medium to severe fog and spring frost. Fruit crops include litchi, mango, guava, loquat, ber and citrus *etc*. Areas under subtropical zone are Jammu, lower and foot hill areas of Himachal Pradesh and Uttarakhand, Haryana, Punjab, Rajasthan, Delhi, Gujrat, Madhya Pradesh, Bihar, Western Parts of West Bengal and Jharkhand.Tropical fruits are grown in areas where winter is absent or mild and frost free, summer is very hot. Fruit crops are pineapple, banana, cashewnut, mango, citrus, grape, Mangosteen, Durian, Rambutan and pomegranate etc. Areas under tropical zone are Tamil Naddu, Andhra Pradesh, Karnataka, Maharastra, Goa, Kerala, Odisha, Eastern parts of West Bengal, Tripura, Assam, Manipur and Meghalaya, Mizoram and some parts of Arunachal Pradesh and Nagaland.

Vegetables Crops

Horticultural grouping of vegetables are mainly based on similarity in nature of characteristics/culture for cultivation or belonging to same botanical family.

1. Cole crops: Cabbage, Cauliflower, Chinese Cabbage, Khol Rabi, Broccoli, Brussels Sprout, KnolKhol.
2. Cucurbits: Cucumber, Squash, Melon, Pumpkin, Gourds, Gherkins.
3. Solanaceous: Tomato, Brinjal, Capsicum, Chilli.
4. Legumes: Peas, Cowpea, Broad Bean, French Bean.
5. Pot herb/Greens: Spinach, Chard, Chocho, Mustard.
6. Salad crops: Lettuce, Celery, Parsley, Endive.
7. Perennial: Asparagus, Drumstick, Parwal.
8. Bulb and corm crops: Onion, Garlic, Leek, Chive, Sallot.
9. Root crops: Radish, Carrot, Beet, Turnip.
10. Tuber crops: Potato, Taro, Cassava, Elephant Foot Yam, Sweet Potato.

Vegetables are also grouped on the basis of season of cultivation as cool and warm season vegetables (Swarup, 2012). Cool season crops are cole crops, kale, onion, garlic, leek, peas, carrot, radish, turnip, beet, asparagus, artichoke, spinach, celery, parsley and potato. Warm season crops are eggplant, cucumber, gourds, pumpkin, tomato, sweet pepper, chilli, beans, okra, melons and cowpea.

Cool season vegetables require vernalization or chilling at low temperatures for flowering. However, heat tolerant varieties are available in cabbage, cauliflower, radish, turnip and carrot suitable for growing in warmer region. In North and East Indian condition,Kharif season vegetables are okra, cucurbits, brinjal, chillies and carrot. Rabi season crops are cole crops, carrot, radish, turnip and beet.

On the basis of photoperiod, long day vegetables are potato, cowpea, lettuce, radish, spinach beet, exotic onion, cabbage, beet, cauliflower, carrot and turnip which require 12 or 13-14 hours of day length for flowering. Short day vegetables are *Dolichos* bean, cowpea, cluster bean, winged bean, sweet potato, pepper, potato, cucumber, Indian onions and pea, whereas, day neutral vegetables are beans, cowpea, tomato, brinjal, okra, cucurbits, french bean, sweet bean, chilli, artichoke and asparagus. Indian onion varieties are mainly short day types which require 10-11 hours for flowering and long day European type onions require 14-16 hours day length. More specifically, beet, carrot, radish and turnip are day neutral for formation of underground part. Vegetables such as lettuce, carrot, potato, sweet potato, tomato and sugar beet are C_3 and *Amarenthus* and globe artichoke are C_2 crops.

Weather and Crop Interaction for Better Growth and Production

Fruit Crops

Temperate fruit crops require chilling temperature during winter dormancy period for specific hours and fulfillment of sufficient chilling hours at 6-9^0 C for 600 hours (Peach) to 1500 hours (apple) ensures higher proportion of bud break and flower retention. High chilling apple varieties namely Red Delicious, Golden Delicious etc. require 1300-1600 Hours, whereas, mid to low chilling apple varieties namely Starkrimson, Mollies Delicious, Shlomit, Maayam etc. require 600-1000 hours. High chilling peach varieties namely Contender, Reliance, Red Haven require 800-1000 hours, whereas, low to mid chilling varieties namely Flordasun, Flordaking, Flordacrest, SaharanpurPrabhat*etc.* require 200-600 hours. The threat arises when there is fluctuation in winter temperature and precipitation. In the recent years, changing climatic patterns in the entire North Western Himalayan region has disturbed the fruit orchards in terms of plant growth, flowering, fruit set and final production. Majority of the orchards have become unproductive due to decline in production potential and are severely affected with diseases and insect-pests. Low and mid hill areas or even some high hill areas have become unfit for high chilling varieties of fruit crops and are being shifted towards higher altitudes. Temperate fruit growers in the low-mid altitude zone are opting for low chilling varieties or diversifying with vegetable crops. Changing climate also affecting the pollinators and pollination, thereby, resulting into poor fruit set. Tropical and subtropical fruitsshow very less degree of tolerance to winter and spring frost. Most tropical crops require warm and humid environment and show intolerance below 10°C.Favourable temperature in tropical areas is around 27-35°C, however, warmest months being only a few degrees higher than the colder months. Temperature differences under tropical conditions between night and day, at any given time, is greater than those between winter and summer.Little variation in day length is also recorded.On the other hand, subtropical conditions experience hotter summers and cooler winters with somewhat winter frost. Relative humidity is lower and day length differences become greater with increased latitude(Samson, 1986 and Nakasona and Paul, 1998).

Another important factor is heat units during fruit growth and development. Accumulation of heat units is required for bud initiation, bud development, bud break, flowering and fruit growth and maturity. Heat requirement varies depending upon the crop and variety. Heat units are also used for fruit maturity indices. Growing degree days (GDD) or heat units is the total heat required for plants or any organism to develop or grow form one stage to another in its life or growth cycle. Heat units are accumulated over time and between minimum and maximum temperature thresholds. Daily maximum (T_{max}) and minimum (T_{min}) temperatures are recorded and compared to the base temperature (T_{base}) which is usually 10°C. GDD is calculated as$[\{(T_{max} + T_{min})/2\}-T_{base}]$. Plants phenological events and appearance of insects and diseases has a close relationship. On the basis of GDD prediction of insect population or pathogens can be done for better pest management.

Mango

It is the leader of all the fruit crops and grown in tropical and sub-tropical areas. Indian climate is most favourable for mango cultivation and can be grown from sea level to 1000 mamsl. Altitude plays an important role in flowering and it has been observed that for every 12 m increase in altitude, there is a delay by 4 days in flowering. Thus, under Indian tropical and sub-tropical conditions, early flowering in mango takes place in the Southern India, followed by Eastern and Northern India. Temperature plays a vital role in the timing of flowering and ripening of fruits. Even in the North Indian conditions, flowering in Bihar, West Bengal and Eastern UP takes place comparatively earlier then other places of North India due to rise in temperature early in the season. Fruit ripening is early in South India, then in Eastern India and thereafter, in the Northern parts of India. In south India, flowering starts during the month of September in early cultivars and end by November in all other cultivars. Optimum temperature around 26° C (23-26°C) is favourable for mango. However, change in climatic factors over the years have influenced the flowering time and in Bangaluru conditions, it has been observed that flowering is taking place as late as in the month of January in place of November-December in many areas.In the subtropic, temperature 5-10°C favours synchronous flowering of mango. Whereas, higher temperature in the range of 10-18°C favours asynchronous flowering. In Tropics, cool temperature followed by rise in temperature causes poor fruit set (Rajan, 2008). Annual rainfall at around 75-100 cm is favourable with a wide range of variation under Indian conditions from 25-250 cm in different areas. However, absence of rainfall during flowering is beneficial; otherwise, rainy days during bloom period is detrimental for pollination and favours anthracnose and powdery mildew. Exposure of the fruits to high sun light intensity during hot periods (above 40° C) for longer duration favours spongy tissues in western India. Continuous overcast and rainfall during fruit maturity period favours fruit fly damage. Frost damage is another threat to mango cultivation especially in northern India which is depended upon the intensity of frost, age of the tree, stage of plant growth and soil moisture as well as tree canopy. Death of new shoots, oozing of gums, leaf scorching and bark splitting are expressions of frost damage. If sub-zero or even 1°C temperature prevails for longer duration then young trees and new shoots are severely damaged. All these factors also augment to the alternate bearing in mango. Protection from hail storm and strong wind velocity is also necessary.

Litchi

Cultivation of litchi is restricted to a narrow climatic zone comprising from Tripura, West Bengal, Bihar, Uttar Pradesh, Tarai and Doon areas of Uttarakhand, some plain areas of Himachal Pradesh and Jammu. Favourable climatic conditions are short, somewhat dry, frost free and cool winter with maximum temperaturerange of 20-22°C and minimum temperature in the range of 5-10°C, whereas, summer should be long and hot with temperature 25-30°C. Temperature below 20°C is favourable for higher flowering. More specifically, dry weather for at least 2 months and accumulation of total 200 hours below 13°C coupled with cooler winter temperature in the range of 0-13°C favoursflower bud differentiation, profuse flowering and better yield. Areas where winter temperature goes above 25°C are not suitable for litchi cultivation, whereas, freezing temperature below -1°C causes serious damage to young trees and shoots and dry chilling wind kills inflorescence. Annual rainfall in the range of 120 cm is good for better plant and fruit growth. Temperature influences the flowering time, bloom intensity and duration, sex ratio, fruit set, fruit growth and quality and ripening time. Temperature around 18°C during early flower development stage favours higher proportion of female flowers and around 23°C lowers it. High temperature in the range of 30-40°C prevails in most of the north Indian Litchi belts during fruit growth and development. However, favourable range is 20-35°C (optimum 30-32°C). Fruit growth declines below 20°C. Effective pollination takes place at 19-22°C, while, pollen tube elongation is restricted when temperature goes below 15°C. In the temperature range of 15-27° C, higher yield can be expected if at least 10% ovules are fertilized. However, female flower abscission occurs when temperature goes above 33°C for longer period. High temperature, atmospheric dryness (low humidity), windy and moisture stress for longer duration during fruit development and maturity favourssun burning,reduce fruit weight and ultimately over all fruit quality and sudden rainfall thereafter causes fruit cracking.Critical period is April, May and June for litchiand intermittent pre-monsoon rain is beneficial for proper fruit growth and quality(Singh and Singh, 2001). Annual rainfall in the range of 110-150 cm is good for litchi cultivation. Litchi growing areas experiences heavy rainfall during July to September. Heavy rain during fruit maturity increases humidity and incidence of diseases and fruit borer becomes a problem.In the foot hill and plainarras of northern Himalayan states temperature below 1°C in the night, wind free bright day time sunshine followed by sudden drop in temperature from evening hours are congenial for frost which causes shoot and flower bud damage.

Guava

Cultivation of guava is most suitable in areas which experiences relatively frost free cooler winter (10°C in night) and dry weather. Altitude range around 100 m amsl is most suitable, however, it can be grown even upto 1500 m amsl. Hot summer with temperature range of 23-30°C is good for fruit set and development, however, below 23°C and above 27°C reduces fruit set. Rainfall in the range of 100-200 cm is suitable mostly during July-September.Alternation of dry and wet condition is ideal rainfall pattern for good guava culture. It can also tolerate very high temperature as high as 40°C and drought spell prevails in North India during summer (April-June). However, prolong drought and low humidity reduces fruit set and enhances post-set fruit drop. Guava is also sensitive to winter frost and causes damage to the young plants, flowers and new shoots.Sufficient soil moisture ensures better vegetative growth, flowering and fruit development. Accumulation of insufficient heat unit during winter extends fruit maturity period, but shortens the harvesting period. Low temperature in winter causes defoliation. Guava flowering time differs in different parts of India. Flowering in guava takes place in the month of April-May, September-October and February-March in eastern India, June-July, October-November, April-May in Western India, April-May and August-September in North India, April-May, August-September, October-November in South India. Spring season (February-March) flowering is called Ambebahar and fruits are harvested in the month of July-September (rainy season). Fruits are Insipid, watery, taste and quality is poor. Monsoon season (June-July) flowering is called Mrigbahar and fruit is harvested in November-January (winter season). Fruit quality is very excellent and commercially profitable. Autumn-winter flowering (October) is called Hasthabahar and fruits are harvested in February-April. Fruit quality is good, however, yield is less. Flowering is discouraged in other seasons so that profuse flowering is induced in Mrigbahar.Chemicals namely NAA (50-100 mg/l) or 2,4-D (50-100 mg/l) or DNOC (5-10 mg/l) or urea (10%) is sprayed at 40-50%flowering stage to deblossom in April-Mayso that yield of quality fruits is increased in winter.Rainy season fruits are infested by fruit fly. Irrigation is not applied during the dry period in the months of February-mid May and hot summer to facilitate defoliation in April-May to force the plants into rest.Tree basins are prepared by digging in June, manures and light irrigation is applied immediately, then, intensity and quantity of irrigation is increased. Profuse flowering is induced after 20-25 days. Red colour development on the peal is mainly dependent on cool nights.

Grape

Grapes are grown in temperate, sub-tropical, mild tropical and hot tropical climate. In temperate climate, grapes are cultivated in dry temperate climate of Kinnaur, Himachal Pradesh and wet temperate climate in Kashmir. However, good quality grapes are grown in sub-tropical, hot tropical and mild tropical climatic regions in India. In north Indian plains suitable areas are Delhi, Bhatinda, Gurdaspur and Ludhiana districts of Punjab, Meerut area of U.P., Hissar and Jind areas of Haryana. During winter vines enter into dormancy and bud break takes place in first week of March. Suitable varieties are those which can complete berry maturity within 90-95 days to avoid rain damage in June. Perlette is most preferred variety and single pruning in December and single harvesting is done. In the mild tropical climate, Bangaluru and Kolar areas of Karnataka, Chittor in Andhra Pradesh, Coimbatore, Theni and Madurai areas of Tamil Nadu experience milder climate with temperature not exceeding 36°C and minimum around 12°C. Main varieties are Anab-e-Shahi, Bangalore Blue, Gulabi and Bhokri. Varieties resistant to rain damage or whose bud differentiation is not hampered under cloudy sky or rain are pruned in any time of the year and harvesting can be done five times in every two years. Otherwise, two pruning and two harvesting schedule is followed, except Thompson Seedless. The prime grape growing zone is hot tropical region comprising Nasik, Sangli, Solapur, Pune, Latur and Osmanabad in Maharashtra, Hyderabad, Ranga Reddy, Mahbubnagar, Anatapur and Medak areas of Andhra Pradesh, Bijapur, Bagalkot, Belgaum and Gulberga region of northern Karnataka. Double pruning is practiced and single harvesting is done. Temperature is in the range of 8-42°C. Vine growth is suffered due to drought and salinity. Pink blush on green berries develops if temperature falls below 8°C. Anab-e-Shahi, Thomson Seedless, Tas-A-Ganesh, Sonaka, Sharad Seedless and Flame Beauty are main varieties. Canes are pruned to single node spurs in March-May to facilitate cane development and these canes are forward pruned in October-November for fruiting. Due to unfavourable weather conditions pruning before October and after November is not possible. Extreme hot temperature is not favorable for growing wine grapes.

Vegetable Crops

Temperature specifically diurnal and optimum temperature plays an important role in selection of seasonal vegetable. Optimum temperature between 10-24°C is suitable for most of the vegetables. In general, cool season vegetables grow well below 18°C and warm season vegetables above 18°C (Swarup, 2012). Many vegetables are susceptible to frost or chilling injury at low temperature (below 10°C). North Indian plains experience severe winter fog and spring frost which cause damage to the vegetables. This damage is more critical at flowering and anthesis stage and crops such as tomato, potato and brinjalare more sensitive. Cabbage, Onion, Garlic, Spinach, Asparagus, Celery and Parsley are relatively frost tolerant. High temperature during flowering, fruit set and fruit development effect fruit quality and production. Pungency in onion and radish is more due to prevalence of prolong high temperature during fruit development. Leaf burning and sun scald are major in tomato and leafy vegetables when solar radiation in very intense. Verbalization, *i.e.* exposure to specific low temperature is very important for cool season vegetables for flowering.Low temperature (below 7^{0}C) for 2-8 hours is essentially (qualitative or obligate) required for crops such as beet, cabbage, kale, brussels sprouts, carrot, onion, leek and parsley. Preferential (quantitative) vernalization requiring vegetables are cauliflower, broccoli, radish, turnip, Chinese cabbage and spinach.Whereas, crops such as beans, peas, potato, tomato, brinjal, chillies, capsicum, cucurbits, bittergourd, okra and asparagus do not require vernalization. Light intensity and duration in terms of photoperiod is also important for many vegetable crops. Cucurbits, peas, beans, potato and sweet potato require high levels of light. Low levels of light are required for onion, carrot, cole crops, lettuce, spinach and celery. Humidity is related to precipitation and temperature which regulate transpiration, plant growth and incidence of diseases and insect-pests.

Cabbage

Cabbage is cool and moist season vegetables, and is cultivated during Autumn to Spring in subtropical and tropical zones of India, whereas, under temperate climate suitable time is Spring to Autumn. Summer season varieties are also available for Indian Plains. Favourable temperature is 15-25°C for head formation of cabbage. Growth slows down below 0°C and above 25°C. Cabbage plants when exposed to 10-12°C for prolonged period induces seed stalk, however, can withstand minimum temperature as low as 4-5°C for short duration and show fairly good hardiness to frost. Under hot and dry conditions yield and quality is poor. Seed germination and seedling growth is good at 12-18°C. Suitable varieties for hills are Pusa Drum Head, Pride of India, PusaMukta, Early Wonder, September, Quisto, and Eclips, whereas, for plains Golden Acre, Maha Rani, Large Solid and Late Drum are suitable. Seed sowing time in hills are January-February, July-August and September-October, and in plains August-November is the main sowing time. In some selected parts of South and Western India sowing can be done round the year.Heat tolerant and tropical; hybrids has facilitated good cabbage production in South Indian conditions. Type or variety also determines the sowing time. Bolting is very less in early varieties when sown early for harvesting in winter, whereas, If late varieties are sown early, heads do not matureuntil spring. Pointed headed or late types bolt less and round headed or early types bolt more readily (Ghosh and Madhavi, 1998). Early maturing varieties namely Golden Acre, Pride of India, Copenhagen Market, Early Drumhead and PusaMukta, PusaAgeti, Sri Ganesh

Goland Early Drum Head become ready for harvest within 60-70 days after transplanting. Mid season varieties namely All Head Early, Wisconsin, All Green, September and PusaMukta become ready for harvesting within 80-90 days after transplanting, whereas, late varieties namely Pusa Drum Head, Danish Ball Head, Late Flat Dutch , Late Large Drum Head and Late K-1take 90-120 days as they require long winter for good quality head.

Cauliflower

Cauliflower cannot withstand extreme of temperatures. In general, 8-25°C is the favourablerange with optimum range of 15-20°C. Early varieties require higher temperature and longer days (Ghosh and Madhavi, 1998). Initial temperature for better growth of young seedling is 23°C and for successive growth 17-20°C is favourable. Winter or late cauliflowers should be grown at 14°C initially and then, optimum temperature for better growth is 15-20°C. More rapid plant growth of late varieties takes place at 23°C. Dry and low humidity is not conducive for cauliflower cultivation. Seed germination is good when soil temperature range is 12-16°C. Proper season and variety selection is very important in case of cauliflower. Indian or tropical types form curds at 20-25°C, heat tolerant, early and require cold treatment for 6 weeks at 10-13°C instead of vernalization. Annual temperate types (*Erfurt* or Snowball type) form curds at 10-16°C (or even at 5-20°C), intolerant to heat, late and require vernalization for 8-10 weeks at 7°C (Fageria*et al.*, 2002). Optimum temperature requirement for curd formation of different Indian cauliflower groups are 20-27°C for early maturing cultivars (September-October), 16-20°C for mid maturingcultivars (November-December) and 12-16°C for late maturing cultivars (December-January). Indian Cauliflowers are grouped as -*Kunwari* (September-October), *Katiki* (Oct.-Nov.), *Aghani* (November-December), *Poosi* (December) and *Maghi* (January). Early maturing varieties are Pusa Deepali, Early Kanwari, Pant Gobhi-1, Improved Japanese, PusaKatki and Pusa Early Synthetic. Mid season varieties arePusa Synthetic, PusaAghani, PusaShubhra, Pusa Himjyoti, Pusa Hybrid-2 and Pant Gobi-4. Late maturing varieties are Snowball-16, Pusa Snowball-1, Pusa Snowball-2, Pusa Snowball K-1 and Late Dania. PusaKatki and Early Kanwari are suitable for cultivation during rainy season *i.e.* June-July and curds are ready in September-October. Early maturing cultivars are short in plant growth and foliage is bluish green in colour with small to medium and loose curd.Late cauliflowers require long growing period and have larger leaves and plant size. If early season varieties are grown in late, poor quality button curds are produced and disorders like Ricey, leafy, fuzzy, loose and yellow or purple curds are major problems. Late varieties if grown early then, leafy growth occur and curd formation is late. Leaves of cauliflower can withstand snow but curds are damaged. Temperature below the optimum level favours formation of small unmarketable hard or button curds and delays curd maturity. Higher temperature produces excessive leaf at the cost of curd formation, generative phage is retarded, bracts of flower stalks grow stronger and become visible at early stage and maturity of these loose curds if formed, at all, is delayed.

Tomato

Tomato is warm season frost sensitive crop and under sunny weather fruit colour and quality is good. However, high temperature, dry wind and low humidity are detrimental for floral parts and there is failure in fruit set. Most favoutrable temperature range for better plant growth and fruiting is 21-23°C; however, 18-27°C is also suitable for better growth. Optimum temperature for pollen germination is 29°C and less pollen germination occurs when temperature falls below 10°C and goes above 37°C. At very high temperature along with dry wind causes blossoms drop and less fruit set due to drying of stigmatic fluid. Critical temperature regimes for different growth stages of tomato are 16-29°C for seed germination, 21-24°C for vegetative growth, 13-18°C (day) and 19-24°C (night) for fruit set, 21-24°C for lycopene formation and 21-32°C for carotene formation (Anonymous, 2011). Varieties/hybrids which are able to set fruit at low temperature upto 6°C are called 'Cold Set' and those which set fruit upto temperature of 300C is 'Hot Set' (Swarup, 2012). Formation of lycopene declines above 30°C and completely stops at 40°C.Temperature and light intensity influence plant growth, fruit set, fruit pigmentation and nutritional value of tomato. Prolong drought situation when followed by sudden rainfall or heavy irrigation then blossom end rot and growth cracks are major problems (Anonymous, 2000). Moreover, under high rainfall condition tomato cultivation is not profitable. In Western (Maharashtra and Gujarat) and Southern India (Swarup, 2012), seed sowing times are June-July and October-November. However, plants raised from seeds sown in January-February are affected by leaf curl virus in the month of May-June. Sowing time is often advanced in these regionsby sowing seed in April-May, September-October and December-January. In Eastern region, September-October and late November-December is the main seed sowing time. In North-Western plains, sowing time is late October-early December. Plants raised from seed sown in June-July are also affected by leaf curl virus in autumn-winter. In North Western hills of India, sowing time is February-March and March-April. Determinate types are preferred in North Indian plains to adjust the short growing season, whereas, indeterminate types are suitable for hills and Southern region as favourable climate prevails for longer period (Frageria *et al.*, 2012).Early varieties are Rajni, Pusa Ruby, Pusa Early Dwarf, Pusa Red Plum, Sioux, Pusa Sheetal, Hisar Lalima, Hisar Anmol and Punjab Kesari. Mid season varieties are Best of All and Anand Tomato 3. Late season varieties are Marglobe and Punjab Tropic. Varieites Sioux, Roma, Best of All, Marglobe, Solan Gola and Solan Shagun are suitable for hills. Sun scald occurs when fruits are exposed to direct sunlight with high temperature and fruits become whitish yellow with leathery texture. Root rot is soil-borne disease and becomes severe after heavy or continuous rainfall. High temperature and soil moisture causes Blossom End Rot and Puffiness. Fruit cracking may occur due to moisture stress under long dry spell during rainy season if irrigation is not properly applied. Further, long duration exposure of the fruits to strong sunlight as well as sudden fluctuation in day and night temperature also favour fruit cracking. Cat face is another disorder occurs under unfavourable weather conditions causing distortion of growth of pistil cells. Water stagnation during rainy season causes Damping off, Backeye rot and *Fusarium* wilt.

Chilli

Chilliis successfully grown in frost free warm and humid areas with minimum temperature of 10°C and maximum in the range of 20-35°C. Optimum temperature is 20-25°C (avg. 24°C). Capsicum requires somewhat milder climate and above 23.5°C temperature is not favourable. Deficit soil moisture regime, high temperature and low humidity during blossom development and fruit formation are the main causes of bud, blossom and fruit drops. On the other hand, excessive rainfall is detrimental to the crops, because it brings about defoliation and rotting of the plant. As a rainfed crop, it is grown in areas receiving an annual precipitation of 63-75 cm (Anonymous, 2011). Soil temperature 30°C retards root growth, whereas, low night temperature in the range of 8-10°C reduces pollen viability resulting into fruits with lesser seeds. Even fruit set may be restricted, if night temperature falls below 10°C. High temperature enhances capsaicin content. Chillibeing a day neutral plant can be grown throughout the year under suitable climate (Fageria *et al*, 2012). Kharif season is the major season for chilli cultivation in India.However, Rabi and Kharif season is good in North India, whereas, in South India transplanting is done in October-November for Rabi, June-July for Kharif and January-February for summer season crop.Common varieties/hybrids are Punjab Lal, Pusa Sadabahar, Agni, Solan Yellow, Pusa Jwal and Punjab Surkha. Capsicum varieties/hybrids are California Wonder, Yolo Wonder, Bharat, Solan Hybrid-2, Pusa Deepti Sun scald occurs if fruits are exposed to strong sun light. Flower and fruit drop is a problem under short day, high temperature and low humidity at the time of flowering.

Brinjal

Brinjal is summer season crop and sensitive to frost which requires long (approx. 4-5 month) and warm growing condition. Low temperature at night and short summer are not conducive for brinjal cultivation and causes fruit deformities. Temperature in the range of 25-27°C is suitable for better growth and production with day temperature below 35°C and night temperature above 16°C. Temperature below 13°C affects plant growth and fruit set. Rainfall 60-100 cm is good for brinjal as it requires high moisture. However, it cannot withstand heavy rainfall during flowering and fruit set. In North India two sowing time is there, first in June-July for autumn harvesting and then in November for spring-summer crop.

Whereas in Southern region, the main sowing time is July-August, however, brinjal can be grown round the year. In hills, seeds are sown in March-April. Early varieties are Pusa Purple Long and Hisar Pragati. Pant Samrat and Punjab Barsatiare good for rainy season. Punjab Moti and Pusa Krantiare suitable for spring and Autumn. Punjab Sadabahar Baingan is suitable for cultivation throughout the year. Punjab Bahar is suitable for spring season.

Potato

Potato is grown in mild climate in temperate, subtropical and tropical zone in India. Suitable areas are where maximum day temperature is not above 35°C and night temperature is not above 20°C (Anonymous, 2013). Optimum temperature range is 15-25°C. Temperature below 21°C is good for tuberization and tuber development; however, ideal range is 16-20°C. Germination is good at 25°C and for vegetative growth, optimum temperature is 20°C. Long photoperiod condition is favourable for haulm growth and delay in tuberization. On the other hand, short photoperiodic condition reduces haulm growth and favours early tuber initiation and maturity is period is also reduced. In general mild temperature during early growth and cool temperature for tuber development is good. Higher temperature is not conducive for tuber formation and even tuber growth is stopped due to rise in respiration rate and the synthesized carbohydrate is consumed at faster rate if temperature goes beyond 30°C (Anonymous, 2000). As a result of which tubers are not in a position to store it for better development.Plants become elongated with wrinkled leaves and silky in appearance as the temperature rises. Such conditions along with long cloudy days, rains and humidity favour various diseases. Potato varieties are grouped on the basis of length of vegetative period and early varieties take around 90 days, mid varieties take 90-120 days and late season varieties take more than 120 days. Soil temperature is also critical for potato cultivation. Small tubers or segments of tubers are planted at 10 cm depth and temperature of the soil should be 8-10°C. Higher temperature damages the tender tubers, whereas, low temperature retards sprouting. If the mean daily atmospheric temperature remains at 10°C for longer time, then sprouting takes place in 27 days, whereas, very early sprouting occurs if temperature goes up to 27°C. Even mild frost with temperature of -1°C to -2°C is detrimental to leaves. Rainfall of 15-20 cm or its equivalent irrigation is good for vegetative growth. Soil moisture level of 65-80 at field capacity is good for potato crop. Heavy rain at sowing or even during sprouting is detrimental for potato. Alternation of dry and rainy spell results into over growth of plants and formation of offshoots from the tubers. Exposure of the tubers to direct sunlight during tuber development or in the storage results into green pigmentation of the surface called greening due to over synthesis of chlorophyll. Green potatoes undergo a chemical change and form solanine, a poisonous alkaloid chemical. However, chlorophyll as such is not harmful, and rather it is an indication of presence of alkaloid.The important potato varieties grown widely in India are Kufri Chandramukhi, Kufri Bahar, Kufri Alankar, Kufri

Sakti, Kufri Badsha and Kufri Satlif for plains, and Kufri Jyoti, Kufri Kundan, Kufri Neelmani and Kufri Giriraj for hills. Kufri Chipsona I and Kufri Chipsona II are the two released varieties for producing potatoes for processing. However Central Potato Research Institute, Shimla, has recently released two more potato processing varieties namely Kufri Chipsona III and Kufri Himsona (Pandey, 2007). Planting season is well defined for different region. In the north western hills, May-June, mid March-April, January-February, Mid February-March and mid February-mid January are the planting time in very high hills, mid hills, north central high hills and north eastern high hills, respectively (Fageria *et al.*, 2012). In Darjeeling and Shillong higher hills, February-March and in Southern hills, March-April (summer season crop), August-September (autumn season crop) and January-February (winter season crop) are planting times. Different panting time in northeastern lower hills and valley areas are January-February and August-September. In north western plains, mid September (early crop), mid October (autumn season crop) and January (spring season crop) are the planting times. In Eastern region comprising West Bengal and Odisha, planting time is early November (for winter season). Mid October is the planting time for north central region for winter crop. High soil temperature above 30°C during plant growth and tuber maturity and storage at high temperature causes Black or Hollow heart.

Cucurbits

Cucurbits are mainly warm season crop successfully cultivated in subtropical and tropical areas. In temperate climate, open field cultivation is not profitable, however, can be grown in the mid to low hill conditions where temperature becomes comparatively higher in the spring and summer. All types of cucurbits grow well under temperature range of 15-35°C, however, day temperature of 23-29°C and night temperature of 15-21°C is favourable for proper growth. Low temperature even at around 10°C slows the plant growth rate and below 5°C plants are damaged and growth completely stops. If frost prevails then plants are killed. Higher temperature in the range of 18-30°C favours faster plant growth. Therefore, it is better to sow the seeds when soil temperature is around 29-32°C for higher and faster seed germination, however, optimum temperature range for seed germinates is 15-35°C. Low temperature and short day condition promote female flowering ratio, whereas, high temperature (35°C or more) and long day condition promote only male flowering (Anonymous, 2009). Early seedlings may be raised in polybags under poly house conditions to avoid outside low temperature. Early transplanting in the open field is done when outside temperature becomes favourable for growth. Pollination is effected under low temperature. Prolonged cool and cloudy or moist weather during flowering period reduce bee activity, and result in poor pollination of flowers and low fruit-set. Important varieties are: 1. Cucumber: Japanese Long Green, PusaSanyog, Straight Eighth, China and Poinsette, 2. Bottle Gourd: Arka Bahar, Pusa Summer, Prolific Long, Prolific Round and Pusa Meghdoot, 3. Bitter

Gourd: Arka Harit, Arka Sumeet, Arka Sujat, Pusa Do Mausmi and Pusa Vishesh, 4. Rigde Gourd: Pusa Nasdar, Pusa Sadabahar and Satputia, 5. Pumpkin: Arka Chandan, Pusa Vishwash and Pusa Vikash, 6. Summer Squash: Pusa Alankar and Early yellow Prolific, 7. Watermelon: Arka Jyoti, Arka Manic, Sugar Baby and Pusa Bedana, 8. Mushmelon: Arka Jeet, Arka Rajhans, Hisar Madhur, Durgapur Madhu and Hara Madhu.

Onion

Onion is successfully grown in mild climate. Somewhat hardy to frost but cannot thrive well in excessive rainfall areas. A temperature range of 13-24°C is suitable for onion. Optimum temperature is 15-21°C for vegetative growth and during bulbing stage 20-25°C is good. The requirement of day length depends on the variety. Varieties of *Kharif* onion require day length of 10-11 hrs. and *Rabi* onions require relatively higher temperature and 12-13 hrs. of day length. Long day varieties will not form bulb in short day, however, short day varieties can form bulb in long day condition. Seedling growth is better in 20-25°C. Relative humidity of 70% is good. Very low temperature in the initial stages may hamper proper plant growth and favours bolting. Varieties suitable for cultivation in Indian plains are short day type. Bulb sprouting in storage can occur when harvesting is not done in dry period and if, *Kharif* onions are harvested before time. On the other hand, bulb of late harvested *Rabi* onions also sprout. Bolting is also another problem caused due to early transplanting of *Rabi* crop in the month of August or late transplanting of *Kharif* crop (at the end of December-January). *Rabi* onions also bolt at 20-25°C. Early maturity and smaller sized bulb of *Rabi* crop are caused due to sudden rise in temperature. Some of the common varieties are Pusa White Round, Pusa Madhavi, Pusa White Flat, Pusa Ratnar, Pusa Red, Arka Pitamber, Arka Pragati, Arka Bindu, Arka Niketan, Arka Kalyan, Agrifound Light Red, Agrifound Light Red, EatlyGrano, brown Spanish, Nasik Red (N-53) and Punjab Naroda.

Garlic

Garlic require cool and humid climate during plant growth and somewhat dry condition during bulb maturity. In North Western Himalayan region fall planting of cloves are done in fall *i.e.* September-October, so that cloves can undergo vernalization (exposure to cold treatment) under natural condition for about 6-8 weeks at around 4.4°C. This period is very much critical for inducement of bulbing. In Madhya Pradesh, Maharashtra, Karnataka and Andhra Pradesh planting time is August-October, in Gujarat, West Bengal and Odisha planting time is October to November and in some places in the hills planting of cloves may be done in March-April. Otherwise, cloves fail to sprout. Planting of cloves should be at 3-4 inches 7.5-10 cm deep to avoid winter injury and heaving. Planting is done well in time so that there is enough time for root growth and some sprouting of leaves under adequate moisture and low temperature. If more growth of leaves take place than leaves suffer winter injury. However, green leaves can tolerate snow. In the late winter or Springvegetative top growth is accelerated as the temperature starts to warm up. Leaves become flat, dark green on flashy stems reaching a height of 30cm or more. Leaf growth is favoured under short day condition and cold temperature, whereas, long day condition (12 hr) is required for bulbing. Treatment of the dormant cloves or young plants at 20°C for 1-2 months hastens bulbing. Too much exposure to the low temperature may result into formation of small bulb at the leaf axils. Improved varieties are Agrifound White, Agrifound Parbati, Punjab Garlic, Yamuna Safed and Sweta. Agrifound Parbati is long day type suitable for North Indian Hills and others are short day type suitable for plains.

Disease-insect and Weather Interaction

Climate and seasonal weather play an pivotal role in disease and pest incidence. Atmosphearic environmental parameters namely light, temperature, snow, rain, atmospheric humidity, dew, cloudiness, sunshine, wind, evaporation, and atmospheric pressure affect the occurrence of disease and insects. But temperature and moisture apparently are the most important factors for most of the diseases and insect incidences.Soil environment regulates heat and moisture around the root zone and are responsible for soil born diseases and insects.The effect of weather on occurrence of disease and insect is the consequence of its action on the susceptible plant (the host); on the parasitic organism (the pathogen that causes the disease); and on the relation between host and parasite which breaks the threshold level of resistance or tolerance.Mango powdery mildew (*Oidiummangiferae*) appears under cloudy weather, rains or morning mists accompanied by cooler nights duringflowering. Disease occurrence may take place in temperature range of 10-31°C and relative humidity of 60-90% (Nelson, 2008). Optimum temperature for infection is 23°C accompanied by high humidity and if abruptly followed by reduction in humidity. Conidia can be spread by wind. Best temperature range for conidia germination is 20-25°C. On the other hand, it has been also reported that conidia can germinate even in the absence of rain or water (Johnson, 1994). The favourable environment for mango anthracnose (*Colletotrichumgloeosporiodes*) is high humidity, frequent rains and temperature range of 24-32°C. Light and continuous rains or heavy dew associated with high humidity and low temperature favour the development of grape downy mildew (*Plasmoparaviticola*). Rain and dew are highly favourable for the spread of grape anthracnose (*Elsinoeampelina*). Increase in maximum and minimum relative humidity and decrease in temperature accompanied by rain during August to September is critical for disease spread (Pampanagouda, *et al*., 2005). Bacterial leaf spot of grape (*Xanthomonascampestris*) is more prevalent during June-August and again in February-March. Temperature range of 25-30°C and relative humidity of 80-90% is favourable for the development of the disease.

Grapepowdery mildew (*Uncinulanecator*) develops under warm and dry conditions. Shade or diffused light also helps in the development of this disease. Banana panama wilt (*Fusariumoxysporum*) is a soil-borne fungal disease and gets entry in the plant body through roots. It is most serious in poorly drained soil.Warm soil temperature, poor drainage, light soils and high soil moisture are congenial for the spread of the disease. Banana leaf spot, leaf streak or sigatoka disease (*Mycosphaerella fijiensis*) spreads under rainfall, dew and temperature is the determinant factor. Conditions favouring massinfection are most common during the rainy season with temperature above 21°C.Black sigatoka (*Mycosphaerella fijiensis*) conidia are mainly water-born for short distance spread, while ascospores are carried by wind to more remote places (Orlandini, 2011). Conidia Formationis favoured at e"70% RH for 32 hours alone or 27-30°C air temperature and e" 70% RH for 24 hours. Ascospore formation occurs at e"70% relative humidity and 27-30°C for 48 hours or e"70% relative humidity and temperature above 27-30°C for more than 48 hour. At the beginning of the rain most of the acospores are released. Infection takes place when leaf is wet or relative humidity is > 90%. Generally, infection is completed in less than 12 hours of moist conditions when optimum temperature prevails. However, it can take 15-24 hours if temperature is not high.In case of pathogen spores of potato and tomato early blight (*Alternariasolani*) disease germinate and produce mycelium at optimum temperature of 23°C (critical range 24-29°C) and 80% humidity. Potato Early Blight becomes more prevalent under warm moist weather. If temperature range is3-26°C (optimum range 17-22°C) and humidity above 90-100% then asexual reproduction of late blight pathogen (*Phytophthorainfastance*) occurs under wet condition (Nelson, 2008). The ideal conditions for infection and spread of late blight disease is day time temperature between 15-21°C, night temperature between 10-15°C, and relative humidity around 100%.

Whiteflies are a serious pest for many vegetable crops especially in polyhouse and as vector for virus. The development time of this insect from egg to adult ranges from 15-70 days depending upon temperature range10 to 32°C and optimum temperature is 27°C (Orlandini, 2011). Lowtemperature increases egg mortality. Under summer conditions adult longevity is less (6-15 days) and during winter period female can live for several months. Brinjal fruit borer infestation is more at low temperature 17-27°C and relatively medium to high humidity 74% (Khuhro, *et al.*, 2011). The severity of infection is positively correlated with maximum humidity and rainfall (Verma *et al.*, 2009). In case of mango mealy bug lower temperature and humidity during November to March is favourable (Anonymous, 2012). Whereas, shady and crowded tree canopy and high humidity with moderately high temperature during February-April and July is congenial for mango hopper. Fruit fly in mango is prevalent at high temperature coupled with high humidity prevailing during May-July months. Mango shoot borer in more under moderate temperature and humidity prevailing during September-October.

Horticulture Under Changing Climatic Scenario

Global climate is changing and has started to show the effect on natural biodiversity, agriculture, horticulture, forestry, fishery and livestock. There is a significant variation in the mean state of climate or its variability which has been recorded over an extended period, globally. An increase in global temperature by 0.76°C has been reported since 1850 (Singh, 2008). Twelve years period during 1995 to 2006 ranked among the warmest years. Inter-Governmental Panel on Climate Change (IPCC) has well documented that global atmospheric concentrations of carbon dioxide, methane and nitrous oxide and greenhouse gases have increased since 1750 (Aggarwal, 2008). Apart from temperature, change is noticed in the rainfall pattern in the form of pre- and post- monsoon as well as during monsoon, winter precipitation including snowfall, unwarranted drought, untimely glacial melting, flood, high solar radiation, tropical cyclones, heat waves, depletion of ground water sources and decline in water table and increase in arid and semi arid areas. Rise in temperature will reduce the crop duration, increase respiration and alter photosynthetic partitioning, and most significantly the plant phenology such as bud initiation and differentiation, flowering, fruit set, fruit drop, fruit growth and development and ripening. Incidence and resurgence of diseases and insect-pests are dependent on temperature and humidity and any change in these parameters will affect the crops by increasing the pathogens and insects population. Many varieties will be susceptible to the diseases and pests. Increase in carbon dioxide level will benefit several crops, however, rise in average temperature and change in other critical parameter will be detrimental for crop growth and development. Many areas, which are famous for specific fruit crop such as apple, mango or litchi or grape or citrus *etc.* growing, will become unsuitable for quality fruit growing if temperature rises even by 0.7-1°C.

Forecasting for Yield and Disease-insect Occurrence

Horticultural crops being high value and perishable in nature require specialized management and handling system. Forecasting for yield and incidence of insect-pest and diseases are the two main objectives to predict the estimated production target and for better post harvest produce management. Weather and climate based information before the cropping season and during the cropping season is helpful for advance planning and crop management.

Reliable and timely forecasts are useful:

i. For proper, foresighted and informed planning.

ii. To counter the uncertainties such as weather, production, policies and prices *etc.*

iii. Horticulture is highly input and cost intensive and requires judicious use of all critical inputs and timely adoption of plant protection measures as prophylactic and for control measures.

iv. Resurgence and emergence of pests and diseases pose a great threat under the changed climatic scenario and unscientific cultivation practices.

v. Season and agro-climatic zone based weather forecasting becomes very essential to adopt specific crop and variety, agro-techniques and plant protection measures etc.

vi. Forecasting on the basis of various aspects has become very much important for various policy decisions relating to storage, distribution, pricing, marketing, import-export etc. and for long term planning and projections of production.

vii. Thus, a sound forecasting approach can assure the farming community about safe guard of their crop from natural calamities, weather vagaries, insect-pest and diseases, and for future planning on the basis of expected production to meet the food, nutrition and and livelihood requirement for ever increasing population.

There are several forecasting models developed for different crop specific to district, agroclimatic zone and state (Agrawal and Mehta, 2007). Weekly or fortnightly weather data along with crop inputs and crop related diseases and insect-pests areassessed and analyzed statistically using regression analysis, complex polynomials through GMDH technique and Artificial Neural Network technique.Peach leaf curl disease, caused by *Taphrinadeformans*(Berk.) Tul., is widespread in all peach growing areas around the world and in India, it is also prevalent in Himachal Pradesh, Jammu and Kashmir, Uttarakand and other low chilling peach growing area in North Indian Plains. An Israeli model forecasting leaf curl disease on*deformans*was validated in the Emilia-Romagna region of northern Italy (Giosue, *et al.*, 2000). When the peach trees are susceptible to infection, the model uses mathematical functions to calculate the risk of infection on the basis of weather conditions (daily rainfall greater than 10 mm, and maximum air temperature greater than 5°C), and it forecasts periods of possible symptom appearance based on the length of incubation. Peach trees became susceptible to infection between the end of January and mid March, when the first leaf buds attained phenological stage C, *i.e.* appearance of leaf apex. The trees remained susceptible for at least 9 weeks: the lastinfection occurred in mid-May. Citrus yield forecasting has been used since 1918 in Florida by U.S. Department of Agriculture (Anonymous, 2003) based on survey opinions of crop observers and statisticians. Four parameters used in the forecast are (1) number of bearingtrees, (2) number of fruit per tree, (3) fruit size, and (4) fruit loss from droppage. The first two of these parameters have the greatest influence on the forecast. The objective data including an early season limb count survey to establish actual fruit set, supplemented with monthly in-season measurements of fruit size and observations of fruit droppage. Florida citrus production over several years is estimated by collecting number of boxes per year. Tree census is done every year by aerial photography. Weather based yield forecasting models have been developed for 12 major horticultural crops in California based on 20 years weather and yield data (Lobell, *et al.*, 2006). Mishra *et al.*, (2004) used weather data of Kakori and Malihabad Dashehari mango belt (Uttar Pradesh) from 1987-1997 to develop a logistic regression model for forewarning system of mango powdery mildew. Different statistical forecasting models (Ramasubramanian, *et al.*, 2005) used are regression models, time series models and probabilistic models which are commonly used.

References

Agarwal, P. K. 2008. Climate Change: Implications for Indian Agriculture. Impact assessment of climate change for research priority planning in horticultural crops. CPRI, ICAR, Shimla, H.P. pp. 1-10.

Agrawal, R. and Mehta, S.C. 2007. Weather Based Forecasting of Crop Yields, Pests and Diseases - IASRI Models, J. Ind. Soc. Agril. Stati.ft. 61(2):255-263.

Anonymous, 2002. Climate change associated effects on grape and wine quality and production

Anonymous, 2003. Forecasting Florida's Citrus Production. Florida Agricultural Statistics Service, 1222 Woodward Street, Orlando, Florida 32803.

Anonymous, 2009. Pump and Squash production. Fact Sheet Agdex 256, order 00-031, Ministry of Agriculture and ZFood, Ontario.

Anonymous, 2010. IPM- schedule on mango. www. nhm.nic.in/Horticulture/IPM.

Anonymous, 2012. Status report on hydrology of arid zones of India. National Institute of Hydrology, Rorkee, Uttarkhand.

Anonymous. 2000. Potato: climate and soil. IKisan.com.

Anonymous. 2000. Tomato: climate and soil. IKisan.com.

Anonymous. 2011a. Tomato crop guide. Haifa Chemicals ltd. www.haifa-group.com

Anonymous. 2011b. Climate and Soil Requirement for Chilli or Capsicum Cultivation. www. Agriinfo.in.

Anonymous, 2013. Potato agro- techniques. www.nhrdf.com/

Fageria, M. S., Choudhary, B. R. and Dhaka, R. S. 2012. Vegetable Crops: production technology, Vol. II. Kalyani Publisher, New Delhi. Pp. 283.

Food and Agriculture Organization of the United Nations. Regional office for Asia and the Pacific Bangkok, Thailand, March 2002.

Ghosh, S.P. and D.L. Madhavi. 1998. Cabbage. In: D.K. Salunkhe and S.S. Kadam (eds) Handbook of Vegetable Science and Technology, Marcel Dekker, NY, pp. 299-321.

Giosue, S., Spada, G., Rossi, V., Carli, G. and Ponti, I. 2000. Forecasting infections of the leaf curl disease on peaches caused by Taphrinadeformans. European Journal of Plant Pathology 106: 563–571.

Khuhro, S. N., Kanher, N. A., Shar, A. H., Mangi, S., Khuhro, R.D. and Narejo, M.U. 2011. Extent of infestation by brinjal fruit borer LeucinodesorbonalisGuen. Sarhad J. Agric. 27(3): 467-469.

Lobell, D. B., Cahill, K. N. and Christopher B. 2006. Weather-based yield forecasts developed for 12 California crops. California Agriculture. 60(4): 211-215.

Mishra, A. K., Prakash, O. and Ramasubramanian, V. 2004. Forewarning powdery mildew caused by *Oidiummangiferae* in mango (*Mangiferaindiaca*) using logistic regression models. Indian journal of Agricultural Sciences. 74(2): 84-87.

Nakasone, H. Y. and Robert E. P. 1998. Tropical Fruits.Cab International. New York.

Nelson, S.C. 2008. Late Blight of Tomato (*Phytophthorainfestans*). Cooperative Extension Service. College of Tropical Agriculture and Human Resources. University of Hawai. Plant Disease PD-45. pp. 1-10.

Nelson, S.C. 2008. Mango Powdery Mildew. Cooperative Extension Service. College of Tropical Agriculture and Human Resources. University of Hawai. Plant Disease PD-46. pp. 1-6.

Orlandini, S. 2011.Selected pest and disease analysis for case study of modelling DIPSA-NIFI, piazzaledellecascine 18, Firenze (IT). CAMI Workshop, 4-5 April 2011.

Pampanagouda, B., Benagi, V.I. and Naraund, V.B. 2005. Role of weather factors in the disease development and spread of anthracnose of grape. Agric. Sci. Digest, 25 (4):311-312.

Pandey, S.K. 2007. Horticulture: Vegetable science- Vegetables, Tubers & Spice Crops. Central Potato Research Institute, Shimla.

Rajan, S. 2008. Impact assessment of climate change in mango and guava research. Impact assessment of climate change for research priority planning in horticultural crops. CPRI, ICAR, Shimla, H.P. pp 36-42.

Ramasubramanian, V., Sharma M.K. and Walia, S.S. 2005. Statistical models for forewarning incidence of major pests of paddy. (Abstract of the paper). 8th Annual conference of society of statistics, computers and application in GVISH, Amravati from 28-30th November, 2005. Statistics and application. 41(1): 81. (www.isari.res.in/ebook).

Samson, J. A. 1986. Tropical Fruits. 2nd ed. Longman, New York.

Singh, H. P. 2008. Impact of climate change on horticulture. Impact assessment of climate change for research priority planning in horticultural crops. CPRI, ICAR, Shimla, H.P. pp. 11-20.

Singh, H. P. and Singh, B. 2001. Lychee Production in India. Lychee Production in the Asia-Pacific Region. FAO Regional office for Asia and The Pacific, Bangkok, Thiland. Corporate Document Repository. 55-77.

Swarup, B. 2012. Vegetable Science and Technology in India. Kalyani Publisher, New Delhi. pp. 656.

Varma, S., Anandhi P. and Singh, R. Kr. 2009. Seasonal incidence and management of brinjal shoot and fruit borer, Leucinodesorbonalis. Journal of Entomological Research. 33(4): 323-329.

❑❑❑

16

Climatic Assessment for Alternate Land Use Plan of Flood Affected Areas: An Observation

Britan Rahman, Pranab Dutta and Dipak Nath

Abstract

Three villages were selected randomly representing the flood affected areas of Dhemaji district of Assam. Analysed data on climatic parameters like rainfall(mm), temperature(°C), potential evapotranspiration (mm), relative humidity (%) of last 10 years collected authentic sources and depth of ground water table was collected from the well water depth from the three villages for twelve months showed that the length of growing period was more than 300 days and during July to September the ground water table is above the soil surface i.e. there is flood in this period, but during January to March, it is very low and water becomes inadequate for growing crops. Other climatic parameters are most suitable for both Kharif and Rabi crops, but annual flooding affects the crop during July to September. From the study it was concluded that short duration varieties of different crops like that of paddy, pulses, oilseeds etc. may be grown at pre and post flood condition.

Keywords: Dhemaji, Flood, Length of growing period.

Introduction

Flood is a regular and annual feature in Assam. Large-scale sand/silt deposition after each wave of flood in the past few years had destroyed cropland in several districts of Assam. It is mainly due to rise of the river Brahmaputra and its tributaries during rainy season. If the problem is not addressed on war footing, the phenomenon would spell disaster for the state. In recent years, the problem has been acute in Dhemaji district in North Bank Plain Zone of Assam. This district is worst affected in terms of frequency and duration of flood. After the recession of floodwater, it was noticed that actual kharif season elapses and a large area remain fallow as entire cropping system of the district is rainfed.

Interpretation of remote sensing data, toposheets and ground truth checking showed that the district was severely affected by flood from the river Brahmaputra and its tributaries *viz.,* Jiadhal, Sisi, Gainadi, Moridhol, Gutung, Dihing etc. An area of about 14,879 ha (Planimeter calculated) was under sand cover. The affected area constitute about 5.0 per cent of the total geographical area and about 15 per cent of the total cultivated area of the district (Vadivelu *et al.,* 2001). How can these lands be used was the question in the minds of everyone. But, till date scientific approaches for managing this region were meagre. At present, the immediate need is to assess the suitable climatic parameters so that short duration varieties of different crops can be accommodated at pre and post flood conditions with the available soil water, to improve the economy and for survival of the people of the affected areas.

Considering the importance of bringing these flood affected areas under pre and post flood crops the present investigation on climate assessment to know the length of growing period has been proposed.

Materials and Methods

Selection of Villages: Based on the sand/silt deposited map of Dhemaji district of Assam prepared by NBSS and LUP, NE Regional Centre, Jorhat in 1999, three villages from different parts of the district were selected for the present investigation. The villages were selected in such a way that they represent the whole flood affected areas of the district. The selected villages are Lalung, Samarajan and Dafalagaon. Each of the villages was visited and surveyed using Participatory Rural Appraisal (PRA) technique for collection of relevant information.

Collection of Climatic Parameters: Relevant climatic parameters (temperature, rainfall, relative humidity and potential evapotranspiration) for last 10 years (1996-2005) were collected from Regional Agricultural Research Station, AAU. North Lakhimpur and District Agriculture Office, Dhemaji. The climatic parameters of these two stations represent the whole of the district.

Determination of Length of Growing Period: The length of growing period or moisture availability period for growth of crop is the period (in days) when precipitation (P) exceeds 50 per cent of Potential Evapotranspiration (PET) as conditioned by the threshold value of soil temperature (5°C or more) plus the time required to evapotranspire an assumed 100 mm of water from the stored moisture from the deep soil profile.

The length of growing period was calculated for Dhemaji district using climatic method as given by Higgins and Kassam (1981). As per the climatic method the growing season starts when precipitation is equal to or greater than 0.5 PET and ends with the utilization of stored soil moisture once rainfall falls below 0.5 PET.

The long term (10 years) rainfall and PET data were used for this purpose. The rainfall and PET distribution patterns were plotted over months and moisture availability period was determined. Depending on the balance between rainfall and PET, the moisture availability period was divided into five sub-classes *viz.*, Humid I, Moist I, Humid II, Moist II and moderately dry period. A sub-class when rainfall is lesser than the PET is considered as "Moist I or Moist II", when rainfall exceeds the PET as "Humid period" and when rainfall is less than 0.5 PET plus time taken to use the stored soil moisture (up to 100 mm) as 'Moderately dry period' sub-class.

Determination of ground water table: The ground water table depth was taken from the well water depth. On 15th of each month data were taken for 12 months during 2006-2007 in the three selected villages.

Results and Discussion

Location: Dhemaji district of Assam occupying the northeastern part of the state lies between 27°5¢ to 27°57¢ N latitude and 94°15¢ to 95°38¢ E longitude. It is situated on the North Bank Plain Zone of Assam with a total geographical area of 3237 sq.km. To its north lies the East-Siang and West-Siang districts of Arunachal Pradesh and the river Brahmaputra surrounds the eastern and southern side of the district and Lakhimpur district is in the west.

Geology: The district is predominantly covered by Quaternary surfacial deposit. Upper Tertiary sediments are exposed only along the foot-hills. Recent river deposits consisting of sand, fine sand, silt and clay are found along the courses of river as well as the present river channels as point bars and channel bars.

Climate: The climate of the district is *per-humid* characterized by high rainfall, mild summers and winters and falls under cool to warm *per-humid* thermic agro ecological sub-zone.

Rainfall: The annual rainfall of the district ranges from 2066.8 mm to 3532.9 mm with an average of 2882.14 mm (Table 1). Analysis of 10 years data showed that district receives maximum rainfall from June to September i.e. 2010.75 mm, which is 69.8 per cent of the total annual rainfall. The rainfall from December to February is very low i.e. 89.42 mm, which is 3.1 per cent of the total annual rainfall. March to May receives around 609.89 mm rainfall, which is 21.1 per cent of total annual rainfall and October to November receives around 172.08 mm rainfall, which is 6.0 per cent of total annual rainfall. The Southwest monsoon starts in April and continues till October. The rainfall pattern is almost similar in the entire district.

Table 1: The seasonal and annual rainfall in the Dhemaji district during 1996-2005

Year	Rainfall (mm)*				Annual Rainfall (mm)
	Dec-Feb	Mar-May	June-Sept	Oct-Nov	
1996	107.5 (3.5)	794.2 (25.8)	2085.6 (67.9)	84.5 (2.8)	3071.8
1997	131.1 (4.5)	574.4 (19.8)	2013.0 (69.6)	175.8 (6.1)	2894.3
1998	49.9 (1.7)	697.3 (23.2)	2146.2 (71.5)	107.7 (3.6)	3001.1
1999	19.7 (0.7)	419.3 (15.0)	2014.1 (72.2)	336.6 (12.1)	2789.7
2000	151.3 (5.7)	511.3 (19.3)	1900.1 (71.6)	91.1 (3.4)	2653.8
2001	164.5 (5.1)	574.2 (17.6)	2402.7 (73.9)	110.5 (3.4)	3251.9
2002	136.1 (3.8)	810.3 (22.9)	2450.8 (69.4)	135.7 (3.9)	3532.9
2003	11.9 (0.4)	880.3 (30.1)	1623.3 (55.5)	408.6 (14.0)	2924.1
2004	89.0 (3.4)	461.2 (17.5)	2019.2 (76.6)	65.6 (2.5)	2635.0
2005	33.2 (1.6)	376.4 (18.2)	1452.5 (70.3)	204.7 (9.9)	2066.8
Mean	89.42 (3.1)	609.89 (21.1)	2010.75 (69.8)	172.08 (6.0)	2882.14

*Figures in parenthesis indicates percentage of annual rainfall

Temperature: The average monthly maximum and minimum temperature of last 10 years recorded in the district are given in Table 2. The average monthly maximum temperature ranges from 25.0°C to 38.0°C and average monthly minimum temperature ranges from 11.3°C to 25.0°C and average monthly temperature ranges from 18.17°C to 31.5°C. The minimum temperature from November to February is less than 14°C.

Table 2: Climatic parameters of Dhemaji district (1996-2005)

Month	Temperature (°C)			Relative humidity (%)	Rainfall (mm)	PET (mm)	0.5 PET (mm)
	Max.	Min.	Average				
Jan.	25.0	11.3	18.17	80.0	34.0	47.7	23.85
Feb.	26.7	12.7	19.67	76.0	43.5	60.2	30.10
Mar.	32.0	16.7	24.33	70.5	106.2	99.5	49.75
Apr.	34.3	18.7	26.50	75.5	168.9	124.7	63.35
May	36.0	20.3	27.83	73.5	345.4	150.1	75.05
Jun	36.7	22.7	29.83	81.5	631.2	102.2	51.10
July	38.0	25.0	31.50	86.5	416.6	150.4	75.20
Aug.	37.3	24.7	31.00	82.0	542.3	107.2	53.60
Sep.	36.7	23.7	30.02	88.0	399.5	93.6	46.80
Oct.	33.7	20.3	27.00	81.5	149.7	98.2	49.10
Nov.	31.0	13.3	22.17	78.0	35.3	66.3	33.15
Dec.	27.3	11.7	19.50	75.5	9.4	46.4	23.20

Source: RARS, NLP and District Agril. Office, Dhemaji.

Potential Evapo-transpiration (PET): The average monthly PET of last 10 years of the district are given in Table 2. The monthly PET varies from 46.4 mm to 150.4 mm. The monthly PET values are less than 50 mm during December and January.

Relative Humidity: The average monthly relative humidity of last 10 years of the district is presented in Table 2. The average monthly relative humidity varies between 70.5 to 88.0 per cent. RH is maximum during June, July, August, September and October.

Fluctuation of Ground Water Table: During July to September the ground water table is above the soil surface i.e. there is flood in this period, but during January to March, it is very low and water becomes inadequate for growing crops. All the three villages have almost similar type of ground water table. The depth of ground water table was recorded from August 2006 to July 2007 (Table 3). The depth was found to be very low during January, February and March and the depth narrowed down towards the end of June. During July, August and September these villages experience inundation, which was directly related with rainfall. It was found that the total rainfall during that period was maximum (1358.4 mm). So from October, the ground water table started receeding depending on the rainfall and reached a depth of 5.0 m, 6.5 m and 6.0 m in Lalung, Samarajan and Dafalagaon respectively on January, which remained almost similar till March. Besides, the depth of ground water was also related with evapotranspiration. It was found that the PET in November to December was minimum. Thus, high rainfall during June, July, August and September in Dhemaji causes flooding in the three villages and the changes in water depth results from the changes in rainfall and evapotranspiration. So during October to December and April to June the ground water table remains within a depth of 4 m and January to March it fluctuates between 4 - 6.5 m in all the three villages.

Table 3: The ground water table fluctuation of the three villages during 2006-2007

Month	Water table depth (m)		
	Lalung	Samarajan	Dafalagaon
August	-0.20	-0.25	-0.15
September	-0.10	-0.20	-0.10
October	+0.25	+0.50	+0.25
November	+2.0	+2.5	+2.0
December	+3.5	+4.0	+3.5
January	+4.5	+5.0	+5.0
February	+5.0	+6.5	+6.0
March	+4.0	+6.5	+6.0
April	+3.5	+4.0	+4.0
May	+2.0	+1.5	+2.0
June	+0.25	+0.30	-0.10
July	-0.25	-0.25	-0.40

Length of growing period: The growing period as per the model exceeds 300 days for Dhemaji district of Assam (Fig. 1). It was observed that after the 1st week of March the rainfall was in excess of PET and therefore, it recharged the soil moisture and gave surplus water from the 2nd week of October, however, the rainfall was lesser than PET, and therefore, the water stress conditions prevailed at this site. The analysis of meteorological data revealed that maximum rainfall received during May, June, July, August and September (Table 1). The district receives rainfall almost throughout the entire year even though short dry spell may take place in some part of the growing period. The area is known to receive adequate quantities of rainfall by the pre-monsoon (March – May), Monsoon (June – September) and less rainfall in the post-monsoon (October – November). The period when rainfall is very low occurs during November to February (< 30 mm per month).

Two to three crops can be accommodated in the district according to this model. Moist I is the period at the beginning of the rainy season when the PET is high, even with low crop canopy due to high temperature and longer day length. Most of the crop establishment operations are done in this period. Moist II is the period at the end of the rainy season when inspite of the high crop canopy, the PET becomes low due to lowering of the temperature and shortening of the day length. Most of the crop harvesting and processing operations are done in this period. The moisture availability pattern when plotted over a seasonal crop calendar provide specific periods of water deficit and surplus and moisture utilization and recharge pattern. After the 1st week of November, however, the rainfall was lesser than

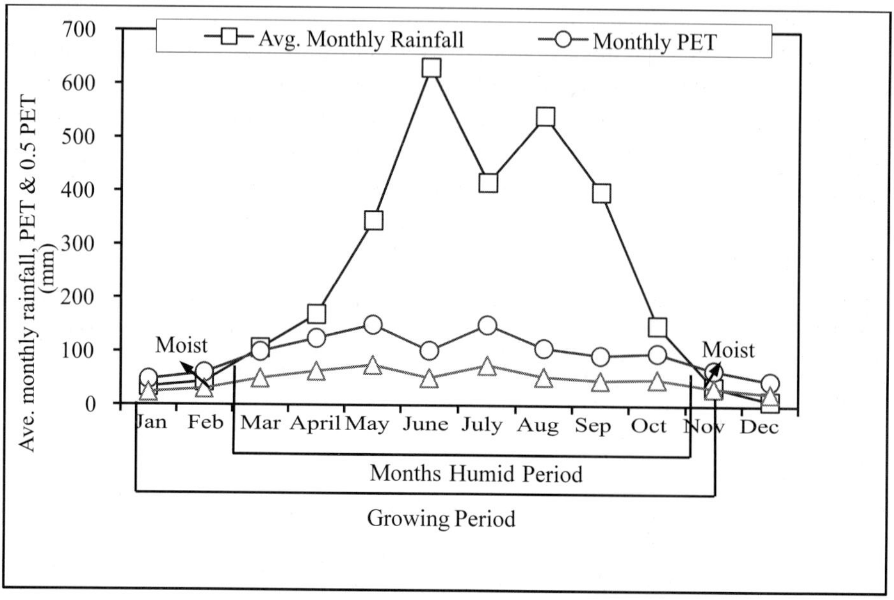

Fig. 1: Length of Growing Period of Dhemaji District

PET, and therefore, the water stress conditions prevailed at this site. From this analysis it is clear that *ahu* rice and other *rabi* crops can be suggested from March, although in the beginning water deficit will be encountered but small amount of rain during that period and the moisture present in the soil will help to establish the crop. From April onward there should not be any problem except flood. The harvest of the *ahu* crop and *rabi* crop is followed by main crop *sali* rice in flood less areas, which is presently followed by the farmers of the district.

Therefore, the farmers in this area, where there is danger of flood may be suggested to go for short duration varieties of rice as double crop in pre and post flood conditions. Cultivation of rice may start in March/April by using a short duration variety (var. Luit, Disang, Kapilee etc) which will be followed by post flood *sali* rice. In shallow water depth rice can also be followed by other crops like black gram, green gram, wheat, mustard, potato and vegetables when the excess water flows out from the field. The areas in which the maximum surface water lies between 30 – 50 cm, can be utilized for growing *ahu* rice followed by *sali* rice. Farmers in these areas may be suggested to grow *boro* rice in the low-lying and swampy areas during November/December to April/ May. When these areas do not receive sufficient rainfall during Jun/July, then some short duration vegetables like okra may be followed after *boro* rice. Though the *boro* rice takes longer duration the yield potential is much higher than *ahu* and *sali* rice. Again, mostly the *sali* rice is affected by flood in these areas; therefore farmers may be advocated to increase the *boro* rice areas.

Conclusion

In future further studies may be initiated to implement these suggestions *viz*. soil management practices along with different crop practices to show the farmer the possibility of growing crops in these areas to sustain their livelihood for the resource poor farmers. So, further studies are needed to find the suitability of these suggestions with actual experimentation in the problem areas.

Acknowledgement

The authors are grateful to Department of Soil Science, Assam Agricultural University, Jorhat-13 and NBSS & LUP, Jorhat for the physical and technical support. The interest and encouragement shown by Dr. T.C. Baruah, Head, Dept of Soil Science, AAU, Jorhat-13 is highly appreciated.

References

Higgins, G.M. and Kassam, A.H. 1981. The FAO agro-ecological zone approach to determination of land potential. Pedologie. XXXI(2): 147-168.

Vadivelu, S.; Baruah, U.; Sarkar, D. and Butte, P.S. 2001. Delineation and characterization of flood deposited sediments in Dhemaji district, Assam. J. Indian Soc. Soil Sci. 49(3): 521-524.

❑❑❑

17

Biodiversity in Mushroom

Jai Prakash Sharma

Abstract

Mushroom has a great potential in India. Its production is gradually increasing at the annual rate of 7%. The total mushroom production in India is 1.13 million tones and on the other hand, mushroom production annually in china is 18.2 million tones. There are a no. 07 wild mushroom available in the world and medicinal properties much as anti Candida, anti tumor (including cervical, uterine, stomach cancer) and antis viral are rotally present. Of the entire wild mushroom tested, T. glolulum has the highest amount of protein, ask, Calcium, phosphorus and iron. A total of 172 wild mushrooms have been reported home NE India and majority of the wild mushroom is of the order Agaricales. There is a need to make a concerted study of the mushroom biodiversity and its use for economic activity.

Keywords: Edible mushroom spices, Yield mushroom, Nutritive and medicinal value of mushroom.

Introduction

Fungi are regarded as being the second largest groups of organisms in the biosphere after the insects .They are one of the most diverse groups of organisms. Among them mushroom is the fruiting body of the macro fungi (Ascomycota and Basidiomycota) which has a short reproductive stage in their life cycle. Wild mushrooms have a profound biological and economical impact and have had a long association with human kind. Wild mushrooms have been consumed by man from ancient times because it is very delicious and important proteinous food supplement and pleasant flavour for livelihood and is very useful for treatment of hyperacidity and constipation. Edible fungi are not only beautiful but play a significant role in the daily life of human beings besides their utilization in industry, agriculture, medicine, food industry, textiles, bioremediation, natural cycling, as biofertilizers and many other ways. Fungal biotechnology has become an integral part of the human welfare (Manoharachary *et al.* 2005). The variety and galaxy of fungi and their natural beauty occupy prime place in the biological world and India has been the cradle for such fungi. Only a fraction of total fungal wealth has been subjected to scientific scrutiny and mycologists have to unravel the unexplored and hidden wealth. One third of fungal diversity of the globe exists in India. Hawks worth (1997) reported about fungal 70,000 species of fungi have been described species since then 270,000 plant species are known and suggested that around 14,000-15,000 species produce fruiting bodies of sufficient size and suitable structure to be considered as macro fungi and the total number of fungal species may approach 1.5 million. Shu-Ting Chang (2006) stated that approximately 14,000 described species of the 1.5 million fungi estimated in the world produce fruiting bodies that are large enough to be considered as mushrooms. nearly 50% i.e. 7000 species are well studied to possess varying degree of edibility, and more than 3000 species spread over in 31 genera are regarded as prime edible characterized until now and only around 5–10% of fungi can be cultured artificially. Thus far, only 200 of them are experimentally cultured, 100 economically cultivated, approximately 60 commercially grown and about 10 have reached an industrial scale (Chang and Miles, 2004). Furthermore, about 1,800 are medicinal ones. The poisonous mushrooms is relatively small (approximately 10%), of these some 30 species are considered to be lethal (Miles and Chang, 1997).Whereas over 2000 species of edible fungi known to man out of a total of 10,000 species of macro fungi reported to be growing in the world (Dhar and Singh, 1999).The people have been eating mushrooms for a long time. They used to go the forests and other wild places and based on their personal knowledge of edible and poisonous mushrooms they used to collect the edible ones and sale in local market and also eat them.

In India, the 'Guchhi' mushroom was a delicacy known to the people prior to the 1950' S.S. Jain (2005) stated that mushroom cultivation in India was started by Dr S.S.Jain (First Mycologist) in early 1950 who has grown *Agaricus* and other species successfully on rotting apple tree twigs and branches, cow dung and wheat straw etc., popularized among the farmers of Himachal Pradesh. Previously, the edible mushrooms were grown in France and Japan. Several reports on wild edible mushrooms have been collected and consumed by people since thousands of years. Archaeological evidences reveal edible species associated with people living 13000 years ago in Chile (Rojas and Mansur, 1995) but in China where the eating of wild fungi was first reliably noted several hundred years before birth the of Christ (FAO, 2004).Hobbs (1995) stated that many cultures have been identified that certain mushrooms could have profound health-promoting benefit. The mushroom industry in 2005 was valued at over $45 billion world market (Shu-Ting Chang, 2006).The mushroom industry can be divided into three main categories: edible mushrooms, medicinal mushroom products, and wild mushrooms. Production of mushrooms worldwide has been steadily increasing; mainly due to contributions from developing countries (Fig.1) such as China (46%), India (3%), and USA (10%).There is also increasing experimentally based evidence to support centuries of observations regarding the nutritional and medicinal benefits of mushrooms. The value of mushrooms has recently been promoted to tremendous levels with medicinal mushrooms trials conducted for HIV/AIDS patients in Africa, generating encouraging results. However, harvests of highly prized edible mycorrhiza mushrooms are continuously decreasing. This has triggered research into devising methods for improved cultivation. It is hoped that there will be even more research into this area, so that larger quantities can be massively harvested through semi cultivation methods. Technological developments in the mushroom industry in general have witnessed increasing production capacities, innovations in cultivation technologies, improvements to final mushroom goods, and utilization of mushrooms' natural qualities for environmental benefits. However, there is always the need to maintain current trends and to continue to seek out new opportunities. The challenge is to recognize opportunities such as increasing consumption capabilities with the increase in world population and to take advantage of this by promoting the consumption of mushrooms.

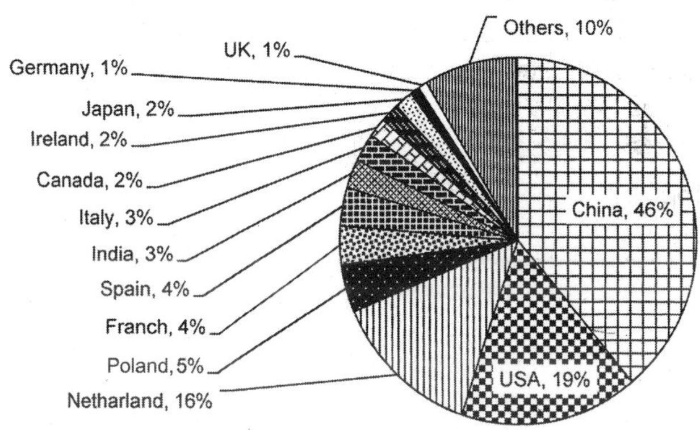

Fig 1: Comparative mushroom production chart of World and India's share

China is the world's largest edible mushroom producer (Table1) (Anonymous, 2003) produces about half of all cultivated mushrooms, and around 2.7 kilograms (6.0 lb) of mushrooms are consumed per person per year by over a billion people and 95% of mushroom production in China is consumed locally, the consumption per capita is likely to be over 10 kg/person/year (Chang, 1999, 2006).This is drastically higher than in US and many European countries where it is around 3 kg/person/year. India produced over 1.13315 lakh tons during 2010-11 (Table2) and exported about 60-70% of it, our per capita consumption is low around 30-40 g/person/year (Wakchaure, 2010).Mushrooms have been exploited commercially in the world. Chhetry and Pfoze (2007) reported 24 different edible fungi collected and identified in local dialects reported that this study focused on 9 locally well known mushrooms in Manipur, India.

Table 1: Total world mushroom production and China's contribution since 1978 (QiTan and Hui Cao, 2010)

Year	Total Production (x1000 tones)	China Production (x1000 tones)	China Contribution (%)
1978	1060.0	60.0	5.7
1983	1453.0	174.5	12.0
1990	3763.0	1083.028.8	
1994	4909.0	2640.0	53.8
1997	6158.4	3918.0	63.6
2002	12250.0	8650.0	70.6
2006	-	14000.0	-
2008	26000.0	18200.0	70.0

Table 2: State wise mushroom production (tons) in India (Wakchaure, 2010)

S. No.	States	Button	Oyster	Milky	Others	Total
1.	Andhra Pradesh	2992	15	15	0	3022
2.	Arunachal Pradesh	20	5	0	1	26
3.	Assam	20	100	5	0	125
4.	Bihar	400	80	0	0	480
5.	Chhattisgarh	0	50	0	0	50
6.	Goa	500	20	0	0	520
7.	Gujarat	0	5	0	0	5
8.	Haryana	7175	0	3	0	7178
9.	Himachal Pradesh	5864	110	17	2	5993
10.	Jammu &Kashmir	565	15	0	0	580
11.	Jharkhand	200	20	0	0	220
12.	Karnataka	0	15	10	0	25
13.	Kerala	0	500	300	0	800
14.	Maharashtra	2725	200	50	0	2975
15.	Madhya Pradesh	10	5	0	0	15
16.	Manipur	0	10	0	50	60
17.	Meghalaya	25	2	0	0	27
18.	Mizoram	0	50	0	0	50
19.	Nagaland	0	75	0	250	325
20.	Orissa	36	810	0	5000	5846
21.	Punjab	58000	2000	0	0	60000
22.	Rajasthan	100	10	0	10	120
23.	Sikkim	1	2	0	0	3
24.	Tamil Nadu	4000	2000	500	0	6500
25.	Tripura	0	100	0	0	100

(*Contd.*)

S. No.	States	Button	Oyster	Milky	Others	Total
26.	Uttarakhand	8000	0	0	0	8000
27.	Uttar Pradesh	7000	0	0	0	7000
28.	West Bengal	50	50	0	0	100
Union Territories						
1.	A&N Island	0	100	0	0	100
2.	Chandigarh	0	0	0	0	0
3.	Dadar & Nagar Haveli	0	0	0	0	0
4.	Daman & Diu	0	0	0	0	0
5.	Delhi	3000	50	20	0	3070
6.	Lakshadweep	0	0	0	0	0
7.	Pondicherry	0	0	0	0	0
	Total	100683	6399	920	5313	113315

In India, the mushroom production systems are of mixed type i.e. both seasonal farming as well as high-techcultivation. Mushroom production in the country started in the 70s and growth rate, both in terms of productivity as well as production, has been phenomenal. In seventies and eighties button mushroom was grown as a seasonal crop in hills, but with the development of the technologies for environmental control and increased understanding of the cropping systems, mushroom production shot up from 5000 tones in 1970 to over 1,20,000 tones in 2011. Button, oyster, milky and paddy straw mushroom is cultivated in different parts of the country as per temperature profile in different seasons. Two-three crops of button mushroom are grown seasonally in temperate regions with minor adjustments of temperature in the growing rooms, while one crop of button mushroom is raised in North western plains of India seasonally. Oyster, paddy straw and milky mushrooms are grown seasonally in the tropical/sub-tropical areas from April to October.The areas where these mushrooms are popularly grown are Orissa, Maharashtra, Tamil Nadu, Kerala, Andhra Pradesh, Karnataka and North Eastern region of India. Many commercial units are located in different regions of our country and produce quality mushrooms for export round the year. The present production of white button mushroom is about 85% of the total production of mushrooms in the country.

Nutritive Value of Wild Edible Fungi

Mushroom is rich in high content of proteins, vitamins, minerals, fibres, trace elements and low/no calories and cholesterol. Many of them have been used in folk medicine for thousands of years. Some of them are neutralceuticals (natural food having potential value in maintaining good health and boosting immune system of the human body) while some can produce potent nutriceuticals (compounds that have medicinal and nutritional attributes and are consumed as medicines in the form of capsules or tablets but not as food).They are the sources of various bioactive substances like, antibacterial, antifungal, antiviral, antiphrastic, antioxidant, anti-inflammatory, ant proliferative, anticancer, antitumor, cytotoxic,DNA damaging, anti-HIV, hypocholesterolemic, antidiabetic, anticoagulant, hepatoprotective, etc. A number of mushrooms form mutual associations (mycorrhizal) with the roots of several forest trees. This beneficial association enhances absorption of water and nutrients amongst trees and is indispensable for the growth and development of these trees and mushrooms. Mushrooms are not of course always beneficial to mankind. Detrimentally, a large number of mushrooms are the wood-rotters causing serious damage to the forest trees hence causing considerable economic loss to timber industry every year. The wild mushrooms are edible; it can be difficult to identify them correctly in the field. Besides, some are toxic, even deadly when eaten (Das, 2009). Fatality due to unwise consumption of wild mushrooms is a common affair amongst regional poor. In India wild mushroom have been reported since its inception from India (Ahlawat *et al,* 2008). The nutritional work of wild edible fungi from Northeast India has been done earlier by Longvah and Deosthale (1998).The Food and Agricultural Organization (FAO) has recommended the use of edible mushrooms as food supplement for protein deficient populations of developing and underdeveloped countries due to the following benefits because it is available in open and unrestricted access to the forests and grasslands, No expenditure is involved for mushroom gathering. Economic benefits in which income generated from the sale of collected mushroom resources goes directly to these inhabitants. As a result, majority of the rural people besides practicing traditional subsistence herding and agriculture also participate in wild mushroom collection. As a nutritional supplement, mushrooms could help diversify an otherwise monotonous diet during the rainy season when there was a paucity of other food resources. There could be several other social and cultural benefits including the healthy bonding between the families and the market place.

Wild Edible Fungi in East Zones of India

The East Zone of India (Fig.2) bordering Nepal, Burma and Bangladesh which includes North Eastern Hill Region viz; Assam, Sikkim, Nagaland, Meghalaya, Manipur, Mizoram, Tripura and Arunachal Pradesh whereas Eastern Zone includes Assam, Bihar, Orissa, Jharkhand, West Bengal, and Chhattisgarh. North Eastern Hill Region is marked by high rainfall have also been reported rich mushroom floras (Verma *et al.* 1985, 1987, 1989, 1995). Nepal is rich in mycodiversity (Rana and Giri, 2006). A total of 150 mushroom species belonging to 37 families and 65 genera were collected from Lukla (2480m above msl) to Panboche (4000m above msl) have been recorded out of which the largest family from Boletaceae and Russulaceae having 18 species followed by Tricholomataceae (16 sps.) and Polyporaceae (9 sps) etc. collected from soil. The appearance, occurrence and dominance were found to be controlled by different factors such as altitude,

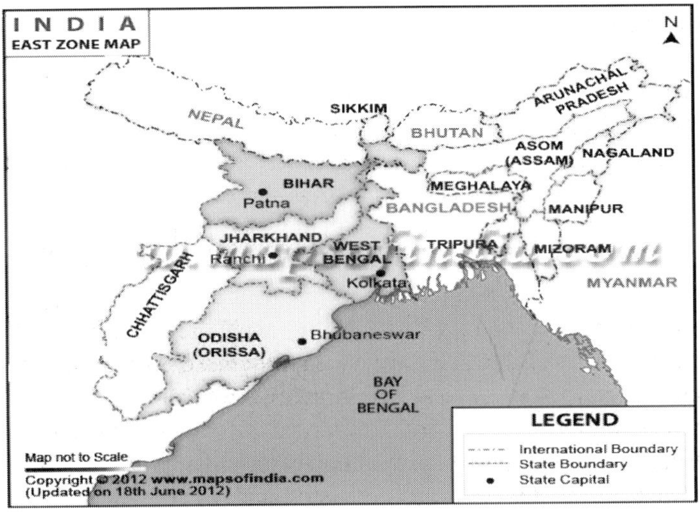

Fig.2: East Zones of India

vegetation, temperature, humidity, etc. the diversity of mushroom species were found the highest (84 sps.) at an altitudinal range of 3500 – 4000m above msl followed by 2500 – 3000m above msl (52 sps.) and 3000 – 3500m above msl (14 sps.).Apart from their importance as a source of food for the regional poor and as mycorrhizal partners of host trees, some are huge sources of bioactive compounds showing medicinal importance and reported 126 wild mushrooms collected from Sikkim.Bhaben *et al.* (2011) reported that more than 13 species of fleshy fungi under 9 genera and six families were identified from Nagaland. No reports are available for artificial cultivation of wild edible fungal flora from India if a technology can be developed which will be beneficial to uplift the economy of people of (Lakhanpal, 1994).Little work on biodiversity on mushroom has been done by from Jharkhand by Sinha (2003). Several workers have reported from Himalayan range mainly by Lakhanpal and his associates (Lakhanpal, 1994, 1997; Sohi and Upadhyay,1986), from Jammu and Kashmir (Kaul and Kachroo, 1974, Abraham and Kachru, 1989; Kaul, 1997, Sagar *et al.* 2005), UP Hills (Harsh and Bist 1985; Singh 1988) Recently a systematic work on wild mushroom diversity from Jammu and Kashmir have been undertaken by Kumar and Sharma (2011). Similarly, the large tract of Indogangetic plains of UP and Bihar have remained almost unattended to, although plains of Punjab (Saini and Atri, 1995; Atri *et al.* 1997) and those of Bengal, (Purkayastha and Chandra 1985) have yielded rich dividends and Assam (Boruah *et al* 1997). Madhya Pradesh (a rich tribal belt) has also received very limited attention (Kumar and Shukla, 1995, Harsh *et al.* 1993, 1996) and different agro-ecological zones from desert ecosystem of Rajasthan (Doshi and Sharma 1997), Vishwakarma *et al.* (2012) reported from Uttaranchal 40 taxa belonging to 11 families were identified. Two families and 2 taxa belonged to class- Ascomycetes viz: Family-Helvellaceae and Morchellaceae, and nine families and 38 taxa to class-

Basidiomycetes viz: Family-Agaricaceae, Amanitaceae, Boletaceae, Cantharellaceae, Coprinaceae, Ganodermataceae, Hydnagiaceae, Lycoperdaceae and Russulaceae. Some macro fungi are also being used as medicine viz: *Morchella esculenta, Hydnum repandum, Agaricus campestris, Cantharellus cibarius, Coprinus comatus and Ganoderma lucidum* from from Uttaranchal. Similarly, in Jharkhand meaning the "forest area" is geographically known as the Chotanagpur plateau, forming the north-eastern portion of peninsular plateau of India which lies between 21 degree 58'10" to 25 degree 16" latitude and 83 degree 22" to 87degree 57" longitude, very few diversity of mushroom have been reported .Six agro climatic zones have been reported and described by Bhatt (2004) .The wild mushroom have been complied and listed in Table 3. Thirty Nine local germplasms of wild mushroom have been reported from several forest area of Ranchi (Sarval forest), West Singhbhum (Bandhgaon forest) and Khunti districts of Jharkhand (Sharma *et al.* 2012) during May to July and August - September months in rainy season of 2002 to 2011. The illustrated photographs of wild mushroom recorded from different states of NE Region were shown in Fig 3.The majority of genera belongs to order Agaricales viz; *Agaricus*, *Termitomycetes* ,*Macrolepiota, Cococybe cyanopus, Calocybe* and *Volvariella* sp. of four families. Among the order agaricales majority of wild six species were of *Termitomycetes(Bada Khukhri)*belongs to family Agaricaceae which are sold at high prices in Ranchi market. Similary *Cococybe cyanopus(Tecnas)* of order Agaricaceae belongs to family Bolbitaceae were the most popular in Ranchi. In addition to above the other mushroom are *Bansh khukhri* (*Dingli*) (*Auricularia* spp.), *Jamun khukhri* (*Boletus* spp.), Paddy straw mushroom (*Volvariella* spp. locally known as *Tormai khukhri* or *Pagla khukhri*), *Balu khukhri, Ganoderma spp.* and *Namak khukhri*.According to survey done by Sharma *et al.* (2002) 10 quintal per season rugra (*Lycoperdon* sp) is marketed and Bada Khukhri (*Macrolapiota* sp) 8 quintal per season and Tecnus (*Termitomycetes* sp.) 2 quintal per day by the tribal and oyster mushroom (*Pleurotus* spp.-red strain locally known as *Kher khukhri* or *Patra*) are sold in Ranchi alone. Some of the important wild mushroom characteristics found in Jharkhand are described below:

Rugra (Lycoperdon/ Scleroderma sp.)

Two species of *Lycoperdon/ Scleroderma* were commonly available which appears first immediately after onset of monsoon (May-June) locally known as *Manie* in oraon and the second one appears in July-August and known as *Chiraya* in oraon. It is found in the sandy upland where stale & burnt leaves are scattered. They are found near the Sal (*Tectona grandis*) trees/ bushes and sold at higher price in urban area. The sporophores of first one are elliptical in shape 0.5 cm to 2.5 cm dia. and brown in colour whereas inner portion are white at young stage and cracked at mature stage while the one are spherical in shape, 1 to 2 cm dia. and white in colour at mature stage and burst. The tribal's like to eat this mushroom very much. The availability of these wild edible fungi have been brought in in telegraph news by Biswas (2010) on July 20 published from Calcutta those lightning strikes, deep cracks form on the surface. Mushrooms grow in these fissures," explained. It's 'raining' *rugra* - the fleshy, button variety of edible mushroom - in mineral-rich Chotanagpur plateau that plays magnet to bolts from the sky.A four-year study in Japan has shown that the yield of button mushroom (*Agaricus bisporus*) and several other mushroom varieties can double if exposed to a ten-millionth second burst of electricity at 50-100,000 volts. No wonder the markets of Ranchi and other towns and cities across the state have been swamped with *rugra,* a tasty, vegetarian alternative to mutton and chicken, ever since monsoon tiptoed into the state on June 18.Mandar homemaker Meena Devi, who troops to Kutchery Chowk to buy her weekly quota of the button variety, claimed that lightning sanitised mushrooms and made them highly nutritious. "So, the rainy season is the best time to have button mushrooms on the menu. It is both delicious and has high nutritive value," she maintained. But nutrition — that too directly from nature — certainly comes for a price. In the local market, white *rugra* sells for anything between Rs 160 and Rs 200 a kg. The black variety is cheaper at Rs 80-100 a kg.Scientists at Central Horticulture Mission in Palandu said that though no extensive research had been conducted in the state on how lightning affected mushroom growth, preliminary studies had confirmed that bolts from the sky made *rugra* sprout from the soil during monsoon. "When lightning strikes, deep cracks form on the surface. Mushrooms grow in these fissures," explained J.P. Sharma, a scientist at Central Horticulture and acting director of the Palandu cradle Ranbir Singh agreed that lightning and mushroom might form an interesting subject for research. "Normally, a fungus does not grow on soil, but on dead or decaying matter such as barks of trees. However, a flash of lightning might disinfect vegetative matter and trigger growth of *rugra*," he said.Prashant Kumar, a professor of horticulture at Birsa Agriculture University, Kanke, pointed out that mushrooms grew only on dead matter that did not contain latex. He conceded that massive R&D was needed to establish the real link between mushroom and lightning.

Bada khukhri (*Termitomycetes* spp. and *Macrolepiota* spp.)

It is found in rainy season (July–October) near termite nests in the forest and garden soil. The stripes are long about 30 cm. and pileus diameter is about 6.5cm. They are excellent in taste like meat and sold at higher prices in the market. People like it too much. It is not given to pregnant ladies as well as small kids as it may prove harmful to them. Tamarind is used in the preparation of dishes mushroom. Powder of *bada khukhri* is used in fish as well as vegetable preparations.

Jamun khukhri (*Boletus* spp.)

It is found near the base of the Jamun (*Syzygium cuminii*) tree during July-October. The fruiting bodies are fleshy; the pileus is greater than 15 cm in diameter, about

600 g in weight, and brown in colour. It is eaten after removing the external layer and its taste is just like liver. Over matured fruiting bodies spoil rapidly and insects' larvae appear.

However, two species of order Agaricales viz; *Amanita* and *Chlorophylum* were poisonous of family agaricaceae and Amanitaceae, respectively. Eight species of *Ganoderma* spp. (medicinal) belongs to order polyporales were preserved. Seven species of *Russula* belongs to order Russulales, Four species of *Boletus* belongs to Bolataceae of order Bolatales were sold at cheap rate. Two species of *Lycoperdon* belongs to order Lycoperales in which *Scleroderma* sp. was the most popular and sold at high prices, *Pleurotus*, *Calocybe* , *Auricularia* , *Polypore,* were seen in the forest .The photographs of above wild mushroom have been shown in Fig3.Considerable biodiversity of mushrooms and potentials of their utilization have been reported earlier by Verma and Upadhyay (1999) from South Bihar and various species of mushrooms are growing in wild as saprophytes, ligninolytic, or mycorrhizic types in the forests of Bihar and they could be utilized for domestic consumption and commercial exploitation. Those genera were to be looked into *Agaricus, Agrocybe, Calocybe, Kuehhneromyces, Lentinus, Macrolepiota, Pholiota, Stropharia, Tricholoma, Termitomyces and Volvariella* spp.Various spp. widespread in their distribution in temperate to tropical conditions could be collected from the forest during rainy season and marketed. The people of South Bihar have to be made aware of these macro fungi.Various members of Gasteromycetes like *Melanogaster* spp. (Phootka) are collected and sold in the market. *Lycoperdon* spp (Rugra), *Malanogaster* spp, *Calvatia* spp. and *Rhizopogon* spp. are worth collecting for consumption. *Terminella* is also grown naturally (Ram and Singh, 1997).It has been reported that the tribal are habitual of mushroom eating in their culture since time immemorial. They collect wild mushroom from the forest and sale them in the market at high prices to cater the need for their survival. As per record 200 kg per week Rugra (*Lycoperdon* sp) is marketed in Ranchi urban (Sinha, 2003). Some of the salient features of the wild species are described below. The mushroom is in food habit and their culture for long time hence it can be popularized easily by training them properly and ensuring the availability of quality spawn. Atri *et al. (*1997) reported Russula Pers. is the most common mushroom of coniferous and broad leaves of the Himalayas and as many as 74 texa spread over 287 collections have been listed. A lot of work has been conducted (Verma, 1997; Verma and Singh, 1981, Verma *et al.* 1985,1987,1989; 1995) in north eastern region and reported that the natural wealth of edible fungi of India if exploited for cultivation, may further augmented the species diversification of cultivated mushroom scenario of the country. The wild mushroom recorded from North Eastern Region of India has been summarized in table 3 according to order, family, scientific name place, season and status of edibility reported by various workers. Das (2009) recorded 126 wild mushrooms collected from Barsey Rhododendron Sanctuary of the state Sikkim are enlisted herewith in table 3 and medicinally important 46 mushrooms are also highlighted with their medicinal properties shown in table 2.The phograpghs of various edible recorded from various places are shown in Fig.3. Out of 126 species 41 were recorded as edible, 17 were noted as poisonous and 68 as inedible or not worth taking. 46 are either used as medicine or potential to give bioactive compounds having medicinal properties. It is observed that most of the species (124) grow during the period of July-August, followed by 95 species during September-October, 66 species during May-June, 17 species during November December, 12 species during March-April and the least number of species i.e. 11 are found in January-February when incessant snowfall and below freezing to freezing temperature threatens the growth of living organisms in this area (Das, 2009). Anonymous (2009) reported that there is good scope for commercial cultivation of wild mushrooms in Mizoram, which offers ideal weather conditions for mushroom cultivation throughout the year. In fact, wild varieties of mushrooms, which are locally known as 'pa', grow abundantly in the State2020 and form one of the traditional cuisines of the Mizos. It is also a common sight in local markets to see wild mushrooms of various shapes and sizes, color and types collected from the forest. Among the indigenous wild mushrooms found in Mizoram, pasi (*Schizophyllum commune*) is the most commonly found mushroom across the State. Pasi is commonly used as food item in Mizoram. It is also found to have medicinal properties including anti candida, anti-tumor (including cervical, uterine, stomach cancer) and anti-viral. It is listed among America's ten best wild medicinal mushrooms and several important pharmaceutical products with proven medical applications have been derived from pasi. With the availability of various species of wild mushroom in Mizoram.

Nutritive Value of Common Wild Edible Mushrooms

The earlier on nutritive value of wild mushroom have been reported by Gbolagade *et al.* (2006) from Southern Nigeria and assessed the nutritive value from young and matured carpophores of twelve common wild edible mushrooms collected from different locations. Of the entire wild mushroom tested, *T. globulus* was the richest containing highest amount of protein, ash, calcium, phosphorus and iron. This was followed in order by *T. microcarpus*, *V. esculenta*, *L. giganteum* and *L. pusilum*. The highest amount of moisture (98.5%) was found in the matured fruit bodies of *L. pusilum* followed by *A. polytricha* (97.1%) and *L. giganteum* (96.9%). *Lycoperdon giganteum* contained greatest amount of soluble sugars and glycogen while *V. esculenta* was the richest in total lipid containing 17.7% of the mushroom dry weight. Generally, the young fruit bodies of all the tested fungal samples were found to be richer than the matured fruit bodies. *T. microcarpus*, *P. atroumbonata*, *L. pusilum* and *V. esculenta* contained abundant amount of K, Na, Mn and Cu respectively. Nutrient composition in wild and cultivated edible mushroom, *Lentinus tuberregium* (Fr.) isolated from the wild and cultivated was analyzed on dry weight basis for their proximate nutrient compositions have been done from Tamil Nadu by Manjunathan and Kaviyarasan(2011)The cultivated mushroom accumulated

Table 3: Diversity of Wild Mushroom recorded in North Eastern Hill and Plateau Region of India

Order	Family	Scientific Name (Number of Species)	Place	Seasons	Edibility	References
Agaricales	Agaricaceae	*Agaricus augustus* Fr.	Sikkim	July to Sept.	Inedible	Das (2010)
	Agaricaceae	*Agaricus xanthoderma* Sikkim Genev.	Sikkim	June toJuly	Poisonous	Das (2010)
	Agaricaceae	*Agaricus silvaticus*	Nagaland	April- Aug.	Edible	Bhuben *et al*. 2012
	Agaricaceae	*Agaricus campestris*	Nagaland	June to Sept.	Edible	Bhuben *et al*. 2012
	Agaricaceae	*Agaricus* sp.(1)	Jharkhand	July - Oct.	Edible	Sharma *et al*. 2012
	Agaricaceae	*Chlorophylum* sp.(1)	Jharkhand	July- Sept.	Poisonous	Sharma *et al*. 2012
	Agaricaceae	*Termitomycetes* spp.	Jharkhand	July- Sept.	Edible	Sharma *et al*. 2012
	Agaricaceae	*Coprinus comatus (O.F. Müll.) Pers*	Sikkim	April to Oct..	Edible in young	Das (2010)
	Agaricaceae	*Lycoperdon perlatum Pers.*	Sikkim	May to Sept.	Edible	Das (2010)
	Agaricaceae	*Lycoperdon pyriforme Schaeff.*	Sikkim	Aug. to Oct.	Edible	Das (2010)
	Agaricaceae	*Lycoperdon* sp.	Bihar	June to Sept.	Edible	Verma and Upadhyaya 1999
	Agaricaceae	*Lycoperdon* sp.(2)	Jharkhand	May to Aug..	Edible	Sharma *et al*. 2012
	Agaricaceae	*Calvatia* spp.	Bihar	June to Sept.	Edible	Verma and Upadhyaya 1999
	Agaricaceae	*Lepiota cristata (Bolton) P. Kumm.*	Sikkim	May to Aug.	Poisonous	Das (2010)
	Agaricaceae	*Lycoperdon elongatum Berk.*	Sikkim	July to Sept.	Inedible	Das (2010)
	Amanitaceae	*Amanita vegenatae*(1)	Jharkhand	July- Sept.	Poisonous	Sharma *et al*. 2012
	Amanitaceae	*Amanita flavoconia* G.F.Atk.	Sikkim	July to Sept.	Poisonous	Das (2010)
	Amanitaceae	*Amanita francheti* (Boud.) Fayod	Sikkim	July to Sept.	Poisonous	Das (2010)
	Amanitaceae	*Amanita fulva* (Schaeff.) Secr.	Sikkim	July to Sept.	Inedible	Das (2010)
	Amanitaceae	*Amanita rubrovolvata* S. Imai	Sikkim	June to Oct.	Inedible	Das (2010)
	Amanitaceae	*Amanita caesarea* (Scop.) Pers.	Sikkim	July to Sept.	Edible	Das (2010)
		Auriculariaceae Auricularia sp.(1)	Jharkhand	July - *Sept*	Medicinal	Sharma *et al*. 2012
	Bolbitiaceae	*Cococybe cyanopus*	Jharkhand	July- Sept.	Edible	Sharma *et al*. 2012
	Cortinariaceae	*Cortinarius iodes Berk. & Curt*	Sikkim	June to Sept.	Inedible	Das (2010)
	Cortinariaceae	*Cortinarius purpurascens (Fr.) Fr.*	Sikkim	May to July	Edible	Das (2010)
	Clavariaceae	*Clavaria fumosa* Pers.	Sikkim	June to Aug.	Inedible	Das (2010)

(*Contd.*)

Order	Family	Scientific Name (Number of Species)	Place	Seasons	Edibility	References
	Clavariaceae	*Clavaria vermicularis* Holmsk.: Fr.	Sikkim	June to Aug.	Edible	Das (2010)
	Clavariaceae	*Clavulinopsis fusiformis*	Sikkim	July to Sept.	Inedible	Das (2010)
	Entolomataceae	*Entoloma lividoalbum (Kühner & Romagn.)*	Sikkim	June to Sept.	Edible in young	Das (2010)
	Entolomataceae	*Clitopilus prunulus (Fr.) P. Kumm.*	Sikkim	July to Sept.	Edible	Das (2010)
	Hygrophoraceae	*Hygrocybe conica (Schaeff.) P. Kumm.*	Sikkim	June to Sept.	Inedible	Das (2010)
	Hygrophoraceae	*Hygrocybe psittacina (Schaeff.) P. Kumm.*	Sikkim	May to Oct.	Inedible	Das (2010)
	Hygrophoraceae	*Lichenomphalia umbelifera* Redhead, Lutzoni, Moncalvo & Vilgalys	Sikkim	July to Sept.	Inedible	Das (2010)
	Hymenogastraceae	*Galerina marginata (Batsch) Kühner*	Sikkim	July to Sept.		Poisonous Das (2010)
	Lepiotaceae	*Macrolepiota sp.*(1)	Jharkhand	June- Aug.	Edible	Sharma *et al.* 2012
	Lepiotaceae	*Macrolepiota*	Bihar	June	Edible	Verma and Aug Upadhyaya 1999
	Lyophyllaceae	*Termitomycetes*	Bihar	June - Aug	Edible	Verma and Upadhyaya 1999
	Lyophyllaceae	*Calocybe*	Bihar	June- Aug	Edible	Verma and Upadhyaya 1999
	Lyophyllaceae	*Calocybe* sp (1)	Jharkhand	July-Aug.	Edible	Sharma *et al.* 2012
	Lyophyllaceae	*Termitomyces eurhizus (Berk.) R.*	Sikkim	July to Aug.	Edible	Das (2010)
	Lyophyllaceae	*Termitomyces medius Heim & Grasse*	Sikkim	July to Aug.	Edible	Das (2010)
	Mycenaceae	*Mycena inclinata (Fr.) Quél.*	Sikkim	July to Aug.	Inedible	Das (2010)
	Mycenaceae	*Mycena leaiana (Berk.) Sacc.*	Sikkim	July to Aug.	Inedible	Das (2010)
	Physalacriaceae	*Armillaria mellea*(Vahl) P. Kumm.	Sikkim	June to Dec.	Edible	Das (2010)
	Physalacriaceae	*Flammulina velutipes (Curtis) Singer*	Sikkim	June to Sept.	Edible	Das (2010)
	Pluteaceae	*Volvariella*	Bihar	June to Sept.	Edible	Verma and Upadhyaya 1999
	Pluteaceae	*Volvariella* sp (1)	Jharkhand	Sept - Oct.	Edible	Sharma *et al.* 2012
	Pluteaceae *V*	*olvariella* sp	Nagaland	June to Sept.	Edible	Bhuben *et al.* 2012
	Pleuotaceae	*Pleurotus* sp (1)	Jharkhand	July - Sept	Edible	Sharma *et al.* 2012

(*Contd.*)

Order	Family	Scientific Name (Number of Species)	Place	Seasons	Edibility	References
	Pleurotaceae	*Pleurotus ostreatus (Jacq.) P. Kumm.*	Sikkim	Aug. to Sept.	Edible	Das (2010)
	Psathyrellaceae	*Coprinopsis atramentaria* (Bull.) Redhead, Vilgalys & Moncalvo	Sikkim	July to Sept.	Edible	Das (2010)
	Psathyrellaceae	*Coprinus disseminatus* (Pers.) J.E. Lange	Sikkim	April to Oct.	Inedible	Das (2010)
	Psathyrellaceae	*Coprinellus micaceus (Bull.) Vilgalys*	Sikkim	July to Sept.	Inedible	Das (2010)
	Psathyrellaceae	*Psathyrella piluliformis (Bull.)*	Sikkim	June to Aug.	Edible	Das (2010)
	Physalacriaceae	*Oudemansiella mucida (Schrad.)*	Sikkim	Aug. to Oct.	Edible	Das (2010)
	Physalacriaceae	*Xerula radicata (Relhan) Dörfelt*	Sikkim	April to Sept	Edible	Das (2010)
	Strophariaceae	*Kuehneromyces mutabilis (Schaeff.) Singer & A.H. Sm*	Sikkim	June to Sept.	Edible	Das (2010)
	Strophariaceae	*Leratiomyces sqaumosus var. thraustus (Kalchbr.) Bridge & Spooner*	Sikkim	Aug. to Sept	Inedible	Das (2010)
	Strophariaceae	*Pholiota squarrosa (Fr.) P. Kumm.*	Sikkim	Aug. to Sept	Edible	Das (2010)
	Strophariaceae	*Agrocybe*	Bihar	June- Sept.	Edible	Verma and Upadhyaya 1999
	Strophariaceae	*Pholiota*	Bihar	June-Sept.	Edible	Verma and Upadhyaya 1999
	Strophariaceae	*Stropharia rugosoannulata*	Bihar	June-Sept.	Edible	Verma and Upadhyaya 1999
	Strophariaceae	*Stropharia semiglobata (Batsch) Quél.*	Sikkim	June to Oct.	Inedible	Das (2010)
	Schizophyllaceae	*Schizophyllum commune Fr.*	Sikkim	April to Sept	Inedible	Das (2010)
	Schizophyllaceae	*Schizophyllum commune*	Mizoram		Edible	Anonymous (2009)
	Schizophyllaceae	*Schozophyllum commune*	Nagaland		Edible	Bhuben *et al.* 2012
	Suillaceae	*Suillus luteus (L.) Roussel.*	Sikkim	June to Sept.	Edible	Das (2010)
	Tricholomataceae	*Trocholoma*	Bihar	June to Sept.	Edible	Verma and Upadhyaya 1999
	Trichomataceae	*Tricholoma virganum*	Nagaland	Mar. - June	Edible	Bhuben *et al.* 2012
	Trichomataceae	*Tricholoma virganum*	Nagaland	June-Sept.	Edible	Bhuben *et al.* 2012
	Trichomataceae	*Termitomycetes eurrhizus*	Nagaland	June-Sept.	Edible	Bhuben *et al.* 2012

(*Contd.*)

Order	Family	Scientific Name (Number of Species)	Place	Seasons	Edibility	References
Aphyllophor ales	Polyporaceae	*Lentinus canctus*	Nagaland	April –Aug.	Edible	Bhuben *et al.* 2012
	Polyporaceae	*Lentinus cladopus*	Nagaland		Edible	Bhuben *et al.* 2012
	Polyporaceae	*Lentinus* sp	Jharkhand	May- July	Edible	Sharma *et al.* 2012
	Polyporaceae	*Pleurotus flabellatus*	Nagaland		Edible	Bhuben *et al.* 2012
	Polyporaceae	*Pleurotus ostreatus*	Nagaland	April - Aug.	Edible	Bhuben *et al.* 2012
	Polyporaceae	*Lentinus* sp	Nagaland	May- July	Edible	Bhuben *et al.* 2012
Auriculariales	Auriculariaceae	*Auricularia delicata*	Nagaland	July-Aug.	Edible	Bhuben *et al.* 2012
	Auriculariaceae	*Auricularia auricula-judae* (Bull.) J.Schröt.	Sikkim	June to Oct.	Edible	Das (2010)
	ncertae sedis	*Pseudohydnum gelatinosum (Scop.) P. Karst.*	Sikkim	July to Sept	Edible	Das (2010)
Boletales	Boletaceae	*Boletus spp.*(4)	Jharkhand	July- Aug.	Edible	Sharma *et al.* 2012
	Boletaceae	*Boletus calopus* Pers.	Sikkim	July to Aug.	Inedible	Das (2010)
	Boletaceae	*Boletus edulis* Bull.	Sikkim	July to Aug.	Edible	Das (2010)
	Boletaceae	*Boletus erythropus*	Sikkim	May to Aug.	Poisonous	Das (2010)
	Boletineae	*Strobilomyces confusus Singer*	Sikkim	June to Oct.	Edible	Das (2010)
	Calostomataceae	*Mitremyces Junghuhnii Schltdl. & Müll.*	Sikkim	May to Oct.	Inedible	Das (2010)
	Paxillaceae	*Paxillus involutus* (Batsch) Fr.	Sikkim	June to Sept.	Poisonous	Das (2010)
	Rhizopogonaceae	*Rhizopogon* spp.	Bihar	June to Sept.	Edible	Verma and Upadhyaya 1999
	Rhizopogonaceae	*Terminella*	Bihar	June to Sept.	Edible	Ram and Singh 1977
	Rhizopogonaceae	*Kuchheromycetes*	Jharkhand	June to Sept.	Edible	Sinha 2003
	Sclerodermataceae	*Scleroderma* sp.(1)	Jharkhand	July - Sept.	Edible	Sharma *et al.* 2012
	Sclerodermataceae	*Sclroderma areolatum Ehrenb. Leopard*	Sikkim	June to July	Poisonous	Das (2010)
	Sclerodermataceae	*Scleroderma cepa Pers.*	Sikkim	June to Aug	Poisonous	Das (2010)
	Sclerodermataceae	*Scleroderma verrucosum (Bull.) Pers.*	Sikkim	June to Aug	Poisonous	Das (2010)
	Suillaceae	*Suillus bovinus (Pers.) Roussel.*	Sikkim	July to Aug.	Inedible	Das (2010)
Cantharellales	Cantharellaceae	*Cantharellus cibarius* Fr.	Sikkim	June to Sept.	Edible	Das (2010)
	Cantharellaceae	*Craterellus cornucopioides (L.) Pers.*	Sikkim	July to Sept.	Edible	Das (2010)
	Cantharellaceae	*Crepidotus applanatus (Pers.) P. Kumm.*	Sikkim	April to July	Inedible	Das (2010)

(*Contd.*)

Order	Family	Scientific Name (Number of Species)	Place	Seasons	Edibility	References
	Clavulinaceae	*Clavulina coralloides* (Holmsk.) J.Schröt.	Sikkim	July to Sept.	Inedible	Das (2010)
	Hydnaceae	*Hydnum repandum L.*	Sikkim	June to Aug.	Edible	Das (2010)
Dacrymycetales	Dacrymycetaceae	*Calocera viscosa* (Pers.)	Sikkim	Aug. to Sept.	Inedible	Das (2010)
	Dacrymycetaceae	*Dacryopinax spathularia (Schwein) G.W.*	Sikkim	June to Sept.	Edible	Das (2010)
Eurotiales	Trichocomaceae	*Trichocoma paradoxa* Jungh.	Sikkim	Throughout the year	Inedible	Das (2010)
Geastrales	Geastraceae	*Gaestrum triplex Jungh.*	Sikkim	Aug. to Oct.	Inedible	Das (2010)
Gomphales	Gomphaceae	*Ramaria stricta (Pers.) Quél.*	Sikkim	July to Sept	Inedible	Das (2010)
Helotiales	Bulgariaceae	*Bulgaria inquinans* (Pers.) Fr.	Sikkim	May to Aug	Inedible	Das (2010)
	Geoglossaceae	*Trichoglossum walteri (Berk.) E.J. Durand*	Sikkim	July to Sept	Inedible	Das (2010)
	Helotiaceae	*Chlorosplenium aeruginascens* (Nyl.) Kanouse	Sikkim	June to Sept	Inedible	Das (2010)
Hypocreales	Hypocreaceae	*Hypomyces chrysospermus* Tul. & C.Tul.	Sikkim	June to Sept	Poisonous	Das (2010)
Hymenochae tales	Hymenochaetaceae	*Porodaedalea pini* (Brot.) Murrill	Sikkim	Aug. to Dec	Inedible	Das (2010)
	Hymenochaetaceae	*Porodaedalea pini (Brot.) Murril*	Sikkim	Throughout	Inedible	Das (2010)
	Hymenoch aetaceae	*Coltricia cinnamomea*	Sikkim	June to Oct.	Inedible	Das (2010)
	Hymenochaetaceae	*Coltricia montagnei (Fr.) Murrill*	Sikkim	July to Oct.	Inedible	Das (2010)
	Hymenochaetaceae	*Inonotus dryadeus (Pers.) Murrill*	Sikkim	Throughout	Inedible	Das (2010)
Leotiales	Leotiaceae	*Leotia lubrica* (Scop.)Pers.	Sikkim	Aug to Oct	Inedible	Das (2010)
Lycoperdonales	Lycoperdonaceae	*Lycoperdon* sp.	Nagaland	June to Sept.	Edible	Bhuben *et al*. 2012
	Lycoperdonaceae	*Lycoperdon* sp.	Jharkhand	July-Sept	Edible	Sharma et al. 2012
	Lycoperdonaceae	*Clavatia gigantean*	Nagaland	July- Aug.	Edible	Bhuben *et al*. 2012
	Lycoperdonaceae	*Vascellum endotephrum (Pat.) Demoulin & Dring*	Sikkim	June to Sept	Inedible	Das (2010)
Pezizales	Helvellaceae	*Gyromitra infula* (Schaeff.) Quél.	Sikkim	June to Sept	Poisonous	Das (2010)
	Helvellaceae	*Helvella crispa* (Scop.) Fr.	Sikkim	June to Jan	Poisonous	Das (2010)
	Helvellaceae	*Helvella elastica* Bull.	Sikkim	Jul to Sept	Inedible	Das (2010)
	Helvellaceae	*Helvella macropus* (Pers.) P. Karst	Sikkim	July to Sept	Inedible	Das (2010)
	Pezizaceae	*Peziza badia* Pers.,	Sikkim	June to Oct	Inedible	Das (2010)
	Pyronemataceae	*Scutellinia scutellata (L.) Lambotte*	Sikkim	June to Nov	Inedible	Das (2010)

(*Contd.*)

Order	Family	Scientific Name (Number of Species)	Place	Seasons	Edibility	References
	Sarcoscyphaceae	*Wynea americana* Thaxt.	Sikkim	June to Oct	Inedible	Das (2010)
	Pyronemataceae	*Aleuria aurantia*	Sikkim	Sept. to Nov	Edible	Das (2010)
Polyporales	Fomitopsidaceae	*Fomitopsis officinalis (Vill.)*	Sikkim	Throughout	Inedible	Das (2010)
	Ganodermataceae	*Ganoderma lucidum (Curtis) P. Karst.*	Sikkim	Throughout.	Inedible	Das (2010)
	Meripilaceae	*Grifola frondosa (Dicks.)*	Sikkim	May to July	Edible	Das (2010)
	Meripilaceae	*Meripilus giganteus (Pers.) P.Karst.*	Sikkim	Aug. to Oct.	Edible	Das (2010)
	Polyporaceae	*Ganoderma* spp.(8)	Jharkhand	July-Sept	Medicinal	Sharma *et al.* 2012
	Polyporaceae	*Polypores* sp(2)	Jharkhand	July-Sept	Medicinal	Sharma *et al.* 2012
	Polyporaceae	*Lentinus* sp.	Bihar	June to Sept.	Edible	Verma and Upadhyaya 1999
	Polyporaceae	*Daedaleopsis confragosa (Bolton) J*	Sikkim	Throughout	Inedible	Das (2010)
	Polyporaceae	*Fomes fomentarius (L.) J.J.*	Sikkim	Throughout	Inedible	Das (2010)
	Polyporaceae	*Laetiporus sulphureus (Bull.) Murr.*	Sikkim	July to Sept..	Edible	Das (2010)
	Polyporaceae	*Lenzites betulina (Fr.) Fr.*	Sikkim	July to Oct.	Inedible	Das (2010)
	Polyporaceae	*Microporus xanthopus (Fr.) Kuntze*	Sikkim	Aug. to Dec.	Inedible	Das (2010)
	Polyporaceae	*Pycnoporus cinnabarinus (Jacq.) P. Karst.*	Sikkim	July to Sept	Edible	Das (2010)
	Polyporaceae	*Trametes gibbosa (Pers.) Fr.*	Sikkim	Throughout	Inedible	Das (2010)
	Polyporaceae	*Trametes versicolor (L.) Lloyd*	Sikkim	May to Feb	Inedible	Das (2010)
	Polyporaceae	*Tremella foliacea Pers.*	Sikkim	May to June	Inedible	Das (2010)
	Polyporaceae	*Tremella mesenterica Retz.*	Sikkim	March to Nov	Edible	Das (2010)
	Polyporaceae	*Trichaptum abietinum (Dicks.) Ryvarden*	Sikkim	May to July	Inedible	Das (2010)
	Sparassidaceae	*Sparassis crispa (Wulfen) Fr.* Sikkim		July to Sept	Edible	Das (2010)
Phallales	Gomphaceae	*Gomphus floccosus (Schwein.)*	Sikkim	May to Sept.	Inedible	Das (2010)
Russulales	Russulaceae	*Lactarius deliciosus* (L. ex Fr.) S.F.Gray	Sikkim	June to Sept.	Edible	Das (2010)
	Russulaceae	*Lactarius piperatus (L.) Pers.*	Sikkim	Aug. to Sept	Edible	Das (2010)
	Russulaceae	*Lactarius volemus (Fr.) Fr*	Sikkim	June to Sept.	Edible	Das (2010)
	Russulaceae	*Lactarius glaucescens*	Sikkim	June to Sept.	Inedible	Das (2010)
	Russulaceae	*Russula* spp. (7)	Jharkhand	July-Sept .	Edible	Sharma *et al.* 2012
	Russulaceae	*Lactarius* sp. (1)	Jharkhand	July-Sept.	Edible	Sharma *et al.* 2012
	Russulaceae	*Lactarius ochrogalactus*	Sikkim	Aug.to Sept.	Inedible	Das (2010)

(*Contd.*)

Order	Family	Scientific Name (Number of Species)	Place	Seasons	Edibility	References
	Russulaceae	*Lactarius yazooensis Hesler & A. H. Sm.*	Sikkim	July to Aug.	Inedible	Das (2010)
	Russulaceae	*Russula compacta Frost*	Sikkim	June to Sept	Inedible	Das (2010)
	Russulaceae	*Russula cyanoxantha (Schaeff.) Fr.*	Sikkim	June to Sept	Inedible	Das (2010)
	Russulaceae	*Russula heterophylla (Fr.)*	Sikkim	June to Sept	Inedible	Das (2010)
	Russulaceae	*Russula natarajanii.*	Sikkim	July to Sept.	Inedible	Das (2010)
	Russulaceae	*Russula nigricans (Bull.) Fr.*	Sikkim	July to Sept	Inedible	Das (2010)
	Russulaceae	*Russula puellaris Fr*	Sikkim	May to Sept	Edible	Das (2010)
	Russulaceae	*Russula virescens (Schaeff.) Fr.*	Sikkim	Aug to Sept	Edible	Das (2010)
	Russulaceae	*Russula xerampelina (Schaeff.) Fr.*	Sikkim	June to Sept	Edible	Das (2010)
	Russulaceae	*Russula senecis S.Imai*	Sikkim	June to Sept	Inedible	Das (2010)
	Stereaceae	*Stereum hirsutum (Willd.) Pers.*	Sikkim	Throughout	Inedible	Das (2010)
Thelephorales	Thelephoraceae	*Sarcodon atroviridis (Morgan)*	Sikkim	July to Sept	Inedible	Das (2010)
Xylariales	Xylariaceae	*Xylaria hypoxylon* (L.) Grev.	Sikkim	May to Dec	Inedible	Das (2010)
	Xylariaceae	*Xylaria polymorpha* (Pers.) Grev.,	Sikkim	July to Nov	Inedible	Das (2010)
	Xylariaceae	*Daldinia concentrica* (Bolton) Cesati & de Notaris	Sikkim	June to Jan	Inedible	Das (2010)

higher concentrations of carbohydrates(58.05%), protein (25%), moisture (12.51%), total ash (5.14%), crude fibre (14.69%), fat (1.54%), potassium(90.8 mg/gm), calcium (87 mg/gm), magnesium (30.4 mg/gm), sodium (37.3mg/gm), iron (6.5 mg/gm), copper (1.0 mg/gm), zinc (4.9 mg/gm), manganese (1.7 mg/gm), energy (338 kcal)., Potassium content was high, followed by calcium, magnesium, sodium respectively. Whereas the wild mushroom proximate value such as carbohydrates (55.8%), protein (25%),moisture (9.4%), total ash (4.7%), crude fibre (3.6%), fat (1.6%), potassium (7.53 mg/gm), calcium (2.66 mg/gm), magnesium (2.45mg/gm), sodium (1.2mg/gm), iron (0.53mg/gm), copper (0.11 mg/gm), zinc (0.41 mg/gm), manganese (0.08 mg/gm), energy (331 kcal).

Cultivated Mushroom Diversity in North Eastern India

Mushroom has a great potential in India and its production is exponentially increasing at an annual rate of 7% (Tiwary, 2004).The commercial mushroom farming exhibited good growth during the last decade of the 21stcentury mushroom production has reached 1.13 million tons in India.

Orissa is the leading state for paddy straw mushroom (*Volvariella vovacea*) production (Table 2 and Fig.4). Himachal Pradesh, Punjab and Haryana are major white button mushroom (*Agaricus bisporus*) growing states in North India with average yield of 20-22 kg per quintal of compost. It is now becoming popular in other states. Several species of edible fungi have been reported from northeastern hills of India (Verma and Singh, 1981). In Eastern zones of India, mushroom production have been reported from Bihar (480t), Nagaland (325t), Jharkhand (220t) and now 67.6 tons from 2001 to 2011-12, Assam (125t),West Bengal and Tripura each (100t), Manipur (60t), Mizoram (50t), Meghalaya (27t) Arunachal Pradesh (26t) and Sikkim (3t). In India button mushroom production (100683t), Oyster (6399t), Milky (920t) and Others *ie.* paddy straw mushroom (5313t) have been reported (Table 2). Oyster mushroom (*Pleurotus* spp.) is very delicious and important proteinous food supplement for livelihood and is very useful for treatment of hyperacidity and constipation. In Bihar, Oyster mushroom was grown in Bhagalpur, Purnia, Hazaribagh and Ranchi districts and production was very scanty before 1992-93 (Ram and Singh, 1997).The production of mushroom have been reported 5 tones during 1992-93 and 10 tones in 1994-95 in Bihar by Sharma and Rai (1994). Now its the total production in undivided Bihar was 31.83 tons from 1986-2001. Bordoloi (2012) reported from Assam that in the tea garden areas, it is often seen that people suffer from malnutrition and deficiency diseases and suggested that mushrooms can make up for this deficit in a large measure.

In addition to wild mushroom, cultivated mushroom are also getting popularity in tribal and urban areas of Jharkhand and adjoining states of eastern region. Of late, eight cultivated mushrooms *viz;* white button mushroom (*Agaricas bisporus)* , summer button *(A. bitorqis)*, oyster mushrooms (*Pleurotus florida, P.sajorcaju, P. ostreatus, P. eous,*) ,blue oyster(*Hypsizygus ulmarius*) and milky mushroom

(*Calocybe indica*) ,Black ear mushroom (*Auricularia polytrica)* and paddy straw mushroom (*Volvariella volvacea*) have become popular and its cultivation technology was standardized by Upadhyay *et al*. (2002) and Sharma and Kumar (2006) and Anonymous (2012).The following biodiversity of twenty seven sub species taxon level have also been evaluated dviz; PF-1, PF-2, PF-3, PF-4 and PF-5 of *Pleurotus florida*, Psc-1, Psc- 2, Psc-.3, Psc- 4 & Psc-5 of *Pleurotus sajor caju* and CI-1, CI-3, CI-6 ,CI-7 and APK-2 of *Calocybe indica* and Vv-1, Vv-2, Vv -4, Vv-6, Vv-7 , Vv-8 ,Vv-9, Vv-10, Vv-11,Vv-12 ,Vv-13 and OE -112 of *Volvariella volvacea* were tested . Shiitake (*Lentinus edodus*) strain SL-5 was tested in Dhanbad. Among these, PF-1 of *P.florida*, PSC - 4 of *P. sajor caju*, CI-7 and APK-2 of *Calocybe indica* and VV-7 , VV-9 and VV-11 were recorded high yielder. Some of the photographs of cultivated above mushroom are shown in Fig.5.

Advantages of Mushroom Cultivation

1. Wastes such as cereal straws are largely burnt by the farmers, which causes air pollution. However, these raw materials can actually be used for the cultivation of mushrooms. This kind of bioconversion exercise can greatly reduce environmental pollution.

2. Mushroom cultivation can be a labour intensive activity. Therefore, it will serve as means of generating employment, particularly for rural women and youths in order to raise their social status. It will also provide additional work for the farmers during winter months when the farming schedule is light.

3. It will provide the people with an additional vegetable of high quality, and enrich the diet with high quality proteins, minerals and vitamins which can be of direct benefit to the human health and fitness. The extractable bioactive compounds from medicinal mushrooms would enhance human's immune systems and improve their quality of life.

4. Mushroom cultivation is a cash crop.The harvested fruiting bodies can be sold in local markets for additional family income or exported for an important source of foreign exchange that will definitely improve the economic standards of the people.

5. Some warm mushrooms, e.g. *Volvariella volvacea* (Straw mushrooms) and *Pleurotus sajor-caju* (Oyster mushrooms) are relatively fast growing organisms and can be harvested in 3 to 4 weeks after spawning. It is a short return agricultural business and can be of immediate benefit to the community.

Mushroom farming is both a science and an art. The science is developed through research the art is perfected through curiosity and practical experience. However, mushroom farming is a business which requires precision. Indeed, it is not as simple as what some people often loosely stipulate. It calls for adherence to precise procedures. If you ignore one critical step, you are inviting trouble, which could lead to a substantially reduced mushroom crop yield.

Fig.3: Diversity of Wild Edible Mushroom North Eastern Hill and Eastern Plateau Region of India *http://www.360flat.com/photo_page/wild-mushroom.asp# (*Contd.*)

Fig.3: Diversity of Wild Edible Mushroom North Eastern Hill and Eastern Plateau Region of India *http://www.360flat.com/photo_page/wild-mushroom. asp (*Contd.*)

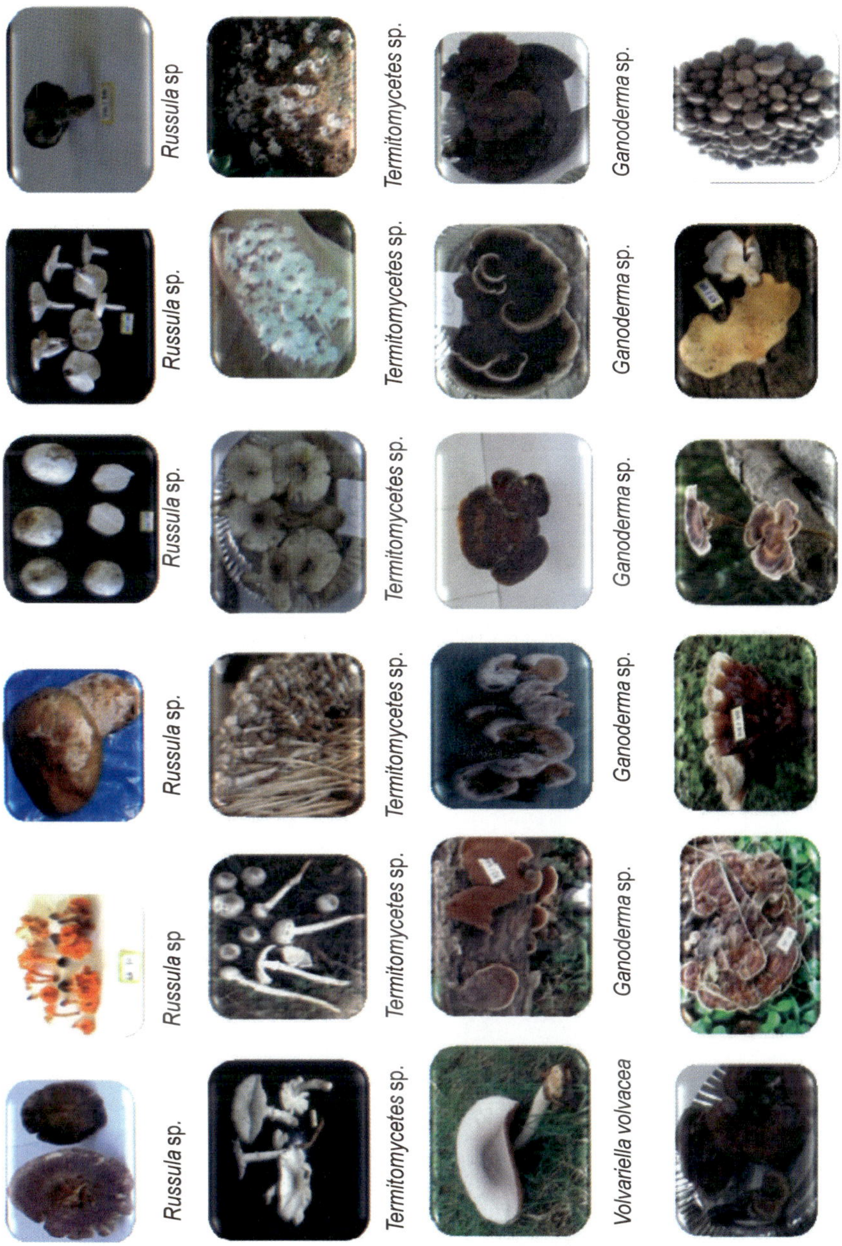

Fig.3: Diversity of Wild Edible Mushroom North Eastern Hill and Eastern Plateau Region of India *http://www.360flat.com/photo_page/wild-mushroom.asp#

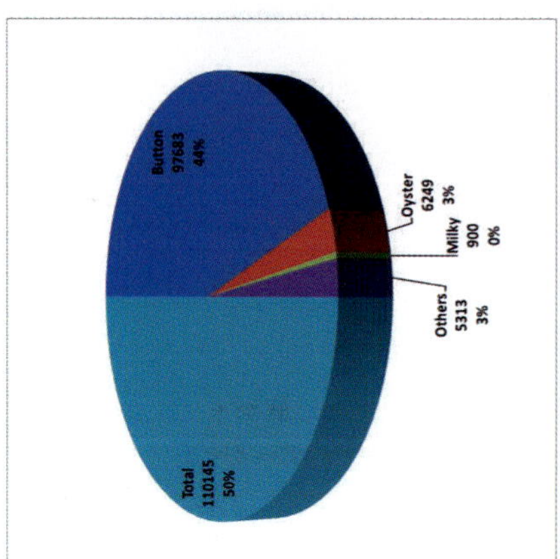

States(28)

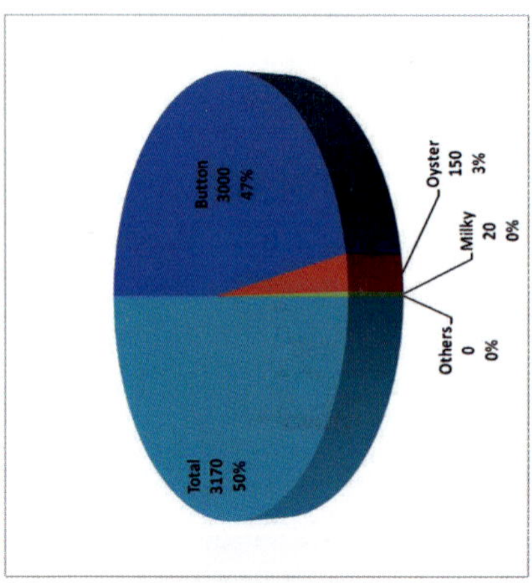

Union Territories (7)

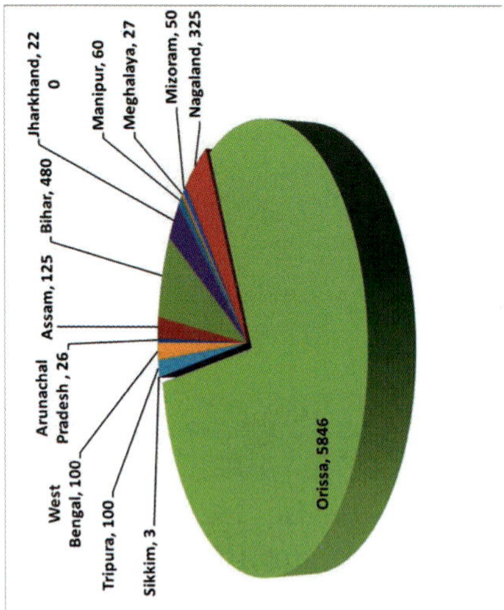

Mushroom Production in East Zone of India (Tones)

Fig.4: Comparative mushroom production and North Eastern in India

Fig. 5: Biodiversity of Cultivated Mushroom

Conclusion

A total of 172 wild mushrooms have been reported from North Eastern region of India and majority of wild mushroom were of order Agaricales (21 families, 69 species). Agaraicaceae was the major family followed by boletales (6 families,14 species), Polyporales, Russulal es (1family, 18species), Pizizales (3 families,12 speci es), Aphyllophorales (2 families, 8 species), Cantharellale 3 families, 5 species), Hymenochaeles (1family, 5species), Lycoperdonales (1family, 4species), Xyalonidales (3 families, 3species), Dacrymycelales (1family 2 species). Thirty nine wild fungi were identified in mushroom diversity from plateau of Jharkhand.Among them, nine species belongs to order agaricales in which six species were *termitomycetes* (*Bada Khukhri)* belongs to family Agaricaceae and two species were found as poisonous viz, *Chlorophylum* sp.and *Amanita vegenatae*.Similarly, *Cococybe cyanopus*(*Tecnas*) of order Agaricales belongs to family Bolbitaceae were popular in Jharkhand. Nine species belongs to Polypoales viz; *Ganoderma* and *Polypore*. Eight species belongs to Russulales.The important species of edible fungi belongs to *Scleroderma*, *Boletus* and *Lycoderma*.Eight cultivated species were successfully cultivated.Among them, P-1 of *P.florida*, Psc-4 of *P. sajor-caju*, CI-7, APK-2 of *C. indica* and Vv-7, Vv-9 and Vv-11 of *V. volvacea* were recorded high yielders.

References

Abraham, S.P. and J.L. Kachroo 1989. Larger fungi from Kashmir, India, VI. The genus Amanita. *Micol. Neotrop. Apl.* 2: 41-51. Agricultural Engineering and Machinery (APCAEM). Convention Centre No. 12, Yumin.

Ahlawat .O.P. Kumar, S., Arumuganathan,T and Tewari, R.P. ()25 Years of AICRP(Mushroom) All India Coordinated Research Project on Mushroom National Research Centre for Mushroom(Indian Council of Agricultural Research) Chambaghat, Solan-173 213 (HP).

Anonymous 2003. China Becomes World's Biggest Edible Mushroom Producer. Allbusiness.com. August 21, 2003. Retrieved 2010-08-04. Road, Chaoyang District, Beijing 100029, P.R. China. www.unapcaem.org.

Anonymous 2009. Mizoram: Export Potential and Prospects. Occasional Paper No. 135. Export-Import Bank of India. Published by Quest Publications. June 2009

Anonymous 2012. Training Manual on Mushroom cultivation Technology. Asian and Pacific Copyright© 2012 UNAPCAEM.

Atri N.S, Saini, S.S. and Saini, M.K. 1997. Studies on genus *Russula* Pers. from North Western Himalaya. *In:* Souvenir and Abstracts. Indian Mushroom Conference 1997 (Abs) p11.

Bhaben, T, Lisha, G, and Chandra, S.G. 2011.Wild edible fungal resources used by ethnic tribes of Nagaland, India. Indian Journal of Traditional Knowledge 10(3):512-515.

Bhatt, R.P. and Lakhanpal, T.N. 1990. Fleshy fungi of North- Western Himalayas V. Indian Phytopathol. 43(2):156-164.

Bhatt, S.C. 2004. The district gazetteers of Jharkhand. Gyan Publication House, New Delhi, p. 354.

Biswas, R. 2010 . Bolt button for bumper mushroom - Edible fungi sprouts on lightning-prone Chotanagpur plateau, swamps markets in capital, elsewhere. Telegraph, Wednesday, July 21, 2010.

Biswas, S.and Singh, N. P. 2004. Productivity of different species of oyster mushroom under Tripura condition: Journal of Hill Research 17(2):74-76

Bordoloi, A.K. 2012.Tips on 'right' mushrooms. The Telgraph, Calcutta, Tuesday, February 21, 2012

Boruah, P. and Singh, R.S. 2001. Edible fungi of medicinal value from the eastern Himalaya region. International Journal of Medicinal Mushrooms, 3: 124.

Boruah, P., Adhikary, R.K., Kalita, P. and Bordoloi, D. 1996. Some edible fungi growing in the forest of East Khasi Hills (Meghalaya). Advances in Forestry Research in India, 14: 214–219.

Boruah, P., Kailta, P., Bordoloi, D., Gogi, P. and Adhikary, R.K. 1997. Some fleshy fungi of ethnobotanic use from north east India. Advances in Forestry Research in India, 16:165-171.

Chang S.T. 2006.The World Mushroom Industry: Trends and Technological Development International Journal of Medicinal Mushrooms 8(4): 297-314.

Chang, S. T. 1999. Global impact of edible and medicinal mushrooms on human welfare in the 21st century: International Journal of Medicinal Mushrooms 1:31-62.

Chang, S.T. Miles, P.G. 2004. Mushrooms- cultivation, nutritional value, medicinal effect and environmental impact. CRC Press, Washington, D.C. pp. 451.

Chhetry, G. K.N. and Pfoze, N. L. 2007. Diversity of wild edible mushrooms and their economic values in Shepoumaramth Naga areas under Senapati district of Manipur. Journal of Mycopathological Research 45(2):249-252.

Das, K. 2009. Mushrooms of Sikkim I: Barsey Rhododendron Sanctuary. Sikkim State Biodiversity Board, Gangtok and Botanical Survey of India, Kolkata. 1-160.

Das, K. 2010. Diversity and conservation of wild mushrooms in Sikkim with special reference to Barsey Rhododendron Sanctuary NeBIO, 1(2):1-13.

Dhar B. L and Singh S.K. 1999. Current status of mushroom genetic resources in India. In Proceedings of the 3rd International Conference on Mushroom biology and Mushroom Products & AMGA's 26th National Industry Conference Sydney. pp 40-46.

Doshi, A. and Sharma, S.S. 1997.Wild mushrooms of Rajasthan. In: Rai D, Verma (eds) Advances of mushroom biology and production. Mushroom Society of India. Solan, India

FAO. (2004). Non wood forest products, wild edible fungi: A global overview of their use and importance. (ed. Boa, E). FAO Publication, Rome, pp. 17-147.

Gbolagade, J., Ajayi, A., Oku, I. and Wankasi, D. 2006. Nutritive Value of Common Wild Edible Mushrooms from Southern Nigeria. Global Journal of Biotechnology & Biochemistry 1(1):16-21.

Harsh, N. S.K., Rai, B.K. and Ayachi, S.S. 1993. Forest fungi and tribal economy- a case study in Baiga tribe of Madhya Pradesh. J. Trop. Forest 9: 270-279.

Harsh, N.S.K., Tiwari, C.K., Rai, B.K. 1996. Forest fungi in the aid of tribal women of Madhya Pradesh. Sustainable Forestry 1: 10-15.

Hawksworth, D.L. 1997. The fascination of fungi: Exploring fungal diversity.*Mycologist* 11: 18-22.

Hobbs, C. 1995. Medicinal Mushrooms: an exploration of tradition, healing, and culture. 2nd edition. Santa Cruz CA, USA. Botanica Press. pp. 252.

Jain, R. K. 2005.The History of Mushroom Cultivation in India. Environment.Bhologi.com

Kaul, T.N. 1997. Introduction to Mushroom Science (Systematics). Oxford and IBH Publishing Co. Pvt. Ltd, New Delhi. P. 198.

Kaul, T.N. and Kachroo, J.L. 1974. Common edible mushrooms of Jammu and Kashmir. Ind. Mush. Sci. 71:26-31

Kumar, S. and Sharma Y. P. 2011. Diversity of wild mushrooms from Jammu and Kashmir (India) Proceedings of the 7th International Conference on Mushroom Biology and Mushroom Products (ICMBMP7)568-577.

Kumar, S.M. and Shukla, C.S. 1995. *In:* Advances in Horticulture Vol. 13 -Mushrooms. Malhotra Publish. House, New Delhi, pp.277-316.

Lakhanpal, T.N. 1994. Prospect of mushroom from the wild, In: Advances in Mushroom Biotechnology, edited by Nair, M.C., Gopalan, C. and Das, L. Scientific Publisher, Jodhpur, India p15-22.

Lakhanpal, T.N. 1997. Diversity of mushroom mycoflora in the North-West Himalaya. *In*: Recent Researches in Ecology, Environmentand Pollution (eds Sati, S. C., Saxena, J. and Dubey, R.C.), Today and Tomorrow's Printers and Publishers, New Delhi, pp 35–68.

Longvah,T.and Deosthale, Y.G. 1998. Compositional and nutritional studies on edible wild mushroom from Northeast India, Food Chem. 63:331-334.

Manjunathan, J. and Kaviyarasan, 2011. Nutrient composition in wild and cultivated edible mushroom, Lentinus tuberregium (Fr.) Tamil Nadu., India International Food Research Journal 18: 809-811 .

Manoharachary, C., Sridhar, K. Singh, R., Adholeya, A., Suryanarayanan, T. S. Rawat, S. and Johri, B.N. 2005. Fungal biodiversity: Distribution, conservation and prospecting of fungi from India. Special Section: Microbial Diversity. Current Science, 89(1):58-71.

Miles, P.G.and Chang, S.T. 1997. Mushroom Biology, Concise Basics and Current Developments. World scientific Publ., London.

Purkayastha, R.P. and Chandra, A. 1985. Manual of Indian Edible Mushroom. Today and Tomorrow's Printers and Publishers, New Delhi.

Qi Tan and Hui Cao 2010. New development of the mushroom industries in China Institute of Edible Fungi., Shanghai. Academy of agricultural sciences, Shanghai 201106, P.R. China. http:wsmbump.org/Bulletin_2_contenent.html

Ram, D. and Singh, A.R. 1997. Mushroom cultivation in Bihar-Present Status and Future Prospects. Souvenir and Abstracts, Indian Mushroom Conference (September 10-12, 1997). Poster 4. 34 p.82.

Rana, P. and Giri, A. 2006. Mushroom diversity in the Sagarmatha national park and its buffer zone area. Banko Janakari. 16(2):17-24.

Rana, P. and Giri, A. 2006. Mushroom diversity in the Sagarmatha national park and its buffer zone area. Banko Janakari, 16 (2):17-24.

Rojas, C.and Mansur, E. 1995. Ecuador: Informaciones generales sobre productos non madereros en Ecuador. *In:* Memoria, consulta de expertos sobre productos forestales no madereros para America Latina *Y el Caribe,* pp. 208-223. Serie Forestal #1. Santiago, Chile, FAO Regional Office for Latin America and the Caribbean.

Sagar, A., Chauhan, A., Sehgal, A.K. 2005. Ethnobotanical study of some wild edible mushrooms of tribal district Kinnaur of Himachal Pradesh. Indian J. Mush. 23:1-8

Saini, S.S. and Atri, N.S. 1995. *In:* Advances in Horticulture Vol. 13 -Mushroom, Malhotra Publishing House, New Delhi, pp. 375-386.

Sarma, T.C., Sarma, I. and Patiri, B. 2010. Wild Edible mushroom used some ethnic Tribes of Western Assam. The Bioscane. An International Journal of Life Science Special Issue, 3: 613-625.

Sharma, J.P. and Kumar, S. 2006. Effect of supplementation on with bran and oil cake on yield of (*Pleurotus* spp.) in Jharkhand. Souvenir and Abstract, National Symposium on Emerging Plant Disease, Diagnosis and Management (Jan. 31- Feb. 2) p 126.

Sharma, J.P., Ajit, Jha, S.Kumar and M. Rai 2002. Jharkand Mei Khumbh Utpadan (Khukhri) Chotanagpur Horticulture, 15-1 9 (1-4) 30-1.

Sharma, J.P., Kumar S. and Verma. R.N. 2011. Present status and future stratifies of mushroom production in Jharkhand. *In:* Diversity and Production of Edible Mushrooms (Eds) S.Kannaiyan, T. Marimuthu and K. Lenin), Associated Publishing Company, pp.184.

Sharma, S.R. and Rai, R.D. 1994. Aiming for the world market. The Hindu Survey of India. pp 191-7.

Shu-Ting Chang 2006. The World Mushroom Industry: Trends and Technological Development International Journal of Medical Sciences, 8(14)10: 297-314.

Sinha, B.K. 2003 Biodiversity and its economics: a case study of Ranchi urban agglomeration. Jharkhand Journal of Development and Management Studies XISS, Ranchi, 1(3):349-360.

Sohi, H.S. and R. C. Upadhyay, 1989. Effect of temperature on mycelial growth of *Pleurotus* species andtheir yield parameters or selected substrates. Mushroom Science, 12(2)49-56.

Sohi, H.S. and Upadhyay, R.C. 1986. Occurrence of *Cladobotryum variospermum* (Link) huges on polyporous fungi under natural condition. Current Science 55:1037-1038.

Tiwari, R.P. 2004. Now in Limelight. The Hindu Survey of Indian Agriculture, pp. 132-33.

Upadhyay, R. C., Verma, R. N., Singh S. K and Yadav M. C. 2002. Effect of organic nitrogen supplementation in Pleurotus species. Mushroom Biology and Mushroom Products. Sánchez *et al.* (eds). 2002UAEM.

Verma, R.N. and T.H. Gourbidhu Singh 1981. Investigation on edible fungi in the North Eastern Hills of India. Mushroom Sci.11: 89-99.

Verma, R.N. and Upadhyay, R.C. 1999. Biodiversity of mushroom and potential of their utilization in tribal belts of Bihar. *In:* National Seminar on Sustainable Horticulture Production in Tribal Region (July 25-26) HARP pp. 140.

Verma, R.N. Singh, G.B. and Singh, S.M. 1995. Mushroom flora of North Eastern Hills. *In:* Advances in Horticulture Mushroom, edited by Chadha, K.L. and Sharma, S.R., Molhotra Publishers House, New Delhi, 329-349

Verma, R.N., 1997. The mushroom Scenario of India –a need for species diversification. *In:* Souvenir and Abstracts.Indian Mushroom Conference 1997 (Abs) p12.

Verma, R.N., Singh, G.B. and Bilgrami K.S. 1987. *In:* Proceedings of the International Conference on Science and Cultivation Technology of Edible fungi, RRL (J&K) 414-421.

Verma, R.N., Singh, G.B. and Singh, S.M. 1985. Meghalaya Sci. Soc. J. 7 and 8: 6-9.

Verma, R.N., Singh, S.M., Singh, G.B. and Bilgrami, K.S. 1989. Curr. Sci. 58:1370 1371.

Vishwakarma M.P., Bhatt R.P and Joshi, S. 2012. Macro fungal diversity in moist temperate forests of Garhwal Himalaya. Indian Journal of Science and Technology 5(1):1928-1932.

Wakchaure, G.C. 2010. Production and Marketing of Mushroom: Global and National Scenario. *In*: Mushrooms cultivation, Marketing and Consumption. (Eds. Singh, M. Vijay, B., Shwet Kamal and Wakchaure, G.C.). Directorate of Mushroom Research (ICAR) Chambaghat, Solan. pp 15-22.

❑❑❑

18

Absorption of Photosynthetic Active Radiation (PAR) and its Effect on Growth Processes and Yield of Wheat Under Different Dates in the Gangetic Plains of Eastern India

Rajib Nath, Mausumi Parya, D. Majumder, Sarika Jena and P.K Chakraborty

Abstract

The Photosynthetically Active Radiation (PAR) measure of radiant power is important in evaluating the effect of light on plant growth. The photosynthetic response correlates better with the number of photons than with energy. This is expected because photosynthesis is a photochemical conversion where each molecule is activated by the absorption of one photon in the primary photochemical process. PAR is defined in terms of photon (quantum) flux, specifically, the number of moles of photons in the radiant energy between 400 nm and 700nm. One mole of photons is 6.0222 x 1023 photons (6.0222 x 1023 is Avagadro's Number). The Phtotssynthetic Photon Flux Density (PPFD), i.e. the photon irridiance, is expressed in moles per squre meter and per second (formerly, Einsteins per squre meter and per second). The absorption of PAR by wheat varieties increased from 7.30 h to 11.30 h followed by a decline at 13.30 h at CRI, tillering and milking stages. This indicates that absorption of PAR at different hours in day seems to play as significant role. Stem, leaf and ear dey weights and LAI were singnificantly related to the absorption of PAR by wheat.

***Keywords**:* Wheat; PAR : Principal Component regression analysis

Introduction

The growth and productivity of a crop are directly related to the absorption of Photosynthetic active radiation (PAR) by the crops. The absorption varies due to variation of crop species, season and time of a day. Wheat is an important winter crop which should be cultivated in large areas of the state with an aim to replace the *boro rice* cultivation which extracts high amount of water from the ground, thus conserving the under ground water. To address this objective it is imperative to identify the sowing time of wheat in accordance with climatic impact on the crop. In this context, studying the PAR absorption pattern by wheat and selection of dates of sowing according to this criterion are seemed to be the most important task of the wheat workers of eastern India. To achieve this objective, a three year study on wheat varieties under different dates of sowing had been carried out to investigate the PAR absorption and its effect on different growth processes and yield of wheat crop.

Methods

The experiment was carried out during winter seasons of 2005-06, 06-07, and 07-08 at BCKV, Mohanpur, Nadia, W.B., India (22.56'N latitude, 88.32'E longitude and an attitude of 9.75m above mean sea level) having topical humid climate with short and mild winter. The average annual rain fall is 1457min, 85% of which is received during June to September, mean monthly temperature range is 10-37°C. The experimental soil is sandy loam and neutral in reaction (PH-6.90) having 0.077% total N, 57.25 Kg/ha available P_2O_5, 221.51Kg /ha available K_2O and 0.78% organic carbon. Five Varieties of wheat (PBW 343, HD 2733, HW 2045, PBW 533 and K9107) were sown on three dates (18th November, 3rd December and18^{th} December) in each block, keeping dates of sowing in the main plot and varieties in the sub-plot in a split plot design having the plot size of $15m^2$.The crop was fertilized with 150Kg N, 60Kg P_2O_5 and 40 Kg K_2O ha^{-1} where N was given in three splits. The direct incident PAR above and below the wheat and canopy was measured by placing the line quantum sensor (APOGEE Logan UT, UK) 25cm above the crop and soil surface across the rows within the wheat canopy keeping the sensor horizontal to the surface. The reflected PAR was measured from the same positions by simply inverting the instrument. The absorbed PAR was calculated from the measured PAR values as follows:

Absorbed PAR (APAR)

=[PAR(o) +RPAR(s)] –[TPAR + RPAR(c)] (Gallo and Doughtry, 1986)

Where,

PAR (o)= the portion of the incident PAR above the canopy

RPAR(s) = Reflected PAR from surface under the wheat canopy (reflected PAR value at bottom level of wheat canopy).

TPAR = Transmitted portion of the PAR through the canopy to the soil surface (incident PAR value at the bottom of the wheat canopy).

RPAR(c) = Reflected PAR from crop (reflected PAR value at the uppermost layer of the wheat canopy).

Percent absorption was calculated by dividing the APAR by PAR (o) and multiplying the factor by 100.

The PAR absorption was measured at CRI, tillering, flowering and milking stages and for diurnal variation; it was measured at 7.30, 9.30, 11.30 and 13.30h on each dates of measurement. Among the growth parameters, leaf area index, leaf and stem dry weights were measured at flowering and milking stages. The yield was estimated from 6m^2 of the area of each plot. Principal component regression analysis (PCRA) was carried out to find out the relationship between absorption of PAR and growth parameters and well as with grain yield.

Results

The result indicated that absorption of PAR increased from 7.30h to 11.30h, followed by a decline at 13.30h at CRI, tillering and milking stages irrespective of dates of sowing and the varieties. Absorption gradually declined with the delay in sowing. Principal component regression analysis showed that stem, leaf, ear dry weights and LAI were significantly related to the absorption of PAR at different growth stages. About 31 to 35% variation in stem weight, 47% variation in leaf weight, 34-35% variations in ear weight and 45 and 74% variation in LAI could be explained through the variation in absorption of PAR at the different growth stages of wheat. Grain yield was significantly related to the absorption of PAR, absorption of PAR at flowering significantly explained the variation in grain yield. Absorption of PAR at different hours in day seems to play a significant role. PCRA suggested that stem, leaf and ear dry weights and LAI were significantly related to the absorption of PAR at 9.30, 11.30 or 13.30h at the different phenophases of wheat.

References

Gallo, K.P. and Daughtry, C.S.T. 1986. Techniques for measuring intercepted and absorbed PAR in corn canopies. Agronomy Journal. 78:752-756

Hippes, L.E. Asrar, G., and Kanemasu, E.T. 1983. Annerning the interception of photosynthetically active radiation in which wheat. Agric. Meterol 28:253-259.

Hogdes. T, and Kanemasu, E.T. 1977. Modelling daily dry matter production of which wheat. Agron. J. 69: 974-978.

❑❑❑

19

Stress Impact Assessment of *Brassica* spp Under Different Dates of Sowing and Irrigation Level

Swaraj Kumar Dutta, Rajib Nath, G. Sounda, D. Majumder and P.K. Chakraborty

Abstract

In order to estimate the irrigation need of the crop, stress degree day index computers from canopy and air temperature could be used. During rabi manon, two varieties (Seeta, Brassica Juncea) and (Benoy, Brassica Campterteris) were grown. The canopy temperature of the crop was measured on 45th, 60th and 75th days after sowing and stress degree day (SDDI) was also measured. About 67, 70 and 51 per cent activation between SDDI and dry matter production could be explained at 45, 60 and 75 DAS of the crop growth period. Reflectance by mustard canopy was 940 measured. Late sown crop reflected less radiation with the advancement of growth. Extinction coefficient was also measured and the increase in the value of extinction coefficient upon irrigation showed canopy development and better utilization of increasing radiation by the crop.

Keywords: Mustard; Stress Degree Day; Reflectance; Extinction coefficient.

Introduction

Anthropogenic activities lead to the initiation of weather variability rather than climate change as envisaged by general public opinion. The most vulnerable variabilities have crept in temperature and rainfall regimes. Shifting of South West Monsoon is on the card which has caused a delay in sowing of winter crop and thus duration of phenophases have been shortened (Parya *et al.*, 2010) which reduced the yield of the crop. Delayed sowing leads to the initiation of moisture stress during the important phenophases. Effect of both the moisture stress and air temperature has been combined in the canopy temperature which may be used as an indicator for explaining the biological response of the crop growing under different environmental situation. Idso *et al* (1977) suggested that the stress degree day index computed from canopy and air temperature might be successively used for judging the irrigation need of the crop. With a view to explore the possibility of using the SDDI and crop reflectance to find out ideal date of sowing and variety of mustard in the Gangetic Plains of West Bengal; the present investigation has been planned.

Materials and Methods

The field experiment was carried out during *Rabi* season at Jaguli Instructional farm, B.C.K.V ($22^0 56'$ N; $88^0 32'$ E) in 2007-08 and 2008-09. The experiment was carried out in split-split plot design with 18 treatment combinations in which 3 dates of sowing D_1 (20th Nov.); D_2 (30th Nov.) & D_3 (10th Dec.) and three irrigation levels I_0 (I_w/CPE=0), I_1 (I_w/CPE=0.5), I_2 (I_w/CPE=1.0) and two varieties V_1 (Seeta; *Brassica juncea*) and V_2 (Binoy; *Brassica campestris)* were considered as main plot, sub plot and sub-sub plot treatments respectively. The canopy temperature of the crop was measured on 45th, 60th and 75th days after sowing (DAS) and the stress degree day (SDDI) was computed following Idso *et al* (1977). The reflectance of PAR from the canopy was measured with the help of Line Quantum Sensor (Apogee Delta T, U.K) and the extinction coefficient was computed as follows:

$I = I_0 . e^{-KL}$,

or, $l_n I/I_0 = -KL$,

or, $K = - I/L [\log e\ I/I_0]$,

$= - I/L [2.303 \log I/I_0]$.

Where, I = Radiation (TSR or PAR) at a particular height of crop canopy

I_0 = Radiation at the top of the canopy.

K = Extinction coefficient

e = Exponential constant

L = Leaf Area Index

Results and Discussion

SDDI and dry matter production: Stress Degree Day Index (SDDI), which is the difference between canopy temperature and air temperature maximum, indicates the extent of moisture stress experienced by a crop. Negative values of SDDI indicates the sufficient availability of moisture to the crop plants resulting in concomitant cooling process in leaf due to higher transpiration. Positive value indicates higher canopy temperature due to less transpiration under moisture stressed condition. When the crop was sown on D_1, dry matter production decreased with increase in SDDI, on 45th DAS about 76% variation in dry matter production could be explained through the variation in SDDI. Throughout the cropping season, I_2 (I_w/CPE =1.0) irrigated crop did not experience any moisture stress while I_1 (I_w/CPE =0.5) irrigated crop experienced stress only at 45 DAS. However the crop under rainfed condition was under stress throughout the season excluding the last phase due to receipt of rainfall. With the advancement in crop age the R^2 values declined at 60 DAS but 80% variation in dry matter production at 75 DAS could be explained due to variation in SDDI. When the crop was sown on D_2 it showed that SDDI significantly explained the dry matter production capability throughout the season. About 67, 70 and 51 percent association between SDDI and dry matter production could be explained at 45, 60 and 75 DAS of the crop growth period respectively. When sown on D_3, the crop showed a very weak and insignificant relationship upto 60 DAS; however on 75 DAS, significant reduction in dry matter production was observed with the increase in SDDI; about 67% variation in dry matter production could be explained through the variation in SDDI.

Reflectance by mustard canopy: The fraction of incident radiation is reflected by the vegetation due to multiple scattering by the canopy elements. The moisture availability also changes the reflectivity to maintain equilibrium between absorption and transpiration by a particular crop (Monteith and Unsworth, 2001). When the crop was sown on D_1, reflectance recorded a polynomial pattern of reduction with the increase in plant height; altogether the association was not so strong with the R^2 values ranging from 0.336 to 0.457. a significant reduction in reflectance was observed with the increase in plant height showing better moisture utilization, when the crop was sown on D_2. The D_3 sown crop recorded a significant rduction in reflectance with the increase in plant height on 75DAS only. The results suggested that the late sown crop reflected less radiation with the advancement of growth. Stanhill (1970) argued that reduction in reflection with the increment of plant height resulted due to better utilization of light by the canopy.

Extinction coefficient: The reduction in the value of extinction coefficient implies that direct solar radiation can penetrate deeper within the canopy (Uchijima, 1976).

When the crop was sown on D_1, under rainfed condition, the extinction coefficient was the lowest; this was also observed under D_2 and D_3 sowing with a few observations. This indicated lower canopy development under rainfed condition. Under irrigated condition, extinction coefficient increased indicating canopy development and better utilization of incoming radiation by the crop. Under late sown condition, extinction coefficient was found to be higher because the crop had to complete its life cycle within a specified period of time.

The results suggested that better canopy development and light utilization would be possible under irrigated condition.

References

Idso, S.B., Jackson, R.D. and Reginato, R.J. 1977. Remote sensing of crop yields. Science. 196: 19-25.

Monteith, J.L. and Unsworth, M. 2001. In : Principles of Environmental Physics. pp 85-88. Butterworth- Heinmann Oxford.

Parya, M.; Nath, R.; Mazumdar, D. and Chakraborty, P.K. 2010. Effect of thermal stress on wheat productivity in West Bengal. J. Agrometeorology 12(2): 217-220.

Stanhill, G. 1970. Some results of helicopter measurements of albedo. Solar Energy. 13:59.

Uchijima, Z. 1976. Microclimate of the rice crop in Climate and Rice. Proceedings of the Symposium on Climate and Rice. The IRRI, Los Banos, Philippines, pp: 115-140.

❑❑❑

Section D: Climatic Influences Over Animal Husbandry and Fisheries

20

Amelioration of Summer Infertility in Livestock

B.S. Prakash

Abstract

Infertility in Livestock is a matter of great concern Anoestrous and repeat breeding in buffalo and bovines are two of the most serious reproductive problems affecting 30-40% of the total cattles and buffalo population on a conservative estimate our country is loosing 20-30 million tonnes of milk due to anestrus and repeat breading in cattle and buffaloes translating to a loss of Rs 40-50,000 crores annually. The used weak symptoms of estrus in the normal breeding season (September to February) become ever weaker during the hot month of summer. The incidences of silent heat is the lowest in December while the break is seen in the hot summer month of April. Summer infertility is also noticedin goat. During summer farmer experience a variety of reproductive problems such as an estrus extended weaning to eastern interiors, poor conception rates, higher unbryo mortality rates and low barrowing rates. There are strategies to ameliorate heat stress induced fertility problems and gonadotropin treatment can prevent seasonal infertility. There is a need to implement several techniques at the field level combining the glance of any detrimental impact on health and environment

Keywords: Infertility in livestock, Ovsynch treatment, Ayurvedic herb, Heat stress in goat and pig.

Introduction

India has vast resources of livestock, which play a vital role in improving the socio-economic conditions of the rural masses in terms of supplementing family incomes and generating gainful employment in the rural sector, particularly, among the landless labourers, small and marginal farmers and women, besides providing cheap nutritious food to millions of people.

Milk is the most important contributor to the output from livestock sector. Although India ranks first in the milk production, but Indian dairy is a classic example of Gandhian philosophy of production by masses rather than mass production. Improved buffalo and zebu production in particular, could significantly enhance the economy and living standards of farmers in India. There are at least 30 different zebu breeds in Indian subcontinent alone: Hariana, Tharparkar, Sahiwal and Red Sindhi are the predominant dairy breeds. The average lactation yield of these breeds is around 1800 litres in a 305 days period. However, most of the cattle in India are of the nondescript types which yield very little milk. In an effort to increase milk production cross breeding of zebu with semen from exotic breeds has been carried out. The average milk production over a 305 days lactation period of milch buffaloes ranges from 1500 – 2500 litres. The adaptation of buffalo and zebu to tropical climates ensures their place in the future of world agriculture. On perusal of Livestock Census of 1992, 1997 and 2003 we observe that there has been a significant decline in cattle population from 204.58 million in 1992 to 198.88 million in 1997 and again to 187.38 million in 2003. Interestingly, there has been a corresponding increase in buffalo population from 89.9 million to 97.9 million indicating a sharp preference for buffalo rearing among the farmers. With an overall achievement of 100 million tonnes of milk in 2006-07 from cattle, buffaloes and goats and a per capita milk availability of 245 g/day, an anticipated 116 million tonnes of milk production in 2011, the Indian dairy scenario is constantly looking ahead & promises to take greater strides in making dairying more remunerative to the farmer. Livestock numbers in India indicate that buffaloes though numbering around 98 million in the country form a third of the total cattle and buffalo livestock population and contribute more than 55% of the milk. There has never been a doubt about the farmers' preference for keeping buffaloes particularly in Northern States of India as the predominant milch animal since it is a triple purpose animal contributing milk, draught and meat as compared to the dual purpose native cattle. The buffalo milk also fetches a good remunerative price to the farmer and the animal can utilize poor quality roughages better than native cattle.

Goat and sheep are also reared by the small, marginal and landless farmer for wool and meat. They contribute greatly to the agrarian economy, especially in the arid/semi-arid and mountainous areas where crop and /or dairy farming are not economical. In addition pig production is also practiced particularly as backyard pig farming systems as part of the mainstream farming in Kerala, Goa, North-Eastern States and to special categories among the socially weaker sections and tribals in Jharkhand solely for meat.

Good reproductive performance is essential for efficient livestock production, so improvement programs should increase reproductive efficiency to the extent that this can be justified economically. However, low fertility in Indian livestock particularly in buffaloes is a matter of serious concern. Some of the main reasons– among other factors – being, ovarian disorders, limited attention to application of the modern techniques for improving the reproductive efficiency, poor exploitation of genetic potential of indigenous animals, and, low absorption of available technology.

Anestrous and repeat breeding in buffaloes and bovines are two of the most serious reproductive problems affecting 30-40% of the total cattle and buffalo population. On a conservative estimate the country is losing 20-30 million tonnes of milk annually on account of anestrus and repeat breeding in cattle and buffaloes which translates to a loss of nearly Rs. 40-50000 crores annually. At a micro level, each missed heat is a missed opportunity. For each heat missed the farmer incurs a loss of milk production of 21 days, in addition to bearing the feeding cost for animal maintenance. This tantamounts to about Rs.3300. *For this kind of money you and I can drive a Ford Ikon for 1 month!*

Artificial insemination (AI), which is a normal practice in cattle, is not as successful in buffalo especially in hot summer months, because of the weakness of oestrus symptoms and the variability of oestrus length, which make oestrus detection very difficult. The usual weak symptoms of estrus in the normal breeding season (September to February) become even weaker during hot months of summer (Prakash 2002). The incidence of silent heat was lowest in December (10.5%) while the peak was seen in the hot summer month of April (70%; Prakash *et al.* 2005; Fig.1).

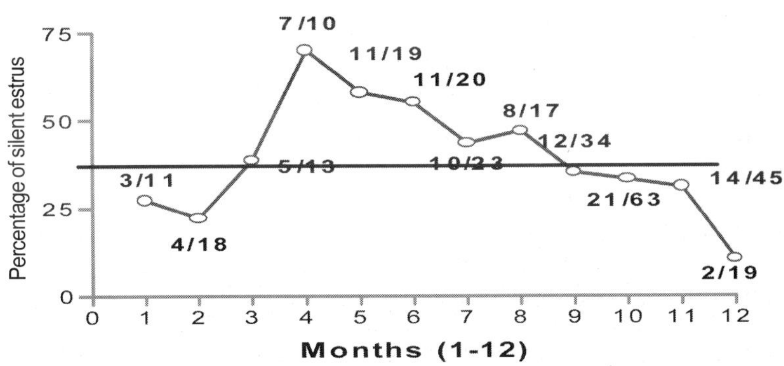

Fig. 1: Incidence of silent estrus in buffaloes during different months

Several studies have attempted to understand the reproductive physiology of buffalo and the factors affecting its behaviour. These have been adequately reviewed (Madan and Prakash, 2007). During the last two decades, considerable attention has been focused on reproductive endocrinology, with the aim of developing models to improve reproductive efficiency, particularly when using controlled breeding techniques. The techniques of ovsynch and heatsynch protocols for estrous synchronization for enhancing fertility in cattle are well documented (Pursley *et al.*, 1995; 1997; Pancarci *et al.*, 2002). The research work conducted at NDRI to ameliorate infertility in buffaloes includes the application of ovsynch and heatsynch estrus synchronization protocols followed by fixed time AI both at farm (Paul and Prakash, 2005; Mohan, Sarkar and Prakash, 2009; Mohan *et al.* 2010) and field levels (Prakash *et al.* 2008; Prakash, 2011); the average success rate for conception being around 50% at fixed time AI and AI at subsequent estrus (both inclusive). Additionally, 30-40 % of anestrus buffaloes become cyclic using these protocols giving a hope to the farmers for getting them pregnant at some stage later.

Buffalo of Mr. Ramchander S/o Sh. Kesi Ram declared pregnant after 14 months of calving at the village Samora using heatsynch

Prolactin Inhibition in Repeat Breeding Buffalo Heifers in the Summer

It is well documented that all kinds of stress lead to increased prolactin release. Circulating prolactin concentrations are 2-6 fold higher throughout the estrous cycle in the low breeding season (summer) compared to those in either the medium (monsoon) or peak breeding seasons in buffaloes (Heranjal *et al.* 1979; Razdan and Kaker, 1980; Pahwa and Pandey, 1984; Galhotra *et al.* 1988, Singh and Madan, 1993). Plasma prolactin concentrations in cycling buffalo heifers were about 40-70 fold higher in summer months in comparison to the concentrations observed in winter months (Roy and Prakash 2007). The extremely high prolactin concentrations in summer months was associated with lower estrogen concentrations and summer anestrus (Singh and Madan, 1993). Under tropical conditions, a large proportion of buffaloes become anestrous during summer months. Progesterone monitoring from blood/milk plasma samples however indicates that most of the buffaloes exhibit cyclicity without overt estrus expression (Kamboj and Prakash, 1993; Prakash 2002) in summer months. Research carried out in the past few decades in several species like human, rats, bitches, ewes, cows and buffaloes have tried to link high prolactin levels in circulation to poor fertility in terms of anovulation/anestrus, suppression of gonadal hormone synthesis and secretion, and poor estrus expression. Keeping the above aspects in view we investigated the influence of prolactin inhibition on endocrine profiles and efficacy of Ovsynch treatment for fertility enhancement in repeat breeding buffaloes in summer months (Roy and Prakash, 2009). Brief results are summarized below:

Inhibition of endogenous prolactin secretion during Ovsynch treatment in buffaloes was brought about by daily administration of norprolac (10 mg, i.m; Novartis, Switzerland) - the latest third generation prolactin inhibitor. The hormones quantified in blood plasma samples collected before, during and after Ovsynch and Ovsynch plus norprolac treatment were LH, prolactin, progesterone, estradiol-17β and total estrogens. The plasma prolactin and progesterone concentrations were negatively correlated during summer estrous cycle, which indicated prolactin-induced suppression of progesterone secretion through poor luteal development. Suppression of norprolac induced prolactin secretion enhanced the preovulatory estradiol-17β and total estrogen concentrations without influencing the preovulatory LH release. The Ovsynch treatment alone reduced the incidence of anestrous from 45% before treatment to only 18% after treatment. However, all anestrous animals (36%) became cyclic when buffaloes were subjected to norprolac plus Ovsynch. Treatment of Ovsynch alone and Ovsynch plus norprolac resulted in 5/11 (45%) and 6/11 (55%) pregnancies respectively after timed AI and subsequent inseminations. It was concluded that ovsynch protocol for estrus synchronization has potential application for improvement of fertility in repeat breeding buffaloes even during extreme summer months which can be further improved through suppression of endogenous prolactin secretion.

Improvement of Immunity through Supplementation of Ayurvedic Herb *Tinospora cordifolia* to Crossbred cattle

The uterus is a unique immunological site as it must maintain immunity against constantly invading pathogens while at the same time remaining tolerant to the fetus which can be considered foreign to the mother. Both innate and acquired immune defense mechanisms are suppressed during pregnancy to avoid fetal rejection by the mother with these mechanisms getting aggravated prepartum and further during early postpartum period. Impairment of the immune status during the periparturient period in cattle increases the risk of uterine infections which are more prevalent during this period compared with other stages of the reproductive cycle. Introduction of cross breeding (the fastest method of improving animal productivity) has certainly improved the milk production in India, but the performance of crossbred cows is affected by high temperature, poor feeding and susceptibility to diseases making them a liability for the resource-poor rural farmer. The overall incidence of mastitis, retention of fetal membranes (RFM), metritis, endometritis and pyometra in crossbred cattle are more in comparison to native cattle.

It is in this context that innovative methods need to be evolved for enhancing the immunity in periparturient period without compromising the health of the developing fetus in order to improve the postpartum reproductive health as well as milk production in the crossbred dairy cow. Medicinal plants have long been used in the Indian system of medicine viz. ayurveda for human beings. Guduchi – botanical name *Tinospora cordifolia* - is a well researched plant for its immunomodulatory activity in terms of immunostimulation. An investigation was therefore undertaken with the objective to study the effect of peripartum dietary supplementation of guduchi on reproductive health, endocrine changes, immunity, calf weight and survival, cyclicity commencement, milk yield and SCC in crossbred cows. Incidence of retention of fetal membranes, endometritis and pyometra were higher in control group of cows in comparison to those recorded in treated group (Mallick and Prakash, 2011). The guduchi supplemented cows exhibited faster uterine involution (28 days *vs*. 42 days) and early commencement of cyclicity (37 days *vs*. 58 days; based on plasma progesterone profiles) in comparison to untreated control group of cows. Mean birth weight of calves from treatment group of cows was significantly higher than that of control group (33.14 kg *vs*. 29.13 kg). Calves born from control group of cows exhibited high mortality (3 out of 7) within 3-4 weeks postpartum in comparison to none among the treatment group. There was however no significant difference in average daily body weight gain of calves in both the groups. There was an improvement in the innate and adaptive immunity in guduchi supplemented cows in terms of higher total leukocyte, lymphocyte, neutrophil count along with increased neutrophil lymphocyte ratio in comparison to untreated cows. Peripartum plasma total antioxidant activity was similar between the two groups. Supplementation of guduchi significantly reduced plasma progesterone concentration prepartum, although there was no significant difference in peripartum plasma total estrogen and PGFM levels (Mallick and Prakash, 2011). A significant increase in milk yield in cows supplemented with guduchi over the untreated control cows was recorded. It was concluded from the study that guduchi supplementation improved the reproductive performance of cows by enhancing the immunity and lowering the prepartum progesterone secretion. Improvement of lactation and milk quality was attributed to increase in GH concentrations, better immunity and certain innate properties of guduchi. A detailed study on a larger population to validate these findings and compare the results both in summer and winter months is in progress in association with the Extension experts.

Summer Infertility in Goats

It is estimated that over 80% of world's goat population is located in tropical, sub-tropical, and arid areas across, Asia, Africa, Central and South America. What makes goats so popular in these areas is there ability to provide high quality food under diverse climatic conditions and resilience to extreme and capricious environment. In small ruminants living at high and mid latitudes, reproductive efficiency appear to be affected by environmental condition throughout the year particularly day length and ambient temperature. Although low summer fertility syndrome is attributed to multiple mechanisms ovulation tends to be delayed in heat stressed goats. There is an increase in percentage of abnormal spermatozoa and decrease in motility with increase duration of heat exposure. To control the sexual activity of females by means of sexually active males is a sustainable breeding technique for goat reproduction in the subtropics called as male effect. There is also an inter-breed difference regarding heat stress induced infertility. Melatonin implants stimulate the sexual behavior of bucks during the non-breeding season. Heat stress during follicular recruitment suppresses subsequent growth to ovulation accompanied by decreased LH receptor level and reduced estradiol synthesis activity in follicles.

Heat Stress Amelioration in Pigs

During summer and early fall, many sow farms experience a variety of reproductive problems, such as, anestrus, extended weaning-to-estrus intervals, poor conception rates, higher embryo mortality rates and low farrowing rates. This phenomenon is commonly referred to as "seasonal infertility" and can also be present in boar studs, where the result is lower semen output and lower semen quality. Most researchers have attributed this seasonal infertility to heat stress and the fact that pigs are inherently seasonal breeders. Major manifestations of seasonal infertility are delayed puberty attainment in gilts, anestrus both in sows and gilts, extended weaning-to-estrus interval, poor conception rate, higher embryo mortality rate, low farrowing rate and total litter sizes are also lower during summer season leading to reduced fecundity and reduced fertility. The primary goal of every operation must be to incorporate environmental strategies to reduce heat stress. Sows and boars both suffer from acute and persistent exposure to elevated ambient temperatures and humidity. As a result, infertility can be relatively short term or, sometimes, a permanent disability from which the animal will never recover. Heat stress to sows and boars occurs when ambient temperatures are outside of the animals' thermoneutral zone, which ranges between 45° and 70° F. In most cases, the effect of heat stress on reproduction has been related to ambient temperatures in excess of 80° F. In studies where temperatures were elevated experimentally for sows and gilts, anestrus increased and conception rates and embryo survival decreased.

There is also evidence that boars exposed to ambient temperatures in excess of 85°F have lower sperm output and poorer sperm motility, and increased morphological abnormalities. Normally, if this exposure is short, recovery will occur in 6 to 7 weeks. It takes 6 to 7 weeks for a new group of sperm cells to mature within the boar's testicle and epididymis, and the immature, developing sperm cells tend to be more sensitive than mature sperm. If a boar is subjected to heat stress sporadically over a two-month period, it is possible that the fertility of his semen could be reduced during this period as well as for the 6 to 7 week period after it has ended.

Strategies to ameliorate heat stress induced fertility problems

- Provision of wallows, shades or cooling systems
- Ensure ventilation systems are in good working order and are providing adequate air movement.
- Carrying out animal handling procedures in cool times of the day
- Plenty of clean, cool, and fresh water is paramount to preventing heat stress
- Adjustment of energy levels and feeding program
- Maximum use of artificial insemination, either as a replacement for natural mating or as an adjunct to natural mating, especially between April and September.
- Implement an early-wean/split-weaning program and avoid multisuckling in summer to reduce the metabolic demand on lactating females and to enhance post-weaning oestrus
- Gonadotropin treatment can prevent seasonal infertility in sow (Taker *et al.*, 2008)
- Establish a long-term programme to reduce the impact of heat stress on the sow herd through genetic selection of animals that can perform well in high temperature.

Although considerable emphasis has been given in this paper on dairy livestock, the issues appear to be common also for other livestock viz. sheep, goats and pigs for meat production. Several techniques have been developed for enhancing production and reproduction in livestock; unfortunately most of these have not yet been adopted at the field level in a big way which is imperative if they are to make an impact towards increasing productivity and thereby enhance farmers' incomes. Perhaps the main reason for research efforts not reaching the farmers doorsteps is the lack of knowledge interaction between farmers, scientists and the industry. It needs to be mentioned here that application of biotechniques for improving reproductive efficiency should be implemented in a sensitive and humane manner keeping in view the primary focus on economizing yield of animals and increasing the number of high genetic merit animals. It should also be borne in mind that this fertility improvement should not be associated with any detrimental impact on health and environment.

References

Galhotra M.M, Kaker M.L, Lohan, I.S, Singal, S.P, and Razdan, M.M. 1988. Thyroid and prolactin status in relation to production in lactating Murrah. *In*: Proceedings of 2nd World Buffalo Congress, 1: 32-34, New Delhi, India.

Heranjal D.D, Sheth A.R, Desai R. and Rao S.S. 1979. Serum gonadotropins and prolactin levels during oestrous cycle in Murrah buffaloes. Indian Journal of Dairy Science, 32:247-249.

Kamboj M. and Prakash B.S. 1993. Relationship of progesterone in plasma and whole milk of buffaloes during cyclicity and early pregnancy. Tropical Animal Health Production, 25:185-92.

Madan, M.L. and Prakash B.S. 2007. Reproductive endocrinology and biotechnology applications among buffaloes. *In*: Reproduction in Domestic Ruminants VI 261-281 (Ed. J. I. Juengel, J. F. Murray and MF Smith. Nottingham University Press, Nottingham, UK).

Mallick, S. and Prakash B.S. 2011. Effects of supplementation of Tinospora cordifolia to crossbred cows peripartum. Anim. Reprod. Sci. 123:5-13.

Mohan K, Kumar, V., Sarkar, M., and Prakash, B.S. 2010. Changes in endogenous estrogens and expression of behaviors associated with estrus during the periovulatory period in Heatsynch treated Murrah buffaloes (*Bubalus bubalis*). Tropical Animal Health and Production 42: 947-952.

Mohan, K., Sarkar, M. and Prakash, B.S. 2009. Efficiency of Heatsynch protocol in estrous synchronization, ovulation and conception of dairy buffaloes (*Bubalus bubalis*). Asian Australasian Journal of Animal Science. 22: 774-780.

Pahwa G.S. and Pandey R.S. 1984. Prolactin in blood plasma and milk of buffalo during oestrous cycle and early pregnancy. Journal of Dairy Science, 67:2001-2005.

Pancarci, S. M, Jordon E.R, Risco C.A, Schouten M.J, Lopes F.L, Moreira M.B.P. and Thatcher W.W. 2002. Use of estradiol cypionate in a presynchronized timed artificial insemination program for lactating dairy cattle. Journal of Dairy Science. 85:122-131.

Paul V. and Prakash, B.S. 2005. Efficacy of the Ovsynch protocol for synchronisation of ovulation and fixed-time artificial insemination in Murrah buffaloes (Bubalus bubalis). Theriogenology, 64:1049-1060.

Prakash B.S. 2002. Influence of environment on animal reproduction. *In*: National workshop on Animal Climate Interaction, pp 33-47, held at Izatnagar, India.

Prakash B.S, Sarkar M, Paul V, Mishra D.P, Mishra A. and Meyer H.H.D. 2005. Postpartum endocrinology and prospects for fertility improvement in the lactating riverine buffalo (*Bubalus bubalis*) and yak (*Poephagus grunniens L.*). Livestock Production Science, 98:13-23.

Prakash B.S, Paul V, Kishore B, Satyapal, Kumar S, Gosain D.K and Markandey J.C 2008. Application of the biotechnique of ovsynch protocol for fertility improvement through oestrous synchronization /induction and set time AI in buffaloes. Proceedings of the National Conference on Distance Education to Animal Farmers with Emphasis on women Self-help Groups, Feb,1-3, Varanasi, Ed. A. Bhaumick, pp 68-72.

Prakash, B.S. 2011. Summer infertility in buffaloes: Problems and Prospects of Improvement using endocrine techniques. *In:* Advanced Training Course on Advances in Applications of Diagnostic Techniques in Veterinary Theriogenology, Centre of Advanced Training in Veterinary Gynaecology and Reproduction, GADVASU, Ludhiana, Feb, 16-Mar 8. 30-38pp.

Pursley J.R, Mee, M.O. and Wiltbank, M.C. 1995. Synchronization of ovulation in dairy cows using PGF_2 alpha and GnRH. Theriogenology, 44:915-923.

Pursley J.R, Wiltbank M.C, Stevenson J.S, Ottobre J.S, Varverick H.A. and Anderson L.L. 1997. Pregnancy rates per artificial insemination for cows and heifers inseminated at a synchronized ovulation or synchronized estrus. Journal of Dairy Science, 80:295-300.

Razdan M.N. and Kaker M.L. 1980. Summer sterility and endocrine profiles of buffaloes. Indian Dairyman, 32:454-459.

Roy K.S. and Prakash B.S. 2007. Seasonal variation and circadian rhythmicity of the prolactin profile during the summer months in repeat breeding Murrah buffalo heifers. Reproduction, Fertility and Development, 19:569-575.

Roy K.S. and Prakash B.S. 2009. Changes in endocrine profiles during ovsynch and ovsynch plus norprolac treatment in Murrah buffalo heifers at hot summer season. Tropical Animal Health and Production, 41: 677-687.

Singh, J.S. and Madan, M.L. 1993. RIA of prolactin as related to circadian changes in buffaloes. Buffalo Journal, 9:159-164.

❑❑❑

21

Poultry Disease with Respect to Climatic Variability and their Suitable Remedial Measures

R.N. Chatterjee and M.R. Reddy

Abstract

Climate change and variability can adversely affect the poultry production in India, because birds can only tolerate narrow temperature ranges. Thermal stress on poultry is major concern. Heat and clod stress can cause immunosuppression and make birds more prone to infectious diseases. Climate change affects the occurrence, distribution and prevalence of poultry diseases. Nutritional strategies aimed to alleviate the negative effects of heat stress by maintaining feed intake, electrolyte and water balance or supplementation of micronutrients such as vitamins and minerals to satisfy the special needs during heat stress have been proven advantageous. Ventilation and cooling methods are essential for birds to survive high summertime temperatures. Tunnel ventilation coupled with evaporative cooling is very effective in keeping birds cool during hot weather. The increasing proportion of poultry production in tropic and subtropical region makes it necessary to reconsider the selection strategy and potential use of naked neck and frizzle genes for improving heat tolerance in chicken. The general principle of disease control including biosecurity, the relevant use of vaccines and medication should be put into practice whatever conditions of climatic stress are present or absent in practical poultry production. Although some of the impacts may happen to greater or lesser extent in short, medium or longer term, it is important to think ahead for future, especially in relation to issues such as building design, environment control, nutritional management disease control to cope with new climate extremes.

Keywords: Poultry production; Impact of temperature and humidity; Diseases spread; Renedict measures.

Introduction

The scientific studies have shown that the global atmospheric concentrations of greenhouse gases (GHGs), have increased markedly as a result of human activities. These increases in GHGs have resulted in warming of the climate system by 0.75°C during last 100 years. The rate of warming has been higher in recent decades. This has, in turn, resulted in increased average temperature of the global ocean, sea level rise, decline in glaciers and snow cover. There is also global trend for increased frequency of droughts, as well as heavy precipitation events over most land areas. Cold days, cold nights and frost have become less frequent, while hot days, hot nights and heat waves have become more frequent. It is projected that by the end of the 21st century rainfall over India will increase by 15 to 40%, and the mean annual temperature will increase between 3°C and 6°C. Such global climatic changes will affect agriculture through their direct and indirect effects on crops, soils, animals and pests and diseases.

Impact of Climate Variability on Poultry

Climate change presents new challenges for the poultry Industry. Chickens, unlike most other animals, do not possess sweat glands to aid in heat loss. The chicken removes excess body heat by *radiation, conduction, convection* and evaporation. When the environmental temperatures are between 28°C and 35°C (82°F and 95°F), radiation, conduction, and convection heat losses are usually adequate to maintain the bird's body temperature. The bird dilates the blood vessels of the skin, wattles, and comb to bring the internal body heat to the skin surface to facilitate conductive, convective, and radiation heat loss. Floor birds will search for cool places in the house and dig into the litter to increase the conductive and convective heat loss. Drooping of the wings promotes convective heat loss by increasing the surface area of the body. Caged birds are more susceptible to heat stress because they are unable to seek a cooler place and there is less conductive heat loss in cages. As the environmental temperature approaches the body temperature of the bird, 41°C (106°F), the efficiency of these heat loss mechanisms diminish. At this point the *evaporation of water* from the respiratory tract becomes the major heat loss mechanism of the bird. When panting fails to prevent the rise in body temperature the bird becomes listless, then comatose, and soon dies. Birds raised from a young age at high environmental temperatures will acclimate to higher environmental temperatures and can maintain good productivity. The nonacclimated flock exposed to rapid increases in environmental temperature (acute heat stress) typically has the greatest loss in production and mortality.

Hot and cold weather can have a severe impact on poultry performance. Production efficiency can be affected long before the temperature reaches a level at which survival becomes a concern. Table 1 is a general guide to the reaction of adult poultry to various temperatures. Heat stress begins when the ambient temperature climbs above 26°C and is readily apparent above 29°C.

Table 1: A guide to the performance of poultry at various temperatures

Below 10°C	Reduction in egg production and weight gains are common with severe negative impact on feed efficiency.
10-21°C	Feed efficiency becomes poorer
21-26°C	Ideal temperature range
26-29°C	Slight reduction in feed consumption can be expected but if nutrient intake is adequate production efficiency is good. Egg size may be smaller and shell quality may suffer as temperature reaches the top of this range.
29-32°C	Feed consumption falls further. Weight gains are lower. Egg size and shell quality deteriorate. Egg production usually suffers. Cooling procedures should be started before this temperature range is reached
32-35°C	Feed consumption continues to drop. There is some danger of heat prostration among layers, especially the heavier birds and those in full production. At these temperatures, cooling procedures must be carried out.
35-38°C	Heat prostration is probable. Emergency measures may be needed. Egg production and feed consumption are severely reduced. Water consumption is very high.
Above 38°C	Emergency measures are needed to cool birds. Survival is the concern at these temperatures.

Impact of Climate Variability on Infectious Diseases

Climate change and extreme weather events also affects the occurrence, distribution and prevalence of infectious poultry diseases. The resulting heat stress leads to immunosuppression and birds are more prone to infections by pathogenic organisms during hot summer days. As a result, India often experience outbreaks of Newcastle disease, fowl cholera during the summer. Climate change would almost certainly alter bird migration, influence the Avian Influenza (AI) virus transmission cycle and directly affect virus survival outside the host. The joint, net effects of these changes are rather unpredictable, but it is likely that AI virus circulation in water bird populations will continue with endless adaptation and evolution. Climate change is a possible factor in the widening spread as it may alter conditions that are involved with disease transmission and persistence including wild bird migration patterns.

Remedial Measures

Housing and Management: The design of poultry houses can greatly help in tiding over the adverse effects of climate change. The building site, orientation, insulation, roof overhang, and equipment design all affect the temperature inside the poultry house. Air movement is particularly important in houses that are ventilated by natural air currents. All poultry houses, but particularly curtain-sided houses, should be positioned so that the roof line runs from east to west. This orientation will keep direct summer sunlight from coming through the sidewall and causing heat to build up within the house. Naturally ventilated houses should not exceed 10m in width for effective cross ventilation. Adequate insulation in the ceiling and sidewalls will pay dividends by reducing the amount of the sun's radiant heat energy that reaches the interior. Installing insulation to the end of a 24-inch roof overhang will prevent solar radiation from penetrating the sidewalls. Insulation also reduces heating costs during winter months. A shiny surface can reflect twice as much solar radiation as a rusty or dark metal roof. Roofs should be kept free of dust and rust. Roof reflectivity can be increased by cleaning and painting the surface with metallic zinc paint by installing an aluminum roof. These practices are particularly effective for buildings that are under insulated. Providing adequate ventilation and stimulating water consumption are essential. A new arrangement for ventilating poultry houses in hot temperatures is tunnel ventilation. This method involves moving air along the building axis from inlets to exhaust fans, providing high airflow velocities. This rapid air movement increases convective heat loss, reducing the effective temperature experienced by the birds. Fogging nozzles and evaporative cooling pads are options that can be used in combination with power ventilation systems and systems and especially with tunnel ventilation. A grass cover on the grounds surrounding the poultry house will reduce the reflection of sunlight into the house. Reducing the number of birds per unit area will reduce the ambient temperature

Nutritional strategies: In addition to heat-stress mortality, economic losses associated with broiler heat stress also occur as a result of lowered growth rate and decreased feed efficiency. Therefore, it is natural for producers to want to stimulate feed consumption in hot weather. However, any management technique which promotes feed consumption or increased activity during the peak hot periods may be counterproductive. During the late afternoon there is a significant rise in body temperature, which, if severe, may kill the bird. A good management strategy for layers to aid in reducing heat stress is to withdraw feed prior to the anticipated time of peak temperature so that it may take an unneeded heat load off the bird. For birds, a period of withdrawal program, it can be beneficial to give supplemental lights during the period of natural darkness. Fasting reduces the heat production from digestion, absorption and metabolism of nutrients. Fasting also has a calming effect. Movement in animals occurs through muscle contraction which generates heat. In hot environments this heat production only adds to the heat load. Therefore, to lessen the heat load, birds should kept as calm as possible. This is especially important during the hottest parts of the day. Once the hottest periods are over and ambient temperature starts to fall, the birds will usually begin consuming feed again.

Heat stress causes birds to decrease feed intake and consequently nutrient intake. Therefore, the dietary nutrient concentrations should be increased. The energy content of the diet, along with other nutrients, should be increased. Increasing fat calories should be considered. Dietary vitamin and mineral concentrations should be re-evaluated. The use of vitamin C, as an anti-stress agent, is often considered during periods of heat stress. Choosing the correct coccidiostat is very important as well as the use of antioxidants and mold inhibitors in stored feed. Protein contributes more to metabolic heat production than do carbohydrate and fat. Therefore, feeding imbalanced diets with regards to amino acids will result in increased metabolic heat production. Amino acid balance in the diet is especially important. Efforts should be made to formulate diets with slightly lower protein levels and to utilize synthetic amino acids, especially methionine and lysine. Birds under heat stress have to make critical life sustaining physiological adjustments. Feed intake is depressed and water intake is increased. Dietary adjustments can help reduce metabolic heat production and maintain nutrient intake. Energy intake and amino acid balance is of extreme importance in heat stress.

Panting is accompanied by an increase in water loss by the lungs. Therefore, more water has to be consumed by birds during hot weather in order to prevent dehydration. Cool drinking water stimulates both feed and water intake. Reducing the body temperature of heat stressed birds is beneficial. When the temperature of drinking water is lower than body temperature it will absorb body heat. Therefore, providing adequate and cool drinking water is extremely important to heat stressed birds. Usually, anything that results in increased water consumption during heat stress will benefit the survival rate. In fact, some researchers have attributed the increased survival rates of heat stressed birds receiving supplemental salts such as potassium bicarbonate, potassium chloride, sodium chloride and ammonium chloride to the increased water consumption which results, not to the salts.

Acclimatization: This may be defined as the ability to tolerate heat stress by gradual contact with high temperature. Young chicks, broilers, layers or breeders can be acclimatized to some degree by daily repeated exposure to a hot climate such as 37°C and a RH of 25% up to 60% for four hours per day or continuous exposure to a relative high temperature for several days. Experimental acclimatization has been established over a period up to 3 weeks without any loss of production but it should be noted that an increase in dietary energy level might have an adverse effect on acclimatization. In parts of the world where there is a gradual increase in ambient temperature with the summer season acclimatization probably occurs naturally. The practice of acclimatization in management may be of value in districts where sudden increases in temperature may be expected. The duration of acclimatization to high temperatures in a flock probably varies with climatic conditions but cool conditions may reduce tolerance to a sudden acute hot spell of weather.

Breeding: There are few options available for mitigating this inevitable adversity through breeding and evolving varieties better adapted to survive. At present, the commercial stocks used in tropical countries are evolved by the breeders doing selection programme in the temperate climate conditions. It would be more appropriate to produce suitable stocks for hot climatic conditions by breeding in tropical climates to reduce environment-genotype interaction. To make genetic improvement, heat tolerance can be incorporated in the selection indices. The research on selection in controlled high temperature has also resulted in developing heat tolerant lines. The heritability of heat tolerance is quite high and rapid progress can be expected. The introgression of genes responsible for heat tolerance would be another strategy to develop heat resistant stocks. Naked neck (Na) gene is one such that reduce the feather cover by 30 to 40%, thereby improves heat dissipation. The effects of this gene under adverse climatic condition in layers and broilers have been well established. The beneficial effect of Na gene in reducing heat stress in fast growing birds exposed to acute heat has been reported. The frizzle gene (F) causes the contour feathers to curve outward away from the body. This has the advantage of easy heat dissipation through convection and radiation heat losses. Another major gene having such application is sex-linked slow feathering (K). The benefits of these feather reducing genes in birds exposed to hot climates have been established. Also some other genes like silky (h), which affects the barbules on the feathers, may improve the effect to dissipate heat. The dominant gene for pea comb (P) also has the feather tract reduction action. The recessive, sex-linked, multiple allelic locus for dermal melanin (id) can improve radiation from skin. There are many reports to prove the individual and interaction effects of these major genes in coping with high environmental temperatures. The experiments with introgressing these genes in commercial lines revealed encouraging results. Therefore, one of the genetic strategies to develop competent lines for climatic change would be the inclusion of these genes in commercial stock. Recent studies revealed that featherless broilers bearing scale-less (sc) gene exhibited economically important advantages when reared under warm conditions.

Health care: The general principle of disease control including biosecurity, the relevant use of vaccines and medication should be put into practice whatever conditions of climatic stress are present or absent in practical poultry production. Certain drugs may be contraindicated in hot climates. An example is nicarbazin which increases metabolic rate and thus reduces the tolerance of chickens to high temperatures.

Conclusions

As a result of climate change, poultry industry is affected by rising temperatures, changes in rainfall patterns or increase of drought. This is directly linked to decreased production and feed ingredient availability and higher incidence of diseases. Poultry need to consider improved management practices to adapt to the changes in climate they are experiencing. Strategies such as introduction of heat tolerant breeds, environmental controlled housing systems, manipulation of nutritional regimes etc are necessary to cope with extreme weather events due to climate change.

References

Amos, T.T. 2006. Analysis of Backyard Poultry Production in Ondo State, Nigeria International Journal of Poultry Science 5(3):247-250.

Arjona, A.A., D.M. Denbow and W.D. Weaver, 1988. Effect of heat stress early in life on mortality in broilers exposed to high environmental temperatures just prior to marketing. Poult. Sci., 67:226-301.

Charles, D.R. 1980. Environment for popultry. Vet. Rec., 106:307-309.

Clemens, J. and H. Ahlgrimm, 2001. Greenhouse gases from animal husbandry: mitigation options. Nutr. Cycl. Agroecosys., 60:287-300.

Cowan, A. and A. Michie, 1978. Environment temperature and broiler performance. The use of diets containing increased amounts of protein. Br. Oyegun, R.O., 1992. Pocket map of llorin. The gateway Poult. Sci. 19:601-1068.

Elijah, O.A. and A. Adedapo 2006. The effect of climate on poultry productivity in llorin Kwara State, Nigeria. International Journal of Poultry Science, 5(11):1061-1068.

Howlider, M.A.R. and S.P. Rose 1987. Temperature and the growth of broiler. Worlds Poult. Sci. J., 45:228-236.

Monteny, G.J. 2003. Strategies for reducing the effects of animal husbandry on climate. In: Interactons between Climate and Animal Production. EAAP Technical Series No. 7. Wageningen, Wageningen Academic Publishers.

Moraes, V.M.B., R.D. Malheiro, V. Bruggeman, A. Colin, K. Tona, P. Van As, O.M. Onagbesan, J. Buyse, E. Decuypere and M. Macari, 2003. Effect of thermal conditioning during embryonic development on aspects of physiological responses of broilers to heat stress. J. Therm Biol., 28:133-140.

Mowbray, R.M. and A.H. Sykes, 1971. Egg production in warm environmental temperature. Br. Poult. Sci., 12: 25-29.

Olanrewaju, H.A., J.L. Purswell, S.D. Collier and S.L. Branton 2010. Effect of Ambient Temperature and Light Intensity on Growth Performance and Carcass Characteristics of Heavy Broiler Chickens at 56 Days of Age. International Journal of Poultry Science 9(8):720-725.

Rosenberg, N.J., 1974. Microclimate: The Biological Environment. New York, John Wiley and Sons.

Shibata, M., 2003. A higher susceptibility of hyperthermic animals to infectious disease. In: Interactions between Climate and Animal Production. EAAP Technical Series No. 7. Wageningen, Wageningen Academic Publishers.

Sluis, W.V. 2007. Intensive Poultry Production. World Poultry Megazine 23(12). Retrieved from www.worldpoultry.net on 15th/01/2013.

❑❑❑

22

Climate Change and Its Impact on Livestock Production Performance

Chander Datt, Ajay Kumar, Monica Puniya, S.S. Kundu and Tapan K. Das

Abstract

Climate change is a change in the distribution of weather over periods of time that range. The effect of climate change is now visible every where including animal farm industry. Projected temperature rise around 2-5°C by 2100 due to climate change is likely estrus, start estrus and decline in reproduction efficiency of livestock species Climate change may have a significant effect on livestock and other animals through is impact on diseases Zoonotic diseases which can be transmitted between vertebrate animals and humans are a matter of great concerns. There is an urgent need for adoption of integrated approaches towards mitigation of advance effects of climate change in animal production. There are nutritional strategies to reduce enteric methane production. Surveillance and monitoring with various adaption strategies are to be brought under purvision to reduce adverse impact of climatic variability in livestock

Keywords: Climate change, Livestock production, Livestock health, Mitigation and adaptation strategies.

Introduction

Livestock sector is a significant contributor towards the rural economy of most of the developing countries including India. Climate change is one of the most serious long term challenges facing livestock owners around the world. Climate change is now universally recognized as a significant global environmental challenge and as a consequence debate over the measures needed to tackle climate change has moved up to the top of the agenda amongst scientific community, civil society, business and government. The United Nations Climate Change Conference - 2010 in Cancan, Mexico highlighted the major challenges and opportunities associated with climate change. Rise in temperature due to climate change is likely to affect livestock production and health (Srivastava, 2010). The contributors towards global warming or climate change include sources like burning of fossil fuel, industries, coal mines, deforestration, agriculture and livestock sector etc. Important green house gases (GHG,s) are CO_2, CH_4, N_2O and hydrofluorocarbons. The GHG,s emission from agriculture sector accounts for 25.5% of global emission and over 60% from anthropogenic sources (FAO, 2009). Indian agriculture sector emitted 334.41 million tones of CO_2 equivalent of which 13.76 million tones is CH_4 and 0.15 million tones is N_2O. Enteric fermentation constituted 63% of the total CO_2 equivalent emission from livestock sector, 21% from rice cultivation. Crop soils emitted 13% of total CO_2 equivalent emission from agriculture. Rest 2.7% of the emission is attributed to livestock manure management and burning of crop residues (INCCA, 2007).

India posses about 13% of the total world livestock population and ranks first. In terms of buffalo, cattle, goat and sheep population, India ranks first, first, second and fourth, respectively and the corresponding population is 55, 16, 20 and 5 % of the world population. Therefore, considering such a huge strength of our livestock and its role in Indian economy, the likely impacts of climate change could be devastating. It is the high time that we should take stock of this alarming situation. Hence, the likely impacts of climate change on livestock production and health have been discussed. Also some measures have been suggested which could positively affect the impending climate changes.

What is Climate Change

Climate change is a change in the statistical distribution of weather over periods of time that range from decades to millions of years. Climate change may be limited to a specific region, or may occur across the whole earth. Climate change is one of the most important global environmental challenges facing humanity with implications for food production, natural ecosystems, freshwater supply, health etc. The earth's climate system has changed on both global and regional scales since the preindustrial era. Further evidences show that most of the warming (of 0.1°C per decade) observed over the last 50 years is attributable to human activities. The Inter-governmental Panel on Climate Change (IPCC) projects predicted that the global mean temperature may increase between 1.4 and 5.8°C by 2100.

The consequences of climate change phenomena are now visible everywhere including in animal farm industry (Lal, 2002) and considered as the serious long-term threat to agriculture (ACIAR, 2007). The livestock systems are susceptible to changes in the severity and distribution of livestock diseases and parasites as potential consequences of global warming. That is why, future of animal production, health, and welfare especially in the Asian countries seems to be largely dependent upon how these changes affect animal production. A lot of concern has been addressed towards the effect of climate change on the physical environment and on human health over the past two decades (FAO, 2007). The majority of our population is rural with agriculture being the main sector of activity. Livestock rearing is the main economic activities available to poor people in developing countries (Steinfeld *et al.,* 2006). Many of the infectious diseases especially those which are vector borne have seasonal patterns and a geographic range that are influenced by environmental variables including biophysical (land cover, landscapes and host abundance, including wildlife) and anthropogenic (socio-political changes, public health, diet and other human behavior) modulate life cycle of pathogen with vectors and reservoirs, resulting in alteration in disease behaviour.

Climate Change and Likely Impacts on Livestock Production

The overuse of natural resources, rapid urbanization, industrialization and economic development has led to increased environmental problems and climate change (Lal *et al*., 2001). Due to existing levels of poverty and population density in the country, even small changes in land area or crop productivity may have serious social consequences (UNEP, 1993).

Large deltas and low lying coastal areas particularly in Asia would be inundated by a rise in sea level (DFID, 2004). Increased precipitation intensity would increase the area prone to flooding throughout these deltas and coastal regions. Conversely, climate change could produce drier conditions in arid and semi arid regions during the summer, leading to more severe droughts. Availability of fresh water is expected to be highly vulnerable to any climate change and tropical cyclones or typhoons would not necessarily increase in frequency but would become more intense (DFID, 2004). Warmer and wetter conditions would lead to increased incidences of heat related infectious disease. Following the increase in frequency and intensity of heat waves, the risk of death and serious illness is expected to increase and will be exacerbated by increased temperature and humidity (McCarthy *et al*., 2001). Livestock growth, reproductive ability and temperature are closely related (Upahyyay *et al*, 2009, Singh, 2010, Chander Datt and Singh, 2008). For example, in case of pigs, if the temperature rises by 1°C above their optimal growth temperature, feed intake decreases by 5% and activity decreases by 7.5%. Pigs subjected to excessively high temperatures are likely to suffer heat stress. It reduces productivity and fertility. It can lower the feed intake of lactating sows, thereby reducing their milk production and resulting in adverse effects on their piglets.

Heat stress also represents a serious problem to dairying especially in tropical countries. When the ambient temperature increases, more heat accumulates than the cow is able to dissipate resulting in heat stress. Heat stress prolongs the oesturs cycle of dairy cow. Signs of oestrus are weakened, the oestrus period shortened, gestation rate decreased and fetal death rate are increased. Thus, heat stress significantly affects reproduction, milk yield and health in dairy cows. In India, The direct heat stress on lactating cows and buffaloes cause a production loss of more than 1.8 million tones of milk. The increase in thermal stress days due to temperature rise has been estimated to cause an additional loss in milk production of 1.6 million tonnes in 2020 accounting about Rs.2365.8 crores at current price levels. Therefore, the annual loss in milk production due to combined thermal stress of summer and global warming on cattle and buffaloes in 2020 are likely to about 3.4 million tones costing more than Rs.5000 crores (Upadhyay *et al.,* 2009) The impact on milk production decline is likely to be higher in crossbreds than in buffalo and indigenous cattle. The temperature variability and rise due to global warming is also likely to affect mature weight of the animals in a negative way and it will also enhance time to attain puberty. Projected temperature rise of 2-5°C by 2100 due to climate change is likely to increase the incidence of silent estrus, short estrus and decline in reproduction efficiency of livestock species.

Over the time, animal populations have acclimatized and adapted to local climates and also are able to cope with a range of weather changes. However, within populations, there is a range of individual sensitivity to extreme weather events. If heat waves increase in frequency and intensity, the risk of death and serious illness would increase. Any increase in frequency of extreme events such as storms, floods, droughts and cyclones would harm human health through a variety of pathways. These natural hazards can cause direct loss of life and injury and affect health indirectly through loss of shelter, population displacement, contamination of water supplies, loss of food production leading to hunger and malnutrition, increased risk of infectious disease epidemics including diarrhoeal and respiratory diseases and damage to infrastructure for provision of health services.

Climate change will have mixed effects on food production globally. Grain production may be severely affected. Livestock can be affected by climate change in two ways: by the quality and quantity of forage from grasslands and supplies of other feeds and by direct effects from higher temperatures and greater temperature extremes. If the quality or supply of forage and feed grains is altered, livestock production may be more affected by the associated changes in pasture and grain prices than by the direct effects of temperature increases.

Under conditions of climatic extremes, there is high probability that forests (which provide fuel, food, timber, fodder *etc.*) will be subject to increased frequency and intensity of stress due to climatic extremes. Several fungal diseases on trees may become more devastating because of the factors like abiotic stresses, such as drought and flooding, are known to predispose trees to several pathogens;

temperature and moisture affect pathogen sporulation and dispersal and changes in climatic conditions are likely to favour certain pathogens; migration of pathogens triggered by climatic change may increase disease incidence or geographical range, when pathogens encounter new hosts and new potential vectors.

Food safety hazards can arise at various stages of the food chain, from primary production to consumption, and climate change may have direct and indirect impacts on their occurrence. There are many pathways through which climate related factors may impact food safety including changes in temperature and precipitation patterns, increased frequency and intensity of extreme weather events, ocean warming and acidification and changes in the transport pathways of complex contaminants.

On one hand, climate affects animal productivity and health. On the other hand livestock also affect climate change. Livestock production systems are being seen as one of the factors contributing to current changes in the environment. An estimated 35% of global greenhouse gas emissions that are derived from agriculture and land use originate from livestock production (McMichael *et al.*, 2007). Livestock currently use almost one third of the world's entire land surface, mostly as permanent pasture, while one third of the world's arable land provides livestock feed (Preston *et al.*, 2006).

Climate Change and Animal Health

Climate changes will have a significant effect on livestock and other animals through its impact on diseases (Grasty, 1999). One of the important factors driving the emergence and spread of diseases in livestock and the transfer of pathogens from animals to humans is climate change. Other factors include changes in the structure of the livestock industry, in breeding and husbandry practices and in international trade in livestock and animal products. Climate change could increase the susceptibility of animals to diseases (For example, exposure to intense cold, droughts, excessive humidity or heat may predispose cattle to complex bacterial syndromes such as mastitis), increase in the range or abundance of vectors/animal reservoirs, prolonged transmission cycles of vectors. Climatic conditions affect animal well-being both directly (through the physical effects of climatic extremes) and indirectly (through influences on the levels of pollution in the air, on the agricultural, marine and freshwater systems that provide food and water and on the vectors and pathogens that cause infectious diseases). The ability of a microbial pathogen to mutate and hence respond to opportunities arising from change is a key factor in considering the potential impact of climate change. Climate change may affect not only the geographical range and abundance of vectors but also the interaction between a pathogen and a vector.

The epidemiology of many diseases are based on transmission through vectors such as ticks, mosquitoes and flies, the developmental stages of which are often heavily dependent on temperature and humidity. Apart from domestic animals such as sheep, goats, and cattle, wild animals are also vulnerable to an extensive range of nematode infections, most of which have developmental stages that are influenced by climate conditions. Any change in climatic conditions could thus affect such agents, either positively or negatively, and therefore, influence livestock health. Environmental conditions such as temperature, humidity and sunlight affect the survival of those pathogens that are able to survive outside the host in the environment. Climate change is expected to affect animal health in a variety of ways, including the occurrence of heat related diseases, extreme weather events, further changes in animal production systems to minimize climate change and environmental damage and the emergence or re-emergence of infectious diseases and also introduction of exotic infectious diseases (Pattanaik and Sharma, 2010).

Rainfall can alter the abundance and type of aquatic habitats available to the mosquito for oviposition. The abundance of midges and mosquitoes may be increased through the generation of vector breeding sites by periods of drought followed by heavy rainfall. Heavy rainfall and flooding can trigger epidemics of mosquito-borne disease in areas where the climate is usually too dry to maintain an abundant mosquito population (Gale and Drew, 2009). The transmission of many diarrhoeal diseases in humans is associated with seasonal rains and flooding.

Livestock diseases are strongly influenced by climate change and transmission of wind borne diseases such as foot and mouth disease and infections transmitted by ticks, flies, mosquitoes, midges and other arthropods may be of great concern with respect to the changing climate (FAO, 2004; Pattanaik and Sharma, 2010). The migration and spread of birds may change and affect the geographical coverage of certain diseases.

It is generally accepted that climate restricts the range of infectious diseases, whereas weather affects the timing and intensity of outbreaks (Epstein, 2004). Thus, changes in temperature and humidity affect the distribution or ecological range of infectious diseases while the frequency and magnitude of outbreaks of disease change with weather extremes such as flooding and droughts. Biological effects of climate changes on animal and plant life include effects on physiology, metabolism or development rate, effects on distribution, effects on timing of life cycle or life history events, adaptation particularly of species with short generation times and rapid population growth rates which may undergo some micro evolution (Hughes, 2000).

There are complex relationships between health and water quality, water quantity, sanitation and hygiene. Heavy rainfall events can transport terrestrial microbiological agents into drinking-water sources resulting in outbreaks of cryptosporidiosis, amoebiasis and other infections. Changes in surface water quality and quantity are likely to affect the incidence of diarrhoeal diseases. A reduction in the availability of clean water also reduces the amount of water available for personal hygiene thus leading to skin infections. Ocean warming, and climate change related acidification and changes in ocean salinity and precipitation also affect the biochemical properties of water, along with water microflora, fisheries distribution, fish metabolic rates, and persistence and patterns of occurrence of pathogenic *Vibrio,*s, harmful algal blooms and chemical contaminants in fish and shellfish. (Tirado and Clarke, 2010)

Certain rodent borne diseases are associated with flooding including leptospirosis, tularaemia and viral haemorrhagic diseases. Other diseases associated with rodents and ticks include plague, Lyme disease, tick borne encephalitis and Hantavirus pulmonary syndrome. Rodent populations have been shown to increase in temperate regions following mild wet winters. These conditions may increase food sources for rodents and promote breeding of fleas and ticks. The assessment anticipates that for at least the first half of the twenty-first century additional UV radiation exposure will augment the severity of sunburn and incidence of skin cancer. Mammalian cellular immunity can be suppressed following heightened exposure to UV radiation. Furthermore, increased UV-B exposure may diminish the host's response to certain vaccinations.

Zoonotic diseases, which can be passed between vertebrate animals and humans, are common throughout the world and constitute an important threat to human health. Ecological change and climatic events can promote epidemic expansion of host and geographic range. Similarly, land use change may also result in new human and animals contacts and the emergence of new disease problems. Globalization and climate change could have an unprecedented worldwide impact on emerging and re-emerging animal diseases. Climate change is disrupting natural ecosystems by providing more suitable environments for infectious diseases allowing disease-causing bacteria, viruses and fungi to move into new areas where they may harm wild life and domestic species as well as humans. Some of the important zoonotic diseases of animals which are most likely to be affected by climate change include pasteurellosis, listeriosis/circling disease, anthrax, infectious necrotic hepatitis/fasciolosis and Rift Valley fever (Singh, 2010). Thus, climate variables are able to affect the prevalence, intensity and geographical distribution of zoonotic helminths, directly influencing free-living larval stages and indirectly influencing mainly invertebrate but also vertebrate hosts.

Climate change and variability also pose a challenge to pest and diseases control measures such as Good Agriculture and Good Veterinary Practices with potential implications for the presence of chemical residues in the food chain. Extreme weather events such as floods and droughts may lead to contamination of soil, agricultural lands, water and food and animal feed with pathogens, chemicals and other hazardous substances, originating from sewage, agriculture and industrial settings. Emergency situations after natural disasters are of special concern for water and food sanitation.

Mitigation Strategies to Combat Climate Change

There is urgent need for adoption of integrated approaches towards mitigation of adverse effects of climate change in animal production. Some of the key points have been given henceforth (Singh, 2010, Srivastava, 2010, Kundu *et al*., 2010, Upadhyay *et al.*, 2010, Goswami, 2010, Kamra and Pawar, 2010, Chander Datt and Singh, 2008)

- Reduction in GHG,s emission: Adoption of cost-effective energy-efficient technologies in electricity generation, transmission, distribution and end-use can reduce costs and local pollution in addition to reduction of GHG,s emissions, shift to renewable energy sources, adoption of forest conservation, reforestation, afforestation and sustainable forest management practices, efficient, fast and reliable public transport systems.
- Nutritional strategies to reduce enteric CH_4 production: For example, balanced feeding, feeding fats rich in MCFA/PUFA contents, use of probiotics, plant extracts, ionophores, halogenated methane analogues etc.
- Other technologies to reduce GHGs pertaining to livestock sector include genetic improvement of animals for desirable traits, biotechnological approaches, animal waste management, grazing management etc.
- Adaptation to climate change: For example, housing designs that enhance summer-time cooling, the" greening" of inner cities, strengthening of coastal buffers and improved control of vector-borne and water-borne diseases.
- Surveillance and monitoring: Surveillance for infectious diseases allows us to know what to expect and to be prepared with the right strategy in place.
- Geographic Information Systems: The system can monitor how our climate is changing and let us know the spread of disease. We can use it to study periods of drought or heavy rainfall which can tell us what pathogens will flourish under which conditions.
- Strengthening veterinary service: Veterinary services including medication and vaccination of animals, epidemiological research, veterinary public health etc. should be strengthened in the areas where diseases are emerging due to climate change.
- Policies and measures: Laws should be implemented to put a price on carbon which could help to accelerate solutions to this crisis. Incentives may be given to the technologies which are cleaner and greener.

The financial, food, fuel and climate crises are individually serious issues, but in combination their impact could be catastrophic for the global economy. Perhaps we stand at a crossroads. Even as we move into a global economic downturn there are growing calls for an accelerated transformation towards low carbon and resource productive economies. To move forward will require new and more efficient solutions, technologies and products alongwith behavioral change. However, we need to create much better international, national and regional structures and systems to adapt to climate change as well those to stimulate, nurture and progress new low carbon ideas through R&D to commercialization and the marketplace. A new system is needed that fosters sustainable, low carbon and resource productive innovation - short, medium and long-term. Designers, scientists, innovators, investors, entrepreneurs, consumers and governments all have a role to play in the change that is needed.

References

ACIAR. 2007. Australian Centre for International Agricultural Research Climate change and agriculture. Australian Government Available at: http://www.aciar.gov.au/node/239

Chander Datt and N.P Singh. 2008. Climate and Livestock Production. In: Climate Change and Food Security (Eds.: M.Datta, N.P. Singh and Er. D. Daschaudari. pp 275-291).

DFID. 2004. Department for International Development Climate Change in Asia. Key sheet. DFID, United Kingdom. Available at: http://www.dfid.gov.uk/pubs/files/climatechange/llasia.pdf.

Epstein P.R. 2004. Climate change and public health emerging infectious diseases. In: Encyclopedia of Energy. 1: 381-392.

FAO. 2004. Food and Agricultural Organization. Impact of climate change on agriculture in Asia and the Pacific In: 27th Food and Agriculture Organization of the United Nations Regional Conference for Asia and the Pacific" held at Beijing, China from 17 to 21 May, 2004.

FAO 2007. Adaptation to climate change in agriculture, forestry and fisheries; perspective, framework and priorities. Departmental Working Group on Climate change FAO, Rome.

FAO, 2009. Submission to UNFCC AWG LCA, Enabling agriculture to contribute to climate change, Food and Agriculture Organisation of United Nations.

Gale, P.T and Drew, A.J. 2009. The effect of climate change on the occurrence and prevalence of livestock diseases in Britain: A review. J. Applied Micro. 106: 1409-1423.

Goswami, S.L. 2010. Biotechnological interventions for coping with impact of climate change in livestock. In compendium of Winter School "Impact, Vulnerability and Adaptation of Indian Livestock to Impending Climate Changes" (Eds.: R.C. Upadhyay, S.V. Singh and Ashutosh) held at N.D.R.I. Karnal from November 9-29, 2010. Pp 13-20.

Grasty S. 1999. Agriculture and climate change. TDRI quarterly Rev. 14: 12-16.

INCCA, 2007. Indian Network for Climate Change Assessment; India: Green House Gas Emissions- 2007, Ministry of Environment and Forests, Govt. Of India.

Kamra, D.N. and Pawar, M.M. 2010. Methane mitigation by plant secondary metabolites- An indigenous approach. In Proc.:"National Symposium on Climate Change and Livestock Productivity in India (Eds.: R.C. Upadhyay, S.V. Singh, Ashutosh, Manju Ashutosh and Anjali Aggarwal) held at N.D.R.I., Karnal from October 7-8, 2010. Pp56-67.

Kundu, S.S., Veena Mani and Chander Datt. 2010. Feeding of cattle and buffaloes during heat stress for maximizing productivity. In compendium of Winter School "Impact, Vulnerability and Adaptation of Indian Livestock to Impending Climate Changes" (Eds.: R.C. Upadhyay, S.V. Singh and Ashutosh) held at N.D.R.I. Karnal from November 9-29, 2010. pp 129-134.

Hughes L. 2000. Biological consequences of global warming are the signal already there? Trends Ecol. Evol. 15: 56-61.

Lal, M. 2002. Regional climate scenarios- Future changes in variability and mean including extreme weather events issue related to agriculture policy in South Asia. In South Asia Workshop on "Adaptation to Climate Change for Agriculture Productivity" May 2002, New Delhi.

Lal, M., Harasawa, H. and Murdiy Arso D. 2001. Asia in climate change, impacts, adaptation and vulnerability. Intergovernmental Panel on Climate change (IPCC). Cambridge University Press.

McCarthy J. J., Canzianin O. F., Leary N.A, Dokken D. J., and White K. S. 2001. Climate change impacts, adaptation and vulnerability. Intergovernmental Panel on Climate change (IPCC), Cambridge University Press.

McMichael A J., Powels J. W., Butler C. D., and Uauy R. 2007. Food livestock production, energy, climate change and health. Lancet, 370: 1253-1263.

Pattanaik, B. and Sharma, G.K. 2010. Impact of climate change on animal health and performance. In Proc.:"National Symposium on Climate Change and Livestock Productivity in India (Eds.: R.C. Upadhyay, S.V. Singh, Ashutosh, Manju Ashutosh and Anjali Aggarwal) held at N.D.R.I., Karnal from October 7-8, 2010. Pp47-55.

Preston B. L., Suppiah R., Macadam I., and Bathols J. 2006. Climate change in the Asia/ Pacific region. A consultancy report prepared for the Climate change and development round table. Commonwealth Scientific and Industrial Research Organization (CSIRO), Australia.

Singh, R.K. 2010. Impact of climate change on pattern of zoonotic disease occurrence and spread in livestock. In Proc.:"National Symposium on Climate Change and Livestock Productivity in India (Eds.: R.C. Upadhyay, S.V. Singh, Ashutosh, Manju Ashutosh and Anjali Aggarwal) held at N.D.R.I., Karnal from October 7-8, 2010. pp38-46.

Srivastava, A.K. 2010. Impact of climate change on Indian livestock productivity. In Proc.: "National Symposium on Climate Change and Livestock Productivity in India (Eds.: R.C. Upadhyay, S.V. Singh, Ashutosh, Manju Ashutosh and Anjali Aggarwal) held at N.D.R.I., Karnal from October7-8, 2010. pp 3-9.

Steinfeld H., Gerber P., Wassenaar T., Castel V., Rosales M. and de Haan C 2006. Livestocks long shadow: environmental issues and options. Food and Agricultural Organization (FAO), Livestock, Environment and Development Initiative (LEAD), Rome.

Tirado, M.C. and Clark, R. 2010. Climate change and food safety: A review. Food Research International.

Upadhyay, R.C., Ashutosh, Singh, S.V., Rita Rani, Syama Asraf and Anil Kumar. 2010. Impact of climate change on milk production of crossbred cows. In Proc.: "National Symposium on Climate Change and Livestock Productivity in India (Eds.: R.C. Upadhyay, S.V. Singh, Ashutosh, Manju Ashutosh and Anjali Aggarwal) held at N.D.R.I., Karnal from October7-8, 2010. pp131-134.

Upadhyay, R.C., Asutosh, Ashok Kumar, Gupta, S.K., Singh, S.V. and Nitika Rani. 2009. Inventory of methane emission from livestock in India. In: Global Climate Change and Indian Agriculture (Ed. P.K. Aggarwal), ICAR, New Delhi, India. pp117-122.

❑❑❑

23

Reproductive Biology of the Catfish *Ompok bimaculatus* (Bloch, 1794) Related to the Affects of Climate Change

Samir Malla, Pritam Goswami and S. Banik

Abstract

The breeding potential of Ompok bimaculatus is declining day by day due to decrease of rainfall and increase of atmospheric temperature. As a result, the population of Ompok bimaculatus *is significantly reducing in the fresh water bodies of Tripura.*

The results of reproductive biology of O. bimaculatus were based on the fish samples collected from different locations of some rivers and Hurijala wetland of Tripura. Total weight of the sampled fish varied from 22 g to 171.50 g and the females had a greater weight and length than males. Total length of the males varied from 16.50 cm to 29.50 cm, while the females ranged from 17.0 cm to 33.0 cm. Female dominance over male was observed in the species. The ratio of males to females was 1:1.58. Significant correlation exists between fish length and weight in both males (R^2 = 0.943) and females (R^2 = 0.827). Monthly variation of maturity stages indicate that O. bimaculatus has spawning period extend from May to August. Length at first maturity for female and male was 17.0 cm and 16.5 cm respectively. Maximum average GSI (%) values obtained for male and female were 2.195 and 15.582 respectively in the month of June (Spawning period) and declined to minimum average value in October to January. The fecundity was varied from 2190 to 41552 eggs/fish individual and was relatively higher during mid-July.

Therefore, shifting of peak fecundity-period from May to mid-July and declining of population of Ompok bimaculatus *in recent period may be the result of increase in atmospheric temperature ($r= 0.967$, $P<0.01$) and decrease in rainfall ($r= 0.988$, $P<0.01$) over a period of two decades from 1990-2010.*

***Keywords*:** Climate change, *Ompok bimaculatus,* Rainfall, Temperature, Conservation.

Introduction

The catfish *Ompok bimaculatus* (Bloch, 1794) is an indigenous freshwater fish belonging to the family Siluridae under the order Siluriformes (Siddiqua *et al.,* 2000). It is highly priced, delicious and nutritious and well preferred especially in the entire East and North East India and Bangladesh because of soft bone and body texture with rich lipo-protein. Once upon a time this species has been commonly available in rivers, lakes, floodplains or beels etc. It has an extensive geographical distribution covering India, Bangladesh, Pakistan, Afghanistan and Burma (Jayaram, 1999, Banik and Malla, 2011).

The population has been declined drastically as a result of several anthropogenic factors like indiscriminate fishing during the breeding season, wide use of pesticide and insecticides from agricultural fields, siltation in habitat, sewage and industrial pollution, extensive habitat degradation and alteration and the recent proliferation of exotic fish species and also due to climatic variability (Bhowmick *et al.,* 2000, Chakrabarti *et al.,* 2009, Banik and Malla, 2010). Due to reduced abundance and restricted distribution, it has been listed as an endangered species as per IUCN criterion (CAMP, 1998). Because of its economic importance, *Ompok* species has attracted interest as a potential species for fish culture, and so a number of studies have been carried out for artificial reproduction and aquaculture purposes (Chakraborty *et al.,* 2007, Rahaman *et al.,* 2008, Vijayakumar, 2010). Its genetic and chromosomal structures have also been investigated (Rishi *et al.,* 1994, Sharma, 2008). Studies related to the wild population of *O. bimaculatus* are generally on growth and feeding behavior (Sivakami, 1982, Taleb *et al.,* 1991, Hanjavanit *et al.,* 2009). However, knowledge on reproductive biology of *Ompok bimaculatus* is very poor.

So, the present studies communicate observations of *Ompok bimaculatus* in relation to different aspects of reproductive biology.

Material and Methods

80 samples (31 males and 49 females) of *O. bimaculatus* were collected at random from different rivers and Hurijala wetland of Tripura (Latitude 23°,50’1" N and Longitude 90°15’00"E at an elevation of 3.8 m above MSL) during February,

2009 to March, 2010. The sex ratio and maturation stage of each specimen was determined microscopically, considering coloration, transparency, superficial vascularisation. Length at first maturity was worked out by plotting the percentages of matured fish (50% Female) against their length. The gonads of fish were taken out very carefully and preserved in 5% buffered formalin. Gravimetric method was followed to determine the fecundity of fish. 0.01 g of each ovary was taken separately from anterior, middle and posterior regions of each ovarian lobe. The Fecundity of the fish was obtained by using the formula given by LeCren,(1951). Gonado-somatic index (GSI) of the male and female fish of the collected samples was determined separately following the equation cited by Parameswarn *et al.* (1974). Ova diameter at different stages of maturity was measured with the help of ocular micrometer. Measurements of ova diameter were taken along the longest axis of the ova (Clark, 1934). Relationship of fecundity with various body dimensions like body length, body weight, ovary length and ovary weight was then log transformed and relationships fitted by least squares regression equation (Bagenal 1978): Log F = log a + b log X, Where - F = Clutch size (fecundity); X = Length/Weight; a = Regression constant and b = Regression coefficient.

Results

Length–Weight relationship and sex ratio

Total length of the sampled fish varied from 16.50 cm (22 g) to 33.0 cm (171.50 g). Of the total individuals, the females were relatively bigger than that of males. Regression equation for male and female fish was shown in figure 1 and 2.

Maturity stages of the gonads

The variation in the maturity stages in female *O. bimaculatus* were seen in nearly all months. Granular structures of the ovaries (Immature) were first seen in late December (Stage I). In January and February, the gender was easily distinguished in naked eye (Stage II) i.e. developing stage. Females with maturing ovaries (Stage

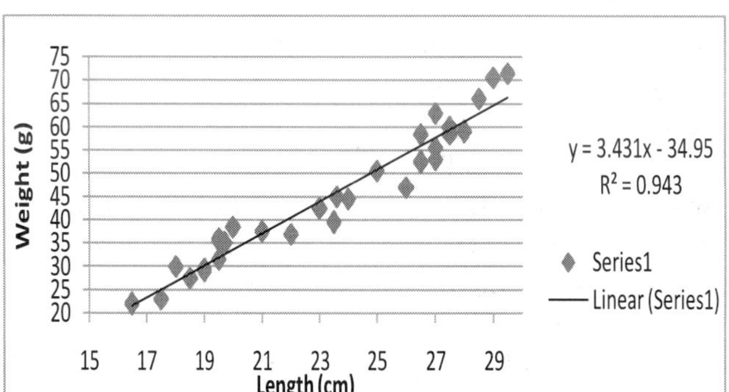

Fig. 1: Regression curve showing the relationship between length and weight in male *O. bimaculatus*

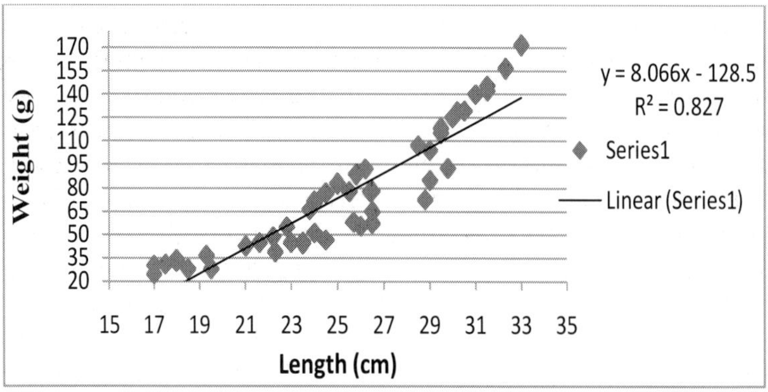

Fig. 2: Regression curve showing the relationship between length and weight in female *O. bimaculatus*

III) appeared during March to May. Mature ovaries (Stage IV) appeared in May-July and the first occurrence of ripe stage (Stage V) was noticed in June. The monthly distribution of the maturity stages confirmed that *O. bimaculatus* began to mature during the winter (December, January and February) and that its sexual development accelerated in spring (March, April and May), spawning mainly between June and August, after which there was a resting period or spent stage during the autumn (VI) (October to December).

Length at first sexual maturity (L_{50})

The percentage distribution of Maturing ovaries and testes (Stage IV and above) in the length group were considered for determination of the minimum size at sexual maturity. The smallest mature male was 16.50 cm in total length (50 % and above) and 22.00 g in weight. The smallest mature female was 17.0 cm (50 % and above) in total length and 24.50 g in weight.

Gonado-Somatic Index (GSI %)

The GSI of females was higher than that of males. In female there were very low GSI (%) observed January to March (mean=4.723) that represented the pre-spawning period. It increased gradually to reach very high value 15.582 in June (Spawning period) and then declined to minimum at 1.54. Contrary to that in male, there were very low GSI (%) (January –March, mean=0.421) which represented the pre-spawning period. It increased gradually to a very high value (2.195 in June, i.e., in spawning period) and then declined to minimum value in September (0.392) (Fig. 3 and 4). Thus it was confirmed with the observations on the monthly fluctuations in the GSI of females and males that *O. bimaculatus* began to spawn in June and continued until August.

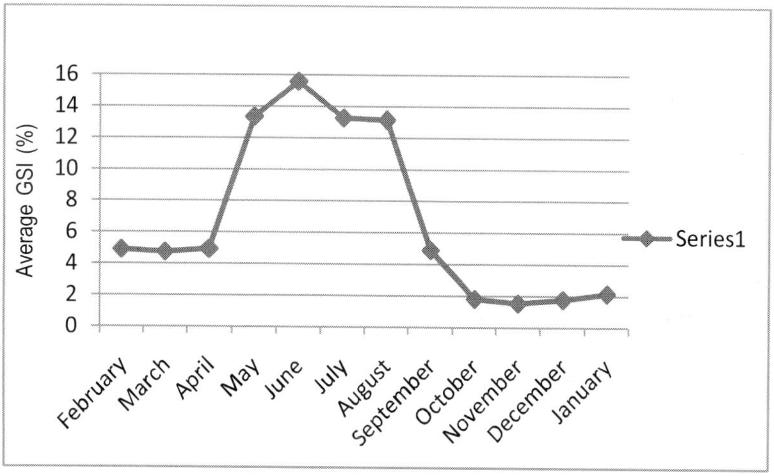

Fig. 3: Monthly variation of Average GSI (%) of female *O. bimaculatus*

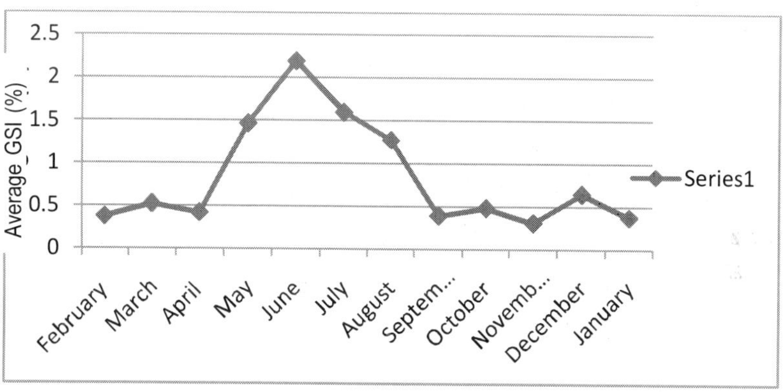

Fig. 4: Monthly variation of average GSI (%) of male *O. bimaculatus*

Fecundity

In the present study, the individual fecundity of fish varied from 2190 to 41552 eggs/fish and the mean fecundity with 95% confidence limits was 19569 ±6,992 eggs/fish for the fishes with 17.0 to 33.0 cm in total length and the body weight ranged from 24.50 to 171.50 g. The monthly variation of mean fecundity was shown in fig.8. Fecundity was high during May- July, which falls sharply during August- October and remain constant until January. Linear relationship of fecundity with total length and body weight of the fish (Figure 5 and 6) were found and the regression models could be expressed as $y = 1237. \times - 2670.$ ($R^2 = 0.721$, P <

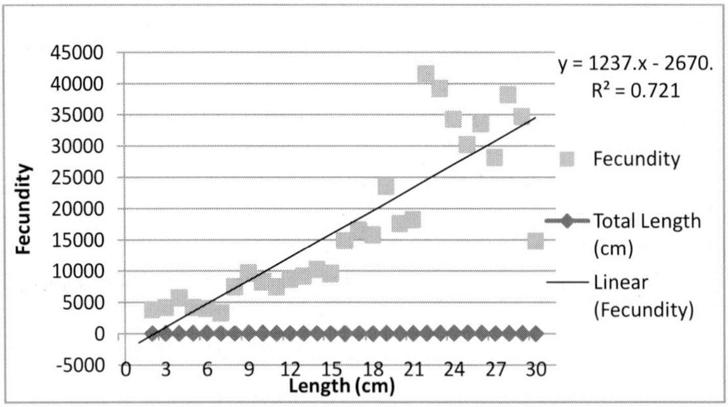

Fig. 5: Relationship between fecundity and length

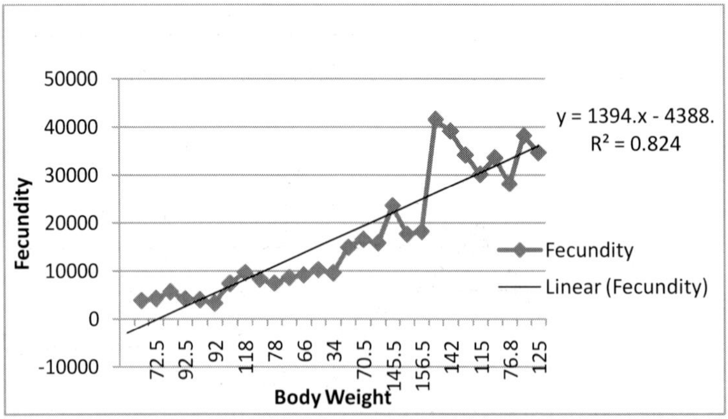

Fig. 6: Relationship between fecundity and weight

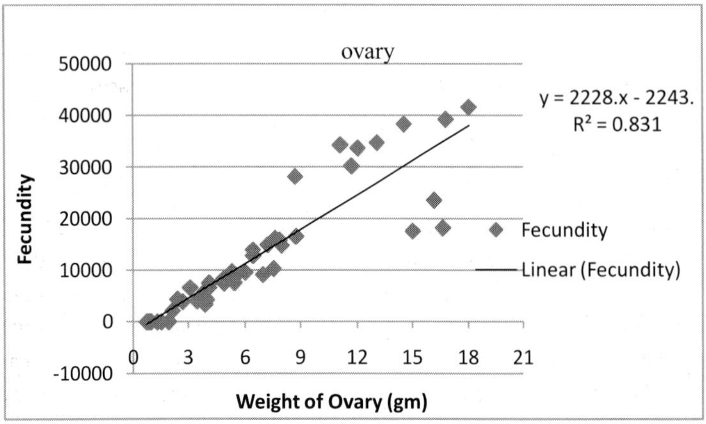

Fig. 7: Relationship between fecundity and weight of ovary

0.001) and y = 1394.x - 4388. ($R^2 = 0.824$, $P < 0.001$) respectively. Fecundity was directly proportional to length and weight of the fish. A significant correlation ($R^2 = 0.831$) was noticed between fecundity and ovary weight (Fig: 7).

Discussion

Based on the results the present observation adds many information on different aspects of reproductive biology of catfish to the existing knowledge ((De laming 1972, DeMartin and Lau 1999, Rimmer and Merrick 1983, Dmitincko, 1970, Gaigher 1977, Abayomi and Arowomo 1996, Parameswaran *et al.,* 1970, Bhowmick 2001, Oymak *et al.,* 2000, Wooton, 1973, Dajoz, 1977, Bagenal 1978, Greenwood 1955, Vanderwaals 1974, Harding 1966, Harding 1966, Healey and Nicol, 1975, Kramer, 1978, Zaret 1980, Ward and Samarakoon 1981, Nawar and Yoakin 1962, Douglas 1979).

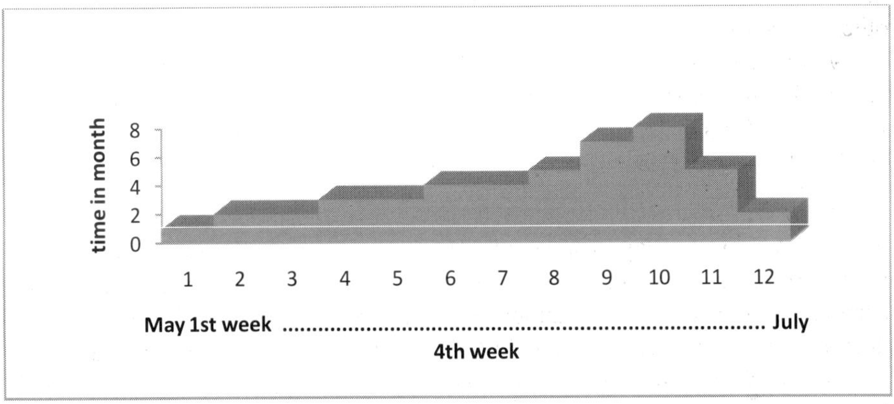

Fig. 8 : Observation of peak-fecundity during 2nd week of July in *Ompok bimaculatus*

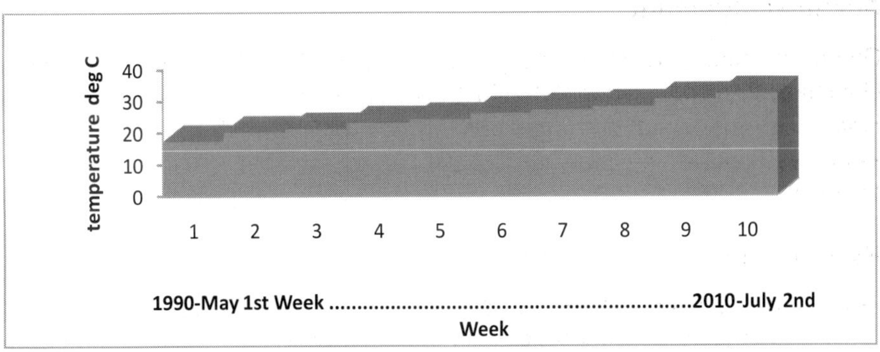

Fig. 9: Shifting of fecundity period in *Ompok bimaculatus* from 1st week of May to 2nd week of July related to increase in temperature during 1990-2010

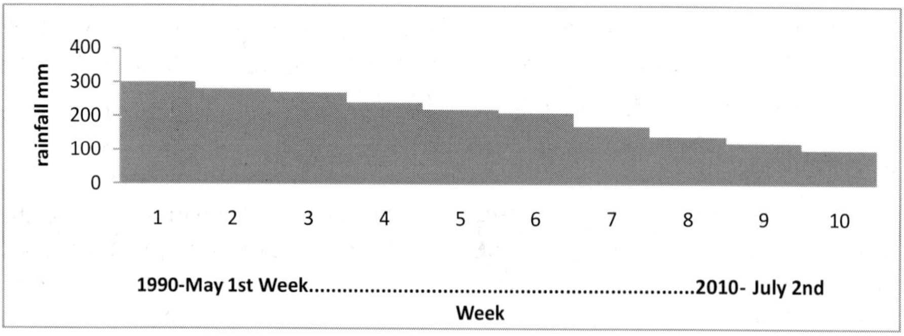

Fig. 10 : Shifting of fecundity period in *Ompok bimaculatus* from 1st week of May to 2nd week of July related to decrease in rainfall during 1990-2010.

Predominance of *Ompok bimaculatus* (female) over male is noticed in the present observations. Similar observations are also being noticed in the population of other catfishes in several studies (Parameswarn *et al.* 1970, 1971, Abayomi and Arawomo 1996, Taiwo and Aransiola 2001). In this context, some of the worth mentioning works are those of Ham (1981), Fagade *et al* (1979), Sivakami, (1982), Bruton, (1979) and Lamai, (1993), Hicklng and Rutenberg (1936), Domagala and Trzebiatowski (1987), Koselov (1984).

Parameswarn *et al.* (1970) reported occurrence of peak spawning in *Ompok bimaculatus* during May to June only. While such observations are being noticed in the present studies during mid July only. The present findings also showing correlation with increase in atmospheric temperature ($r = 0.967$, $P<0.01$) and decrease in rainfall ($r = 0.988$, $P<0.01$) over a period of two decades from 1990-2010. Our earlier observation also depicts similar findings (Banik 1994). So, it may be the probable reason behind declining of population of *Ompok bimaculatus* in recent period.

Acknowledgement

One of the authors (SB) gratefully acknowledges NBFGR, Lucknow and DST, Government of India, New Delhi for financial support on the studies of *Ompok bimaculatus*. The authors thankfully acknowledge the Head, Department of Zoology, Tripura University (a Central University) for providing laboratory facilities.

References

Abayomi, O.S. and G.A.O. Arawomo, 1996. Sex ratio and fecundity of, the clariid catfish (Clarias gariepinus) in Opa reservoir, Ile –Ife, Nigeria. Transactions of the Fisheries Society of Nigeria. 1996: 122-130.

Bagenal, T., 1978. Aspects of fish fecundity. In Ecology of Fresh Water Fish Production. 2nd edn, (Shelby, D. G.,ed.), Blackwell: Oxford, London. pp. 101-135.

Banik, S. 1994. Limnological studies of Gomoti river of Tripura in relation to climate change. In : Proceedings of Vulnerable assessment – climate change, published by Ministry of Environment and Forests, Government of India, New Delhi: 172-182.

Banik, S. and S. Malla, 2010. Rotifer as Growth Promoter for a rare fish Ompok bimaculatus related to Climate change. Proceedings of the workshop on "Aquaculture Biotechnology for Rural Development" at Dept. of Zoology, Tripura University, West Tripura, during 25th to 28th July, 2008 pp. 18-28.

Banik, S. and S. Malla, 2011. Habitat Mapping for Ompok bimaculatus (Bloch) in Gomoti River of Tripura. The 10th Agricultural Science Congress at National Bureau of Fish Genetic Resources (ICAR), Lucknow, U.P., India during 10th to 12th February, 2011 pp. 388.

Banik, S., S. Som, N. Roy, S. Nath and S. Choudhary. 2002. Threatened fish fauna of Tripura: Stratagies for conservation. In: S.Banik ed. Proc. Nat. Sem. On Conserv. Of Aqua. Res. Pp Dept. life Sci. TU. India : 112-119.

Bhowmick, M.L., Mondal, S.C. Chakrabarty, P.P., Das, N.K., Saha, R.N. and S. Ayyapan, 2000. Captive breeding and rearing of Ompok pabda. A.G. Sarkar (Ed) Fish biodiversity of North East India. NBFGR NATP Publ.: 62-68.

Bruton, M.N., 1979. The breeding biology and early development of the catfish (Clarias gariepinus) in Lake Sibaya, South Africa with a review of breeding in species of the subgenius Clariaus (*Pisces: Claridae*). J. of Transdisciplinary environmental studies, 35: 1- 45.

CAMP. 1998. Report of the Workshop on Conservation Assessment and Management Plan (CAMP) for Freshwater Fishes of India, Zoo outreach organization and NBFGR, Lucknow, India, 156 p.

Chakrabarti,. N.M., Chakrabarti, P.P. and S.C., Mondal, 2009. Ompak bimaculatus and Ompak pabda comparative Morphometric and Meristic Study of Embryonic Larval Development. Fishing Chimes, 29(6): 8 – 9.

Chakrabarty, N.M., Mondal, S.C. and P.P., Chakrabarty, 2007. Artificial breeding seed production and rearing of butter fish Ompok pabda – A significant mile stone in technology advancement. Fishing Chimes, 26 (10): 134 – 136.

Clark, F.N. 1934. Maturity of the California Sardine (Sardinella caerulea), determined by ova diameter measurement. Fish. Bull. Sacramento, 42: 1-49.

Dajoz, R., 1977. Introduction to ecology. Hodder and Stoughton, London, pp: 416.

Damagala, J. and T. Trzebiatowski, 1987. Effect of the female size and its eggs diameter on survival and growth of a young stock of trout.Gosp. Rybna. 3: 8-9.

De Vlaming, V.L., 1972. The effect of temperature and photoperiod on reproductive cycling in the estuarine gobiid fish (Gillichthys mirabilis). Fish, Bulletin., 73: 1137-1157.

DeMartini, E.E. and B.B. Lau, 1999. Morphometric criteria for estimating sexual maturity in two snappers, Etelis carbunculus and Pristopomoides sieboldii. Journal of Fish Biology. 97: 449-458.

Demska-Zakes, K. and M. Dlugosz, 1995. Fecundity of vendace from two lakes of Mazurian district. Rybna: 31:37-50.

Dmitincko, E.M., 1970. Reproduction of the sea catfish (Arius thalassinus: Rupel) in the Arabian Res. J. Fish. & Hydrobiol., 3(1): 22-31.

Douglas, C., 1979. Sexual maturity and fecundity of African catfish (Clarias gariepinus) with an observation on spawning behaviour of the nile catfish, (Clarias lazera), Zool. J. of the Limnological society of London., 35: 1-45.

Fagede, S.O., Adebisi, A.A. and Aatanda, A.N. 1984. The breeding cycle of tilapia, Sarotherodon galilaeus in the IITA Lake, Ibadan, Nigeria. Hydrobiologia., 100: 493-500.

Gaigher, L.G., 1977. Reproduction in the catfish (Clarias gariepinus) in the hardap dam. South West Africa. Madoqua, 10: 52-59.

Greenwood, P.H., 1955. Reproduction in the catfish (Clarias mossambicus). Peters. Nature London., 176: 516-518.

Ham, R., 1981. The ecology of six native and two introduced fish species in Enoggera creek system, south east, Queens land. B.Sc (Hons) Thesis. Griffith univ. Brisbane., pp: 157.

Hanjavanit, C. and N. Sangpradub, 2009. A preliminary examination of the gut contents of three species of freshwater fishes, Barbonymus altus (Gunther, 1868), Notopterus notopterus (Pallas, 1769) and Ompok bimaculatus (Bloch, 1794) from Nong Han Kumphawapi, Udon Thani, Thailand. KKU Sc. J. (Supplement) 37: 1-10.

Harding, D., 1966. Lake Kariba: the hydrology and development of fisheries. In Manmade lakes. 2nd edn. (R.H. Lowe Mc 25. Connel ed.). Academic press, London. pp: 123-134.

Healey, M.C. and R.A.W. Nicol, 1975. Fecundity comparisons for various stocks of Lake white Fish (Coreqonus clupeaformis). J. of Fisheries Research, 32: 404-407.

Hickling, C.F & E. Rutenberg. 1936. The ovary as an indicator of spawning period in fishes. Mar. Biol. Assoc. UK 21: 311 – 412.

Jayaram, K. C. 1999. Freshwater fishes of Indian region. Narendra Publishing House, New Delhi, India. pp. 551.

Koselov, B.V. 1984. Ekologia rozmnozenia ryb. Nauka, Moscow, Russia.

Kramer, D.L., 1978. Reproductive seasonality in the fishes of a tropical stream. Ecology., 59: 291-346.

Lamai, S.L., 1993. Aspects of the Applied biology of African catfish (Clarias gariepinus) (Burchell, 1822). Aquacultural techniques and deldrin toxicity. Ph.D.Thesis, Dept. of Applied Zoology, University of Reading, U.K., pp: 204.

LeCren, E.D. 1951. The length-weight relationship and seasonal cycle in gonad weight and condition in the perch (Perca fluviatilis). J. of Animal Ecology 20: 201-219.

Nawar, G. and E.G. Yoakin, 1962. A study on the fecundity of the nile catfish, Clarias lazera, Annuals and magazine of natural history, 135: 345-365.

Oymak, S.A., Solak, K., .nl., E.: Some biological characteristics of Silurus triostegus Heckel, 1846 from Ataturk Dam Lake (Turkey). Turk. J. Zool. 2000; 25: 139-148.

Parameswarn, S., Sevaraj, C. and Radhakrishnan, S. 1970. Observations on the biology, induced breeding and culture possibilities of Ompok bimaculatus (Bloch) in ponds. Proc. Nat. Acad. Sci. India 23(B):145-157.

Parameswarn, S., Sevaraj, C. and S. Radhakrishnan, 1971. Notes on the life history and biology of the cat fish Ompok pabda (Hamilton). J. Zool. Soc. India. 23:137-150.

Parameswarn, S., Sevaraj, C. and Radhakrishnan, S. 1974. Observation on the biology of Labeo gonius (Hamilton). Indian J. Fish., 21: 54-75.

Rahman, M.A., S., Yeasmine and M.N., Khan, 2008. Growth, survival and production of a threatened silurid catfish, Ompok pabda (Hamilton) fingerlings in nursery ponds. Malays. Appl. Biol. 37(1): 47-56.

Rimmer, M.A. and J.R. Merrick, 1983. A review of the reproduction and development in the forktailed catfishes (Ariidae). J. of the Limnol. Society of London., 107: 41-50.

Rishi, S. A.D.K. Thind and K.K. Rishi, 1994. Cytogenitic analysis of Ompak bimaculatus (Siluridae, Teleostei) a species coming endangered. In: Threatened fishes of India (Eds. Dahadrai, P.V., P. Das, and S.R. Verma) Natcon Publ. No. 4: 343-346.

Sharma, R., 2008. Chromosomal studies on Ompak bimaculatus (Bloch) and Ompak pabda (Hamilton-Buchanan, 1822). Indian J. of Animal Sci. 78(2): 234-236.

Siddiqua, K.A., M. S. Islam, M.G., Hossain,and A.T.A., Ahmed, 2000. A histological study of the spermatogenesis in Ompak pabda (Hamilton-Buchanan, 1822). Bangladesh J. of Fish. Res. 4: 185-189.

Sivakami, S., 1982. Length- weight relationship and relative condition in Ompok bimaculatus (Bloch) from Bhavanisagar Reservoir (Tamil Nadu). Geobios (New Report) 1:202-207.

Taiwo, I.O. and M.O. Aransiola, 2001. Length – weight relationship, condition factors and fecundity of two bragrid cat fishes Chriysich thysnigrodigitatus and Chrysichthys walkeri in Asejire Lake. In Transactions of the Fisheries Society of Nigeria. Lagos. 2001pp: 277.

Taleb, M.A., Rahman, M.H. and Hossain, M.A. 1991. Food and feeding habit of small omnivorous fish, Ompok pabda (Hamilton). Proceedings of 15th Annual Bangladesh Sciences Conference. Abstract Section IX, No. 20, p. 17.

Vanderwaals, B.C.W., 1974. Observation on the breeding habits of African catfish Clarias gariepinus (Burchell). Journal of Fish Biology. 48: 6-10.

Vijayakumar, C. 2010. Conservation of an endangered Indian catfish Ompok malabaricus through captive breeding and establishment of captive population. International J. of Biological Technology 1(2): 131-133.

Ward, J.A. and J.I. Samarakoon, 1981. Reproductive tactics of the Asian cichlids of the genus Etrophus in Srilanka. Environmental Biology. 6: 95-103.

Wooton, R.J., 1973. The effect of size of food ration on egg production in the female three-pined stickleback, Gasterosteus aculeatus. J. of Fish Biology. 5: 89-96.

Zaret, T.M., 1980. Life history and growth relationships of Cichla ocelaris, a predatory South American Cichlid. Biotropica, 12: 144-157.

❑❑❑

24

Climate Change and Aquatic Resources: Issues and Challenges

Ratan K. Saha

Abstract

Climate change is a matter of great concern considering aquatic resources and the productivity. The likely impact of Climate Change is noted in the domain both in marine as well as estuarine capture fisheries, fresh water capture fisheries and freshwater aquaculture. Climate change is having a linkage with spread of fish diseases as increasing water temperature will shift the balance in favor of either the host or pathogen changing the frequency and distribution of diseases. There are a no of adaptation and mitigation measures in aquaculture in order to increase the productivity and properly maintain the survivability of many aquaculture species.

Keywords: Global warming, Water resources, Fresh water aquaculture, Coastal area, Brackish water fisheries, Adaptation and mitigation measures.

Introduction

The United Nations Framework Convention on Climate Change defines climate change (CC) as "a change of climate which is attributed directly or indirectly to human activity that alters the composition of the global atmosphere and which is in addition to natural climate variability observed over comparable time periods". Climate change may be limited to a specific region, or may occur across the whole Earth. Fisheries have been in limelight during the past three decades as the fastest growing food production system. Climate is now recognized as a major factor affecting the productivity of key species in world fisheries. The mechanisms that link climate to fish productivity need to be better understood to ensure that natural and greenhouse gas induced climate changes are incorporated into the management of fisheries. The extent of climate-driven changes(e.g., changes in growth, reproductive success, life cycle closure, successful recruitment, changes in suitable habitats and mortality) in commercially and ecologically important fish species will be mediated by the capacity for individual species (or populations) to adapt to changes in important abiotic and biotic factors.

Adaptations could include both changes in the phenology of important life history events (e.g., migration, spawning) and/or physiological changes (e.g., thermal reaction norms of key traits such as growth, increased tolerance to loweredpH/ acidification). The rising ocean acidity makes it more difficult for marine organisms such as shrimps, oysters, or corals to form their shells– a process known as calcification.Thus the entire marine food web is being altered– there are 'cracks in the food chain'. As a result, the distribution, productivity, and species composition of global fish production is changing, generating complex and inter-related impacts on oceans, estuaries, coral reefs, mangroves and sea grass beds that provide habitats and nursery areas for fish. Changing rainfall patterns and water scarcity is impacting on river and lake fisheries and aquaculture production.Aquaculture can be more environmentally damaging than exploiting wild fisheries on a local area basis but has considerably less impact on the global environment on a per kg of production basis.The worldwide capture fisheries have decline sharply in recent decades due to over fishing, pollution, the mismanagement, habitat and coastal zone modification as well as climate change impact.

Climate change is likely to pose threat to water resources and fishery resources of India.This is visible in the form of temperature increase, enhanced anomalies in the rainfall pattern and frequency and severity of extreme events. Agriculture and fisheries sector is largest consumer of water, is highly vulnerable due to its direct dependence on climate parameters. Various problems concur due to climate change such as over heat of pond water, bad water quality, disease prevalence, insufficient water in the ponds, reduce feeding activity and lower growth rate hampered the fish growth and survival in culture, nursery and hatchery level.

Management strategies are being developed to address climate change and fisheries, including risk and vulnerability assessment frameworks, pro-active planning with stakeholders regarding potential impacts and responses and examining existing regulations to identify gaps created by altered species distribution (e.g. unregulated fishing). Overall, fisheries governance systems are needed which are flexible and can quickly adapt to changing ecological and human societal conditions. Adoption of an integrated social-ecological approach that improves the adaptive capacities of ecological and human social systems will help to sustain food security from marine wild capture fisheries.

In order to promote the efficient management of freshwater biodiversity and eventually inverse its decline, there is an urgent need to provide solid estimations of fish species losses under probable climate change and water consumption scenarios. In the culture fisheries, integrated aquaculture-agriculture-animal husbandry-apiculture (A^3 or A^4) system is an important means of diversifying and improving the farm productivity both in wet and dry seasons. In each management mode, the production rate has been found satisfactory and also maintaining favorable ecological conditions.

Causes of Climate Change (CC)

The amount of solar radiation that reaches the Earth is about equal to the amount reflected, maintaining an energy balance. This energy balance is now being altered by two components of the atmosphere. First, greenhouse gases (GHGs) - the most well-known being carbon dioxide (CO_2), methane (CH_4), nitrous oxides (N_2O), sulphur hexafluoride (SF_6), hydrofluorocarbons (HFCs), and perfluorocarbons (PFCs), capture some of the energy reflected from the surface of the earth, keeping it in our atmosphere and warming the planet. India is currently the fourth-largest emitter of greenhouse gas (GHG) emissions.

Second, aerosols - miniscule particles suspended in the air - primarily have a cooling effect on the Earth because they reflect solar radiation back into space before reaching the atmosphere, though aerosols like black carbon produced from burning of biomass and diesel engine exhaust have a warming effect.

Global Warming and Consequences

Global Warming is defined as the increase of the average temperature on Earth. As the Earth is getting hotter, disasters like hurricanes, droughts and floods are getting more frequent.Over the last 100 years, the average temperature of the air near the Earth's surface has risen a little less than 1° Celsius ($0.74 \pm 0.18°C$). It is responsible for the conspicuous increase in storms, floods and raging forest fires we have seen in the last ten years. Out of the 20 warmest years on record, 19 have occurred since 1980. The three hottest years ever observed have all occurred in the last ten years.

In general several effects of global warming, including steady sea level rise, increased cyclonic activity, and changes in ambient temperature and precipitation patterns, have affected or are projected to affect the subcontinent.

As per Intergovernmental Panel on Climate Change (IPCC) (2001) Assessment Reports on ocean systems concluded that global warming will affect the oceans through changes in sea-surface temperature(SST), sea level, ice cover, ocean circulation, and wave climate.The first among the countries to be affected by severe climate change is Bangladesh. Its sea level, temperature and evaporation are increasing, and the changes in precipitation and cross boundary river flows are already beginning to cause drainage congestion. There is a reduction in fresh water availability, disturbance of morphologic processes and a higher intensity of flooding and other such disasters. Bangladesh only contributes 0.1% of the world's emissions yet it has 2.4% of the world's population. In contrast, the UnitedStates makes up about 5 percent of the world's population, yet they produce approximately 25 percent of the pollution that causes global warming (MERIC, 2008).As a whole some impacts of global warming have already become visible in India which are as follows:

- Monsoon rains have become less predictable.
- Glaciers are melting, more floods and droughts occur.
- Mangrove forests are disappearing at an alarming rate. Public health, biodiversity, agricultural production, access to drinking water, and even national security will be affected.
- Increased landslides and flooding are projected to have an impact upon states such as Assam (Dasgupta, 2007).
- Ecological disasters, such as a 1998 coral bleaching event that killed off more than 70% of coralsin the reef ecosystems off Lakshadweep and the Andamans, and was brought on by elevated ocean temperatures tied to global warming, are also projected to become increasingly common(Normile, 2000, UCS,2005, Aggarwal, 2007).
- The corresponding sea level rise at the end of the 21st Century relative to the end of the 20th Century ranges from 0.18 to 0.59m.
- Ongoing sea level rises have already submerged several low-lying islands in the Sundarbans, displacing thousands of people.
- More specifically for fisheries, climate change-related warming may result in:
    - Longer growing seasons and increased rates of biological processes – and often fish production;
    - Greater risk of oxygen depletion;
    - Species shift to more tolerant of warmer and perhaps less- oxygenated waters;
    - Redeployment or re-design and relocation of coastal facilities;
    - Coastal cultures may need to consider the impacts of sea-level rise on facilities and the release of contaminants from nearby waste sites;

- Changes in precipitation, freshwater flows, and lake levels
- Introduction of new disease organisms or exotic or undesired species
- Establishment of compensating mechanisms or intervention strategies
- A longer season for production and maintenance; and
- Modification of aquaculture systems, e.g. keeping them indoors under controlled light, may be needed more often to protect larvae from solar UV-B.

Now, as a whole all the consequences can be discussed in to two groups such as direct and indirect which are as follows:

1. **Direct consequences**
   - Direct biophysical and ecological consequences ofprojected changes.
   - Direct impact on species compositions anddistribution.
   - Direct impact on catch potential.
   - Direct impacts on aquaculture production.
2. **Indirect consequences**
   - Interaction between other sectors (powergeneration, food production).
   - Strong competition for resources (land,freshwater).
   - Mitigation actions of other sectors.
   - Most already present (human driven problems):

Climate change will make them worse.

Effects of CC on Water Resources as a Whole

As a whole due to the climate change all the available reservoirs of water resources are in trouble as because:

- Decreasing rainfall/precipitation
- Decreasing flows in existing river basins as well as significant decrease in open water areas
- Floods and prolonged droughts
- Water logging/drainage congestion
- Sea level rise
- Saline water intrusion
- Altered rainfall patterns and river flows
- Gradual acidification of ocean ecosystem

Effects of Climate Change on Freshwater Resources in Particular

A recent research project (Xenopoulos *et al.*, 2005) has addressed the losses in freshwater biodiversity under probable climate change and water consumption scenarios:

- About half of the investigated rivers worldwide would experience reduced water availability due to both global warming and withdrawal of water for human needs.
- By 2070, in these drying rivers, loss in local fish species would range from less than 4% to more than 22% with a maximum of 75% loss in the most affected rivers.
- Over the past 30 years, freshwater species have declined faster compared to terrestrial or marine species.
- Unfortunately, growing evidence indicate that this trend is likely to persist in the future.
- On one hand, freshwater ecosystems will probably further suffer from invasive species and land use changes.
- Further, freshwater biota is likely to be impacted by the predicted reduction in water availability driven by increased water consumption for human uses and indirectly related to global climate change.
- In order to promote the efficient management of freshwater biodiversity and eventually inverse its decline, there is an urgent need to provide solid estimations of fish species losses under plausible climate change and water consumption scenarios.
- In response to this need, a team of international experts used two scenarios from the Intergovernmental Panel for Climate Change (IPCC) to cover a large range of possible climate change outcomes.
- For the first time, they combined these scenarios with a global hydrological model to estimate possible losses in river water availability due to climate change and trends in water consumption.
- Linking the obtained results to known relationships between fish species and changes in water availability, they investigated the riverine fish richness over the next 70 years in more than 300 worldwide riverine basins.
- In both scenarios, their calculations showed that by 2070, water availability would decrease up to 80% in more than 130 investigated rivers with available fish data.

- About half of these rivers were predicted to lose more than 10% of their fish species when climate change and water consumption impacts were considered.
- A maximum of 75% of local fish losses were calculated for Cauvery, Parnaiba, and San Tiguel rivers.
- The effect of climate change was the most important factor of freshwater fish loss under the scenarios, while anthropogenic water withdrawal contributed much less to species loss (an additional 0–5%).
- However, in regions where substantial irrigation has occurred in the past and is expected to increase (e.g., Southern Asia, and the Middle East), water consumption is particularly important for fish biodiversity decline.
- According to the authors, these forecasts are underestimates of fish losses because many other important drivers were not included in the calculations (e.g. introduced species, reductions in channel-flood-plain connectivity).
- In any event, the scientists argue that along with the international efforts to minimize global warming, reducing water consumption could be an initial conservation strategy to prevent the predicted loss of fishes worldwide.

Effects of Climate Change on Fish/Shellfish Production

Impacts of climate change on fish and shellfish production may be of physiological, ecological and operational as mentioned in the table:

Physiological	Ecological	Operational
1. Growth	1. Habitat degradation	1. Squeezing species diversity
2. Reproduction	2. Disease interactions	2. Shifting fishing site/area
3. Environmental stress	3. Risky marine capture fisheries	
4. Temperature fluctuation	4. Increased fuel consumption	
	5. Unrest ecosystem	

Effects of Climate Change on Freshwater Aquaculture

1. Shrinking down freshwater aquaculture area due to prolonged drought.
2. Delayed monsoon/saline water intrusion upsetting gonadal development of fishes causing:
   - Earlier sexual maturity.
   - Decrease in spawn production.
   - Effects on hatching process-drift in genetic resource.
3. Prolonged drought dwindling culture period reduces desired production.

Effects of Climate Change on Open water Fisheries Resources

- Feeble freshwater run off
- Shifting interface area to upstream shrinking connectivity
- Massive siltation at the open water systems
- Water logging in beels/haors/ox-bow lakes (depressed water bodies)
- Loss of aquatic biodiversity
- Shifting of spawning, feeding and nursery grounds
- Changes in migratory routes
- Low precipitation and habitat degradation disrupting
- Natural breeding of carp in natural water bodies (Ganga, Yamuna, Brahmaputra, etc.)
- Natural Egg production in rivers reduced nearly 75% in last two decades
- Inland capture fisheries declined from 50% to 35% of total fish production during last decades
- River siltation disrupting migratory routes of fish e.g., Hilsa
- The top 10 (225) rivers most at risk (endangered): Ganges, Indus etc.
- WWF assessed the 6 most important threats:

1. dams and infrastructure, 2. excessive water extraction, 3. climte change,
4. invasive species, 5. over-fishing, and 6. pollution.

- Biodiversity shift in Hooghly estuary
    - The salinity regime of the estuary has changed and major portion in its middle stretch has almost changed into a freshwater zone, due to increased freshwater discharge from the Farakka barrage.
    - The biodiversity changes reflected that a sizeable length of Hooghly Estuary has lost its estuarine characteristics.
    - The Barrackpore stretch of the estuary is dominated by freshwater mullet *Sicamugil cascasia.*
    - The brackishwater fish species, *Liza parsia*, *Scatophagus argus*, *Mystus gulio*, *Osteogeneiosus militaris*, *Hemiramphus gaimardi*, are no longer available in large stretches of the estuary.
    - The plankton composition of this estuary, has changed.

  - The once dominant marine diatoms like *Chaetoceros*, *Bacteriastrum* have been replaced by the freshwater plankton species *Scenedesmus*, *Pediastrum* and *Microcystis*.
- Changes in the Fisheries of Ganga and Yamuna
  - The most important hydrological change observed in river Ganga is the sharp increase in sediment load during current decade by almost 20-times at Allahabad and 30 times at Varanasi.
  - This change has resulted in considerable reduction in discharge rates, seriously impacting the breeding and recruitment process of commercially important fishes and significant shift in species composition in favour of miscellaneous and exotic species.
  - Fish landings at Sadiapur and Daraganj centers were estimated at 56.89 tonnes and 29.51 tonnes respectively.
  - The fish catch data from Ganga river system at Allahabad indicated dominance of miscellaneous fish group (54.3%) followed by common carp (24.4%).
  - The landings of major carps and catfishes declined sharply and presently these are contributing only 9.5% in the total landings.

Effects of Climate Change on Coastal Area

The main impacts of climate change on coastal areas are:

- Sea Level rise.
- Sea level rise may destroy the mangrove forest as well as destroy the marine fish nursery ground.
- Reduction of freshwater availability by salinity interference.
- Increasing cyclone frequency.
- Decrease brackish water fish production.
- Water salinity varies from 0-20 ppt.
- Water salinity and its distribution are increasing with the increase of sea level rise.
- Soil salinity in South Western part of Bangladesh is also increasing.
- This salinity interference including sea level rising creates harmful effect on existing fish species.
- Water salinity exceeds the expected salinity level that especially required for freshwater fish production.

- Salinity interference may be of two type of effects as follows:

Salinity Interference

Positive Effect

↓

- Create opportunities for catching and cultivating brackish and marine species

Negative Effect

↓

- Threatened to freshwater fisheries
- Loss of fish biodiversity

Potential species

- *Penaeusvannamei* (Western White Shrimp)
- *Penaeus monodon* (Tiger Shrimp)
- *Chanos chanos* (Milk Fish)
- *Mugil cephalus* (Mullet)
- *Tilapia nilotica* (Tilapia)

Effects of Climate Change on Brackish water System

Saline intrusion

- Rising sea level may substantially diminish critical coastal habitats.
- Effects physiology of fin fish and shell fish.
- In recent years salinity increased up to 20/25 ppt in the coastal belt areas.
- Currently, salinity persistence in the coastal water bodies prolonged up to 5-6 months, which was 23 months earlier.
- Destruction of the sundarbans mangrove forest as well as the fish nursing grounds.
- Shifting interface area to upstream of the river system.
- Natural disaster like Sidr, Ayla, etc. causing saline water logging along with reduced production from shrimp/ fish farming.
- Freshwater fish/ prawn farming hindered due to saline intrusion.
- Sudden fluctuation in temperature and salinity triggering horizontal disease infestation
- Alarming declined in Fisheries biodiversity e.g., mullets, catfish etc. in the Sundarbans region

Effects of Climate Change on Marine Fisheries

- Decrease in natural recruitment pattern due to destruction of nursery ground like the mangroves
- Increased frequency of storm surge disturbing costal aquaculture
- Marine fisheries resources declining throughout the last two decades
- Over 100 important fish and other aquatic species disappeared (FAO, 2009)
- Force fish to migrate in areas/ regions until find suitable habitat.
- Rise in surface temperature bleaches reefs even coral die.
- Occurrence of bloom & less light penetration
- Increases the frequency of El-nino.
- Gradual increase in salinity and ocean acidification along with temperature fluctuation.
- Average tropical sea surface temperature is predicted to increase by 50-80%.
- Increased temperature may affect the distribution pattern of some fish species
- Some of them may be migrate to the higher latitude for cooler place.
- Atmospheric CO_2 level will be high and this will cause the average ocean pH to drop by 0.4-0.5.
- Fish embryos and larvae are more sensitive to pH change than juvenile and adults.
- Eggs of the pelagic fishes might be more susceptible to pH change.
- Increased level of dissolved CO_2 also reduces the pH of animal tissue and it may affect the marine fish physiology.

Effects of Climate Change on Livelihood

Fisheries and aquaculture are receiving increasing attention, not only because they represent an important source of livelihoods and food, but also because of our increasing understanding of aquatic ecosystems. Further, fishery and aquacultureproduction provide direct employment and revenue to an estimated 43.5 million people; mainly fishers but also increasingly fish farmers. Detailed statistics are often not readily available, especially for small-scale fishing activities in developing countries. The general trend is that the number of jobs for fishers is stagnating and that opportunities in aquaculture have been increasing.

However, the livelihood components may be described as natural, financial, physical, human, and social. The findings of a recent study conducted at Bangladesh reveled that the livelihood options were changed through period due to the changes of climate as given in the table below:

Period	Fishing	Agriculture	Others (Boating, grazing, Vegetation collector etc.)
1992	50%	20%	30%
2001	35%	29%	36%
2006	24%	38%	38%

Climate Change and Fish Diseases

- As cold-blooded animals, fish are highly dependent on environmental conditions, especially temperature, to help maintain critical physiological processes such as immune function that can affect whether a fish gets a disease or parasite, how it is affected by it, and how the disease progresses.
- Increasing water temperatures will shift the balance in favour of either the host or pathogen, changing the frequency and distribution of disease.
- A number of endemic diseases (e.g. enteric red mouth, furunculosis, proliferative kidney disease and white spot) will become more prevalent and difficult to control as water temperatures increase.
- Climate change also alters the threat level associated with exotic pathogens.
- Increasing water temperatures and the negative effects of extreme weather events (e.g. storms) are likely to alter the freshwater environment adversely for both wild and farmed fish populations, increasing their susceptibility to disease and the likelihood of disease emergence.
- For wild populations, surveillance and risk mitigation need to be focused on locations where disease emergence, as a result of climate change, is most likely.
- The risk of viral haemorrhagic septicaemia (VHSV), infectious haematopoietic necrosis virus (IHNV) and spring viraemia of carp virus (SVCV) declines as infection generally only establishes when water temperatures are less than 14°C for VHSV and IHNV and 17°C for SCVC.
- If that temperature is exceeded over a period of time, not only may die-offs occur, but also, the increased stress and altered immune function will lead to greater levels of infectious or parasitic diseases which is why global warming is of particular concern.
- Disease is often ignored as a factor affecting wild populations of fish and wildlife because the effects are difficult to observe and quantify.
- The consequences of increased frequency of extreme weather events can be visualized from the below mentioned figure and it is a chain effects:

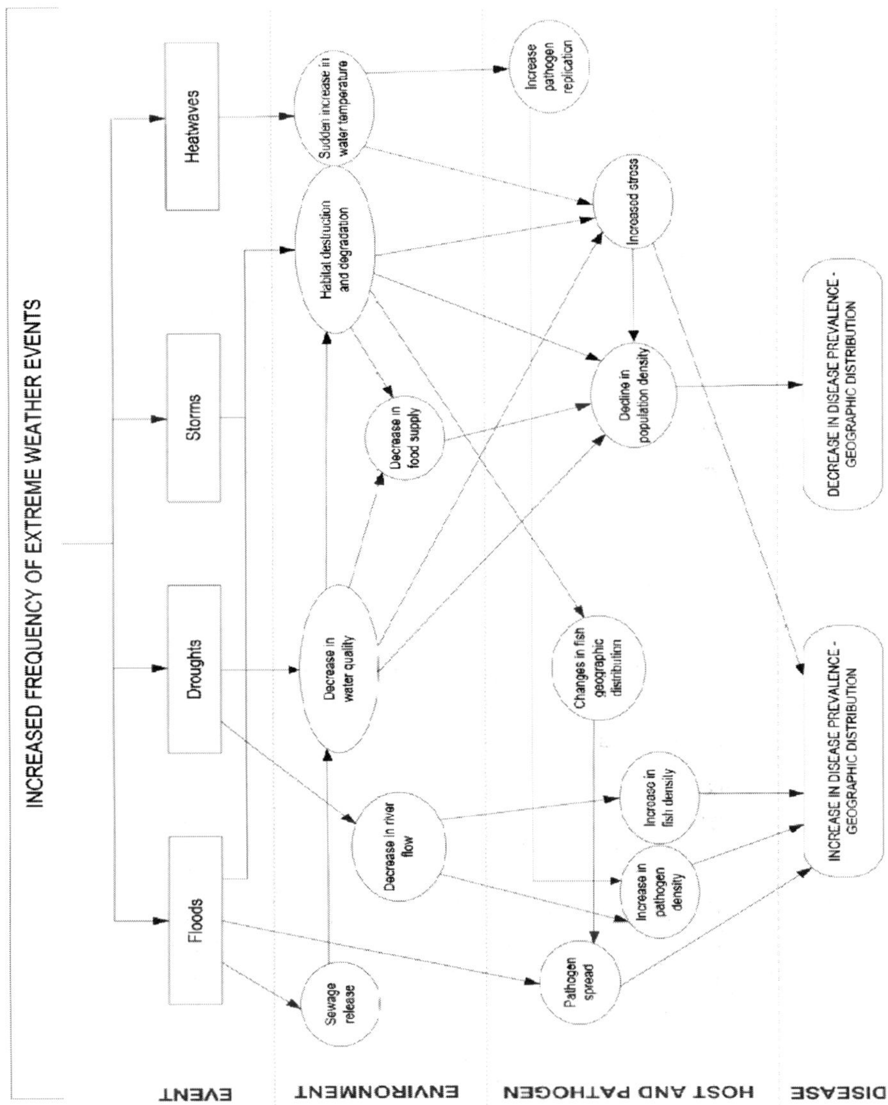
INCREASED FREQUENCY OF EXTREME WEATHER EVENTS
EVENT
ENVIRONMENT
HOST AND PATHOGEN
DISEASE
Floods
Droughts
Storms
Heatwaves
Sewage release
Decrease in water quality
Decrease in food supply
Habitat destruction and degradation
Sudden increase in water temperature
Decrease in river flow
Pathogen spread
Increase in pathogen density
Increase in fish density
Changes in fish geographic distribution
Decline in population density
Increased stress
Increase pathogen replication
INCREASE IN DISEASE PREVALENCE - GEOGRAPHIC DISTRIBUTION
DECREASE IN DISEASE PREVALENCE - GEOGRAPHIC DISTRIBUTION

Table 1: Impact of climate change on parasitic and other diseases of aquatic animals

Host	Disease / Parasite	Response to high temperature	Reference
Large mouth bass (*Micropterussalmoides*)	Red sore disease / bacterium *Aeromonashydrophila*	Susceptibility to the disease increases	Esch and Hazen (1980)
Mosquitofish (*Gambusiaaffinis*)	Asian fish tapeworm (*Bothriocephalusacheilognathi*)	-do-	Granath and Esch (1983)
Trout (*Onchorhynchus* spp.)	Whirling disease / Myxozoan *Myxobolus cerebralis*	-do-	Hiner and Moffitt (2001)
Juvenile cohosalmon (*O. kisutch*)	Blackspot disease/trematode larvae (metacercariae)	Virulence is directly correlated with daily maximum temperature	Cairns *et al.* (2005)
A variety of reeffish	Ciguatera fish poisoning (CFP) caused by bioaccumulation of algal toxins	Increased incidence of CFP due to increased temperature	Tester *et al.* (2010)
Rainbow trout, *Oncorhynchusmykiss*	Infected with *Ichthyophonus* sp.	More rapid onset of disease, higher parasitetoad, more severe host tissue reaction and reduced mean-day-to death at higher temperature	Kocan *et al.* (2009)
Freshwater bryozoans infected with myxozoan, *Tetracapsuloidesbryosalmonae*	Spores released from sacs produced by the parasite during infection of freshwater bryozoans are infective to salmonid fish, causing the devastating Proliferative Kidney Disease (PKD)	Exacerbate PK Doutbreaks and increase the geographic range of PKD as a result of the combined responses of *T. bryosalmonae* and its bryozoan hosts to higher temperatures.	Tops *et al.* (2009)

- As the emergence of disease is linked directly to changes in the ecology of hosts or pathogens, or both (Harvell *et al*., 1999), climate change will have a profound impact on the spread of parasites and disease in aquatic ecosystems (Harvell *et al*., 1999, Harvell *et al*., 2002, Marcogliese, 2008).
- Climate change will affect parasite species directly resulting from the extension of the geographical range of pathogens (Harvell *et al*., 2002).
- Many diseases display greater virulence at higher temperatures that might be the result of reduced resistance of the host due to stress or increased expression of virulence factors/increased transmission of the vectors.
- Some examples have been summarized in table 1 and effects of temperature on parasites, hosts and transmission are given in table 2.

Table 2. General effects of increased temperature on parasite life cycles, their hosts and transmission processes (Marcogliese, 2008)

Effects on parasites	Effects on hosts	Effects on transmission
Faster embryonic development andhatching	Altered feeding	Earlier reproduction in spring
Faster rates of development and maturation	Altered behavior	More generations per year
Decreased longevity of larvae and adults	Altered range	Prolonged transmission in the fall
Increased mortality of all stages	Altered ecology	Reduced host resistance Potential transmission year round

Climate-induced Changes in Fisheries System

Physical	Ecological	Fisheries
• Temperature • Sea level	• Species distributions • Timing of reproduction, migration, abundance	• Productivity • Distribution
• pH • Salinity • Currents • Nutrients • Oxygen • Ice cover • Turbulence & mixing • Wind speed & direction • Storm frequency & intensity • Evaporation • Precipitation • Runoff	• Physiology • Recruitment • Behaviour • Sex ratios • Calcification rates • Upwelling timing & latitude • Habitat loss • Coral bleaching frequency & severity • Disease	• Species availability • Timing • Days at sea • Accidents • Loss of harbours, home • Damage to productive assets • Increased insurance costs • Livelihood diversification • Ability to plan seasonal livelihood activities

Source: Allison *et. al*., 2009

Problems in Aquaculture Activities due to Climate Change- A Case Study at Bangladesh

A unique study has been conducted by the group of researchers at Bangladesh with different stakeholders to assess the extent of problem faced by them due to climate change during various activities in fisheries and aquaculture and summarized as below:

Types of problems	Fish Farmers (n=45)	Hatchery owners (n=15)	Nursery owners (n=15)	Average
Natural hazards	82.22	86.66	80.00	82.96
Manmade hazards	2.22	0.00	6.66	2.66
Both	15.55	13.33	13.33	14.66
Over temperature	62.22	60.00	73.33	64.00
Fluctuating temperature	37.77	40.00	26.66	36.00
Lower temperature	0.00	0.00	0.00	0.00
No rain	66.66	66.66	60.00	65.33
Fluctuating rainfall	33.33	33.33	40.00	34.66
Excess rainfall at a time	0.00	0.00	0.00	0.00

Further, the hatchery and nursery owners also expressed their experiences that various operational activitieswere hampered due to the fluctuation of temperature and rainfall as given in the table:

Hatchery Nursery

SI.No	Types of problems	%average	SI. No	Types of problems	%average
1.	Fish do not ovulate	20.00	1.	Lower growth rate	20.00
2.	Less amount of eggs	33.33	2.	Deformity of fry	20.00
3.	Eggs not fertilized	20.00	3.	Lower survival rate	33.33
4.	Lower hatching rate	13.33	4.	Loss of crops	00.00
5.	Lower survival rate	13.33	5.	Less water available	06.67
			6.	Mass mortality	20.00
	Total	**100**		**Total**	**100**

Likely Impacts of Climate Change of Fisheries and Aquaculture

The likely impact s of climate change in the different sectors of fisheries and aquaculture has been furnished in the table as under to understand the consequences we have to face (adapted from Ali, 1999):

Fisheries & Aquaculture Systems	Likely Impacts
Marine and Estuarine Capture Fisheries	• Affect water temperature and salinity regimes in the upper Bay of Bengal. • Increased water temperature may likely to advance the sexual maturation and the process of and timing of spawning of adult hilsha population. • This may upset timing of spawning migration into the fresh water rivers and estuarine areas. • This may altered the abundance of fish and the fishing ground in fresh water and estuaries. • Other important fisheries based on the populations of Bombay duck, pomfret and Indian salmon, having wider range of habitats may be affected by the increased temperature with reproductive physiology.
Freshwater Capture Fisheries	• Rising sea level may likely affect in reduction fresh water habitat. • Thus strictly freshwater Cyprinids, Anabantids, Channidaeand many other fishes may like suffer. • Climate change may enhance precipitation and flood and increased water flow and though the tidal rivers, streams and creeks • Increased riparian fresh water flow into estuaries and coastal areas will alter the estuarine ecology • Such alteration may affect the habitats of various shrimp and fish making inhospitable and inaccessible
Freshwater Aquaculture	• Pond based aquaculture in coastal area might be affected by sea level rising through submerging the ponds • Rising temperature may affect the hatchery based fish seed production across the country and aquaculture enterprise

Climate Change Adaptation and Mitigation in Fisheries and Aquaculture

The impacts of climate change can be addressed through adaptation and mitigation. The costs and benefits of adaptation are essentially local or national, while the costs of mitigation are essentially national whereas the benefits are global. Some activities generate both mitigation and adaptation benefits, for example, the restoration of mangrove forests can protect shorelines from erosion and provide breeding grounds for fish while also sequestering carbon. Following potential adaptation strategies or measures have been developed through expert consultation during 7-9 April 2008 at Rome (FAO, 2008).

1. Adaptation strategies in fisheries

Adaptation strategies are location and context specific and, hence, difficult to model and predict. There is a need for implementation of adaptive holistic, integrated and participatory approaches to fisheries management, as required for an ecosystem approach. However, some examples of potential adaptation measures in fisheries have been given in the table:

Table 1: Examples of potential adaptation measures in fisheries

Impact of climate change on fisheries	Potential adaptation measures	Responsibility	Reactive/ Anticipatory
Reduced yield	Access higher value markets / shifting targeted species	Public / private	Either
	Increase effort or fishing power*	Private	Either
	Reduce costs to increase efficiency	Private	Either
	Diversify livelihoods	Private	Either
	Exit the fishery	Private	Either
Increased variability of yield	Diversify livelihood portfolio	Private	Either
	Design insurance schemes	Public	Anticipatory
Change in distribution of fisheries	Migration of fishing effort / strategies and processing / distribution facilities	Private / public	Either
Reduced profitability	Exit the fishery	Private	Either
Vulnerability of infrastructure and communities to flooding, sea level and surges	Add new or improved physical defences	Private / public	Anticipatory
	Managed retreat/ accommodation	Private / public	Either
Rehabilitate infrastructure, design disaster response	Private/public	Reactive	
	Integrate coastal management	Public	Anticipatory
	Set up early warning systems, education	Public / private	Anticipatory
Increased dangers of fishing	Set up weather warning system	Public	Anticipatory
	Invest in improved vessel stability / safety / communications	Private	Anticipatory
Influx of new fishers	Support existing local management in stitutions, diversify livelihoods.	Public	Either

* May risk exacerbating over exploitation.

2. Adaptation measures in aquaculture

An ecosystem approach toaquaculture (EAA) management would be a most effective thematic adaptation measure. Adaptations include changing to less carnivorous species, genetic improvements, feed source diversification, better formulation, quality control and management.

The Table below summarizes most relevant specific adaptation measures for aquaculture.

Table 2: Climate change-related impacts and potential adaptation measures in aquaculture

Climatic change element	Impacts on aquaculture or related function	Adaptive measures
Warming	Raise above optimal range of tolerance of farmed species	Use better feeds, more care in handling, selective breeding and genetic improvements for higher temperature tolerance (and other related conditions).
	Increase in growth; higher production.	Increase feed input; adjust harvest and market schedules.
	Increase in eutrophication and upwelling; mortality off armed stock.	Improve planning and siting to conform to CC predictions; establish regular monitoring and emergency procedures.
	Increase virulence of dormant pathogens and expansion of new diseases	Focus management to reduce stress; set up biosecurity measures; monitor to reduce health risks; improve treatments, management strategies; make genetic improvements for higher resistance.
	Limitations on fish meal and fish oil supplies/ price	Identify fish meal and fish oil replacement; develop new forms of feed management, make genetic improvement for alternative feeds; shift to non-carnivorous species; culture bivalves and seaweeds wherever possible.
Sea level rise and other circulation changes	Intrusion of salt water	Shift stenohaline species upstream; introduce marine or euryhaline species in old facilities.
	Loss of agricultural land	Provide alternative livelihoods through aquaculture, building capacity and infrastructure.
	Reduced catches from coastal fisheries, seed stock	Make greater use of hatchery seed; protect nursery habitats; develop/use formulated pellet disruptions, reduced options for aquaculture feeds; income loss to fishers feeds (higher cost but less environmentally degrading); develop alternative livelihoods for suppliers.
	Increase of harmful algalblooms (HABs)	Improve monitoring and early warning systems; change water abstraction points where feasible.
Acidification	Impact on calcareous shell formation/deposition	Adapt production and handling techniques;move production zones.
Water stress and drought conditions mariculture	Limitations for freshwater abstraction	Improve efficacy of water usage; encourage non-consumptive water use in aquaculture, e.g. culture based fisheries; encourage development of where possible.

(*Contd.*)

Climatic change element	Impacts on aquaculture or related function	Adaptive measures
	Change in water-retention period (inland systems reduced, coastal lagoons increased)	Use different/faster growing fish species; increase efficacy of water sharing with primary users, e.g. irrigation of rice paddy; change species in lagoons.
	Reduced availability and period change of wild seedstocks	Shift to artificially propagated seed (extracost); improve seed quality and production efficiency; close the life cycle of more farmed species.
Extreme weather events	Destruction of facilities; loss of stock; loss of business; mass scale escape with the potential to impact on biodiversity	Encourage uptake of individual/cluster insurance; improve siting and design to minimize damage, loss and mass escapes; encourage use of indigenous species to minimize impacts on biodiversity, use non-reproducing stock in farming systems.

3. Adaptation measures in post-harvest, distribution and markets

Both capture fisheries and aquaculture feed into diverse and spatially extensive networks of supplyand trade that connect production with consumers, adding significant value and generatingimportant levels of employment. To some extent, this system can be used to provide an importantmediation and buffering function to increasing variability in supply and source location, but directimpacts will also affect its ability to do so. A range of issues and adaptation measures can beconsidered as given in Table 3.

Table 3: Climate change-related impacts potential adaptation in post-harvest/distribution

Impact on postharvest, distribution/markets	Potential adaptation measures	Responsibility	Reactive/ Anticipatory
Reduce costs to increase efficiency	Source products more widely, change species, add value, reduce losses.	Private	Either
	Develop more flexible location strategies to access materials.	Private/public	Either
	Improve communications and distribution systems	Public/private	Either
	Reduced or more variable yields, supply timing	Private	Either
	Diversify livelihoods	Private	Either
Temperature, precipitation, other effects on post-harvest processes	Change or improve processes and technologies	Private/public	Either
	Improve forecasting, information	Public/private	Either
Vulnerability of infrastructure and communities to extreme events	Add new or improved physical defences, accommodation to change.	Private/public	Either
	Rehabilitate infrastructure, design disaster response	Private/public	Reactive
	Set up early warning systems, education	Private/public	Anticipatory
Trade and market shocks	Diversify markets and products.	Public/private	Either
	Provide information services for anticipation of price or market shocks.	Public/private	Anticipatory

Possible Solutions

Ecosystem approaches to fisheries (EAF) and to aquaculture (EAA) that embed precautionary approach applications within integrated management (IM) across all sectors have the potential to increase ecosystem and community resilience and provide valuable frameworks for dealing with climate change. This would create flexible management systems and support decision-making under uncertainty. However, it would require rapid adjustment of management tools and regulations as necessitated by changed conditions or circumstances. Following are some of the possible solutions:

- Active participation at global and regional level, to ongoing debate and collaboration, to obtain the best possible information of fisheries-related impacts.
- Allocating research funds to analyze local and regional potential changes in resource magnitude and composition and likely socio-economic impacts.
- Sharing information obtained with the sector on potential changes, their scale and possible effects on resources and fisheries.
- Establishing institutional mechanisms to enable or enhance the capacity of fishing interests (fleets and other infrastructures) to move within and across national boundaries as a consequence of changes in resources distribution. This implies developing bilateral agreements.
- Preparing contingency plans for segments of the sector that might not be able to move, particularly for disadvantaged areas and small-scale fishers lacking mobility and alternatives.
- Developing effective national and international scale resource management regimes and associated monitoring systems to facilitate adaptation of exploitation regimes in a shifting environment.
- Strengthening regional fisheries management organization and other mechanisms to deal with cross-border stocks.
- Integrating fisheries management into coastal areas management to ensure that fisheries needs are taken up when dealing with protection of coastal areas from sea level rise, etc.
- Analyzing aquaculture sustainability in an ecoregional context, forecasting changes in productivity or resistance and in required related changes in culture systems, cultured species or delocalization of productive systems.
- Fostering interdisciplinary research, with scientists meeting periodically to exchange information on observations and research results, and meeting with managers to ensure the proper interpretation of results and the relevance of research
- Foreseeing and planning infrastructure adaptations.
- Adoption of Integrated approaches i.e., IFS (Integrated Farming System Models such A^2, A^3, A^4; A=Aquaculture, A=Agriculture, A=Animal Husbandry, A=Apiculture).
- Creation of water structure in the seasonal rivers / rivulets both for irrigation & aqua-bio-resources.
- Last but not the least, the most important aspect is that the responsibility of conservation of aquatic bio-resources rests with ecologists, taxonomists,

fish breeders,geneticists,law enforcingagencies and allother connected directly or indirectly with exploitation of aquatic resources.

Conclusions

From the above discussions and analysis following points may be concluded for our future guidance:

- Overall economy of the country is vulnerable to climate change impact, which could affect our food security and level of poverty by elevating stress on fisheries production.
- But the detailed effects on climate change affecting the fisheries sector is yet to be worked out carefully and methodically.
- It might bring economic hardship or might increase the opportunity to higher fish production through adaptive measure and increase the economic growth.

References

Aggarwal D., and Lal, M., 2007. Vulnerability of the Indian coastline to sea level rise" (PDF). *SURVAS (Flood Hazard Research Centre)* http://www.survas.mdx.ac.uk/pdfs/3dikshas.pdf. Retrieved 2007-04-05.

Ahmed, Ahsan; Koudstall, Rob; Werners, Saskia 2006. Key Risks. Considering Adaptation to Climate Change Towards a Sustainable Development of Bangladesh". http://www.mungo.nl/CC_Bangla.htm. Retrieved 2008-10-18.

Ali, A., 1999. Climate change impacts and adaptation assessment in Bangladesh. Climate Research12:109-116.

Allison, E. H. *et al.*, 2005. Effects of climate change on the sustainability of capture and enhancement fisheries important to the poor: analysis of the vulnerability and adaptability of fisherfolk living in poverty, London, Fisheries Management Science Programme MRAG/DFID, Project no. R4778J. Final Technical Report, 164 pp.

Allison, E.H., *et al.*, 2007. Enhancing the resilience of inland fisheries and aquaculture systems to climate change. Journal of Semi-Arid Tropical Agricultural Research 4 (1).

Allison, E.H., *et al.*, 2009. Vulnerability of national economies to the impacts of climate change on fisheries. Fish and Fisheries10(2):173–96. doi:10.1111/j.1467-2979.2008.00310.x. http://www. imcsnet. org/imcs/ docs/vulnerability_of_fisheries.pdf. bass: implications for red-sore disease, Trans. Am. Fish. Soc., 109: 532-536.

Brander, K. M., 2007. Global fish production and climate change. *Proc. Natl. Acad. Sci. U.S.A.* 104 (50): 19709–14. doi:10.1073/pnas.0702059104. PMC 2148362. PMID 18077405. http://www. pnas.org/ cgi/ pmidlookup?view=long&pmid=18077405.

Cairns, M. A., Ebersole, J. L., Baker, J. P., Wigngton, P. J. Jr., Lavigne, H. R., and Davis, S. M., 2005. Influence of summer stream temperatures on black spot infestation of juvenile coho salmon in the Oregon Coast Range, Trans. Am. Fish. Soc., 134: 1471-1479.

Cheung, W. W. L., *et al.*, 2009. Redistribution of Fish Catch by Climate Change. A Summary of a New Scientific Analysis. Pew Ocean Science Series. http://www.seaaroundus.org/ClimateChange/images/Pew%20OSS%20 Final%20climate%20change%20and%20fisheries.pdf.

Coral Reef Targeted Research, 2008. Climate change: It's now or never to save coral reefs CFTR Advisory Panel 2 Issue 1.

Dasgupta, S., 2007. Warmer Tibet can see Brahmaputra flood Assam. *Times of India* (Times Internet Limited). http://timesofindia.indiatimes.com/NEWS/ India/Warmer_ Tibet_ can_see_Brahmaputra_flood_Assam/articleshow/1556649.cms. Retrieved 2007-03-18.

Doney, S. C., 2006. The Dangers of Ocean Acidification. Scientific American. http:// www.precaution.org/lib/06/ocean_acidification_from_c02_060301.pdf.

Esch, G. W. and Hazen, T. C., 1980. Stress and body condition in a population of largemouth

FAO, 2007. Building adaptive capacity to climate change. Policies to sustain livelihoods and fisheries

FAO, 2008. Report of the FAO Expert Workshop on Climate Change Implications for Fisheries and Aquaculture Rome, Italy, 7–9 April 2008. FAO Fisheries Report No. 870.

FAO, 2009. Climate change implications for fisheries and aquaculture. Overview of current scientific knowledge Fisheries and Aquaculture Technical Paper 530, Rome.

FAO, 2009. The State of World Fisheries and Aquaculture Rome.

Ficke, A. D., Myrick, C. A. and Hansen, L. J., 2007. Potential impacts of global climate change on freshwater fisheries. *Fish Biology and Fisheries*17(4):581–613. doi:10.1007/s 11160-007-9059-5. http://assets.panda.org/downloads/fwfishreport902nov05.pdf.

Granath, W. O. Jr., and Esch, G. W.,1983. Survivorship and parasite- induced host mortality among mosquio fish in a predator- free, North Carolina cooling reservoir, Am. Midland Naturalist, 110: 314-323.

Gray, D. M. and Prowse, T. D., 1993. Snow and floating ice. In: D.R. Maidment (ed.). Handbook of Hydrology, pp. 7.1–7.58. McGraw-Hill.

Handisyde, N., *et al.*, 2006. The Effects of Climate change on World Aquaculture: A global perspective. Department for International Development UK. http://www.aqua.stir.ac.uk/ GISAP/pdfs/ Climate_ full.pdf.

Harrabin, R., 2007. How climate change hits India's poor. BBC News. http://news.bbc.co.uk /2/ hi/south_asia/6319921.stm.

Harvell, C. D., Kim, K., Burkholder, J. M., Colwell, R. R., Epstein, P. R., Grimes, D. J., Hofmann, E. E., Lipp, E. K., and Osterhaus, A. D., Overstreet R. M., 1999. Emerging marine diseases- climate links and anthropogenic factors, Science, 285:1505-1510.

Harvell, C. D., Mitchell, C. E., Ward, J. R., Altizer, S., Dobson, A. P., Ostfeld, R. S. and Samuel, M. D., 2002. Climate warming and disease risks for terrestrial and marine biota, *Science*, 296(5576): 2158-2162.

Hiner, M. and Moffitt, C. M. 2001. Variation in infections of *Myxobolus cerebralis* in fieldexposed cutthroat and rainbow trout in Idaho, J. Aquat. Anim. Hlth, 13: 124-132.

IPCC, 2001. Climate Change 2001: The Scientific Basis. Contribution of Working Group I to the Third Assessment Report of the Intergovernmental Panel on Climate Change. J.T. Houghton, Y. Ding, D.J. Griggs, M. Noguer, P.J. van der Linden, X. Dai, K. Maskell and C.A. Johnson (eds.). Intergovernmental Panel on Climate Change. Cambridge University Press, 881pp.

IPCC, 2007. Summary for Policymakers. In: Change 2007: The Physical Science Basis. Contribution of Working Group I to the Fourth Assessment Report of the Intergovernmental Panel on Climate Change. Solomon, S., D. Qin, M. Manning, Z. Chen, M. Marquis, K. B. Averyt, M. Tignor and H.L. Miller (eds.). Cambridge University Press, Cambridge, United Kingdom and New York, NY, USA. (Hereafter abbreviated to IPCC AR4 – WG1 – SPM) Table SPM-3, page 13.

Klyashtorin L. B., 2001. Climate change and long-term fluctuations of commercial catches: the possibility of forecasting Technical paper 410, FAO fisheries, Rome. ISBN 9789251046951

Kocan, R., Hershberger, P., Sanders, G., and Winton, J., 2009. Effects of temperature on disease progression and swimming stamina in Ichthyophonus-infected rainbow trout, *Oncorhynchus mykiss* (Walbaum), J. Fish Dis., 32(10): 835-43.

Lau, W. K. M., 2005. Aerosols may cause anomalies in the Indian monsoon (php). The Climate and Radiation Branch at NASA's Goddard Space Flight Center. NASA. http://climate.gsfc.nasa.gov/ viewImage. php?id=110. Retrieved 2007-04-17.

Magnuson, J. J., Robertson, D. M., Benson, B. J. Wynne, R.H., Livingstone, D. M., Arai, T., Assel, R.A., Barry, R.G., Card, V., Kuusisto, E., Granin, N.G., Prowse, T.D., Stewart, K. M. and Vuglinski, V. S., 2000. Historical trends in lake and river ice cover in the Northern Hemisphere. Science, 289:1743–1746.

Marcogliese, D. J., 2008. The impact of climate change on the parasites and infectious diseases of aquatic animals, Rev. sci. tech. Off. int. Epiz., 27(2):467-484.

MERIC, 2008. Bangladesh. http://www.ded.mo.gov/ researchandplanning/ indicators/ international/ cty5380.stm.

Mohanty ,B., Mohanty, S., Sahoo, J. and Sharma, A. 2010. Climate Change: Impacts on Fisheries and Aquaculture, Climate Change and Variability, Suzanne Simard (Ed.), ISBN: 978-953-307-144-2, InTech, Available from: ttp://www.intechopen.com/ books / climate-change-and-variability/climate-changeimpacts-on-fisheries-and-aquaculture

Normile, D., 2000. Some coral bouncing back from El Niño. *Science* 288 (5468): 941–942.doi: 10.1126/ science. 288.5468.941a.PMID 10841705. http://www.scienceonline.org/cgi/ content/summary/288/ 5468/ 941a. Retrieved 2007-04-05.

PACFA, 2009. Fisheries and Aquaculture in a Changing Climate. http://www.fao.org/climatechange/ 17789-1-0.pdf.

Pandian P. P., 2011. Status of efforts in addressing potential impacts of climate change on fisheries and aquaculture in India", Pandian P. P. Dy. Commissioner (Fisheries), Govt of India, Ministry of Agriculture, Dept. of Animal Husbandry, Dairying & Fisheries, New Delhi 30.05.2011, Hits- 245

Pant, G. B., 2003. Long-term climate variability and change over monsoon Asia (PDF). Journal of the Indian Geophysical Union7(3):125-134. http://www.igu.in/7-3/2GBpant.pdf. Retrieved 2007-03-24.

Sarwar G.M. 2005. Impacts of Sea Level Rise on the Coastal Zone of Bangladesh. Lund University. http://www.lumes.lu.se/database/alumni/04.05/theses/golamsarwar.pdf. [Masters thesis Lay summary].

Sethi, N., 2007. Global warming: Mumbai to face the heat. Times of India. http://timesofindia. indiatimes.com/ NEWS/India/Global_warming_Mumbai_to_face_the_heat/articleshow/ msid-1556662,curpg-1.cms. Retrieved 2007-03-18.

Sriskanthan, G. and Funge-Smith, S. J., 2011.The potential impact of climate change on fisheries and aquaculture in the Asian region. FAO Regional Office for Asia and the Pacific, Bangkok, Thailand. RAP Publication 2011/16, 41 pp.

Tester, P. A., Feldman, R. L., Nau, A. W., Kibler, S. R., and Wayne Litaker, R., 2010. Ciguatera fish poisoning and sea surface temperatures in the Caribbean Sea and the West Indies. Toxicon.

Toman, M. A., Chakravorty, G. 2003. India and Global Climate Change: Perspectives on Economics and Policy from a Developing Country, Resources for the Future Press.

Tops, S., Hartikainen, H. L., and Okamura, B., 2009. The effects of infection by *Tetracapsuloides bryosalmonae* (Myxozoa) and temperature on *Fredericella sultana* (Bryozoa). Int J. Parasitol., 39(9): 1003-1010.

UNDP. India and Climate Change Impacts. http://www.undp.org.in/index.php?option=com_ content&view= article&id=297&Itemid=466.

Union of Concerned Scientists, 2005. Early Warning Signs: Coral Reef Bleaching. http:// www.ucsusa.org/global_warming/science/early-warning-signs-of-global-warming-coral-reef-bleaching.html. Retrieved 2007-04-05.

United Nations Framework Convention on Climate Change, 1994. http://unfccc.int/essential_background/convention/background/items/1349.php.

World Bank – Climate Change Adaptation (website)

World Bank, 2006. Aquaculture: Changing the Face of the Waters: Meeting the Promise and Challenge of Sustainable Aquaculture.

World Bank, 2009. World Development Report 2010: Development and Climate Change. Chapter 3.

World Fish Center, 2008. The Millennium Development Goals: Fishing for a Future: Reducing poverty and hunger by improving fisheries and aquaculture.

www.intechopen.com

Xenopoulos, M. A., Lodge, D. M., Alcamo, J., Marker, M., Schulze, K. and Van Vuuren, P., 2005. Scenarios of freshwater fish extinctions from climate change and water withdrawal. Global Change Biology11(10):1557-1564.

❑❑❑

25

Impact of Climate Change on Fish Diseases and Immune Response and Some Control Methods for Common Diseases for Mitigation

Lopamudra Sahoo, C. Debnath, J. Parhi and M. Datta

Abstract

Aquaculture is facing the serious effects of climate change in the form of scarcity of water and shift in dissolved oxygen level leading to outbreak of diseases. In a study conducted in Farm facilities of ICAR, Tripura it was seen that rise in temperature do have significant effect on hematological and biochemical parameters of fish which are basic index of good health of any organism. Many different diseases occur in the farmers' pond and which can be controlled. So some basic diseases and their controlling strategies have been pointed out that can lead to better management of fish diseases in pond condition.

Keywords: Climate Change, Fish, Immune response, Diseases, Mitigation.

Introduction

Aquaculture is a booming sector addressing food security and economic growth for India. This sector essentially requires good quality water, but water is becoming a scarce commodity leading to availability of water as a serious concern for any farmer. While the positive benefits of climate change to aquaculture in the form of longer growing season is being seen as a possible advantage, at the same time decrease in availability of water continue to harrow the farmers. Disease per se is an indirect effect of climate change on aquaculture but which can lead to huge economic loss if left unaddressed. Increased water temperature leads to other associated physical changes in pond environment such as shift in dissolved oxygen level. These have been linked to more frequent algal blooms and increase in intensity and frequency of disease outbreaks.

Immune system of any organism combats against the disease causing agent warding off the infectious agent. To see the effect of climate change in view of temperature on the immune response, a study was conducted on the effect of seasonal variation of temperature and other water quality parameters on fish diseases.The work was started from beginning of summer of 2012 and data were recorded till November. The months were broadly divided into 3 seasons: Summer (March-June), rainy (July-September) and winter (October-November). In each month 10 fishes were sampled from ICAR farm facilities for the study.The water samples were collected from the different FSR ponds of ICAR at weekly intervals for analysis of different physic-chemical parameters such as water temperature, pH, dissolved oxygen and total alkalinity following the protocols of APHA. The levels of different hematological parameters like total erythrocyte and total leukocyte count were evaluated in this study. It was seen that TEC (total erythrocyte count) in summer season was significantly different from the TEC of rainy and winter season. Biochemical reaction in fish body could have quickened in an environment with optimum temperature range and increased oxygen supply leading to increased RBC count and hemoglobin concentration. Low temperature may reduce RBC products.In our study on total leukocyte count we found that seasonal variation had no significant effect on Total leukocyte count. But it was seen that TLC in summer season was higher than TLC in winter and rainy season. Higher WBC indicates higher disease resistance and in our study higher WBC in summer seasons could be explained that *L.rohita* exhibited better immunity in warmer months when the temperature was most optimum for its growth and activity than in colder months. In our study on seasonal variation of temperature on serum proteins we found significant effect of temperature on serum proteins. Dietary protein can affect the serum protein as well and with the same feed being provided throughout all the seasons, optimum activity leading to catabolism of proteins and thereby a significant reduction in serum proteins in summer season could be explained because of this in our study. In our study on serum albumin, significant difference in albumin values was found in all the seasons. Increase in albumin in rainy season may be due to the fact that induction of albumin synthesis in spawning time. At present it is seen that temperature is being beneficial enhancing the immune status of fish. But it is very early to say so. The study is going on and after some continuous data recording it will be possible to arrive at a definite conclusion.

Concept of Fish Health Management in Aquaculture

"Causes of diseases are reasonable and discoverable causes are removable"

Just as in land animals or in agricultural crops fishes and shellfishes living in the aquatic environment are susceptible to diseases whenever certain abnormal biological, physiological or environmental changes occur. The problem of diseases has assumed greater significance in the context of aquaculture of these organisms in different culture systems. The fish function as hosts to a wide spectrum of parasites and as intermediate hosts to a number of parasites of man and animals.

Environment

Poor water quality can lead to outbreak of diseases in culture ponds. In other words it can be said that the quality of water indicates the status of health of fish living in it. Hence the first step should be maintenance of best quality environment in the culture unit. To maintain ambient aquatic environment the following basic steps must be followed:

Removal of unwanted aquatic weeds: Unwanted aquatic weeds should be removed because it reduces the pond productivity. Removal can be done manually, chemically and biologically.

Liming: Liming helps in maintaining the pH of pond water and acting as barrier for many types of pathogens. Liming should be done periodically based on the pH of water.

Fertilization: In Integrated farming system, organic manures like cow dung, poultry manure are used commonly in fish ponds. But precaution should be maintained not to dump excess manure in the pond that can result in algal blooms thereby leading to depletion of oxygen and causing stress to fishes.

Fishes

While selecting candidate species for integrated farming system, care should be taken for maintaining appropriate stocking density, species combination etc. One of the criteria for selection in Integrated farming system is adaptability to crowded condition and relative resistance to diseases.

Indian Scenario

Fishes suffer from a wide variety of diseases of infectious and non- infectious etiology in their natural environment. Fish mortalities are due to bacteria, parasites, fungus, nutritional deficiencies, metabolic disorders and faulty environmental conditions like oxygen deficiency, high or very low |water temperature, chemical pollutants, overcrowding, mal-nutrition, algal blooms etc. However, fish mortality is not the only criterion to evaluate the effect of fish disease. Even the morbidity, which leads to weight losses and poor growth in surviving fish, contributes substantial losses to the farmers. Since last decade there is an increased realization of the necessity to avoid infectious agents to maintain healthy stocks to enhance production.

Important Bacterial Diseases of Fishes

1. Fin rot and tail rot

It is found in both young and adult fishes. The disease is contagious and is capable of causing immense damage. It is caused due to a mixed infection of *A. hydrophila*and *Pseudomonas fluorescens.*

Control measures: Bath treatment in 1:2000 copper sulphate for 1-2 minutes or painting of concentrated copper sulphate solution in the affected fish.

2. Dropsy

It is another important fish disease in India where rohu, catla and mrigal are affected mostly in the late winter. The epidermis and body cavities get filled with fluid and scales protrude out from their pockets leading to severe anaemic condition. Causative agent isolated in infectious dropsy of Catla catla are mixed infection of *A. hydrophila*and myxosporidian parasite and also *A. hydrophila*in conjunction with malnutrition in Rohu.

Control Measures: Treat the affected fishes fully using terramycin in the feed.

3. Eye disease of Catla

Causative organism identified as *A.liquifaciens*infecting the eyes of catla. The corneas of the eye gets vascularized leading to opacity and complete necrosis resulting in the death of the fish.

4. Corneal opacity in silver carp

Caused due to Gram positive bacterium, *Staphylococcus aureus*causing mass mortality.Similar eye disease in *Channamarulius*is also caused by *Staphylococcus.*

Treatment: Chloromycetin bath @8-10 mg/litre of water has been found effective in controlling the disease at an early stage. Disinfecting the environment with $KMnO_4$ at a dose of 0.1 0 ppm followed by liming @ 300 ppm check the disease.

5. Gill hyperplasia syndrome in common carp

Myxobacterial organisms have been attributed to a condition known as gill hyperplasia syndrome in common carp *(Cyprinuscarpio)*

6. Ulcerative disease

In catla bilateral ulcerations of the opercula and the head are observed. In several cases the ulcerations have been reported to have penetrated the opercular bones and cranium. *A. hydrophila*has been isolated in most of the cases, although several other bacterial forms were also present as secondary invaders. It is beyond doubt that in the ulcerative disease of fish these organisms might be playing an important role not only in necrotizing the muscle tissue but also damaging the internal organs viz. Kidney, liver and spleen. The causative organism has been identified as *Flexibactercolumnaris.* Topical application of potassium permanganate or short bath in 500 ppm of $KMnO_4$ has been found to be very effective in completely curing the disease.

7. Edwardsiellosis

Edwardsiellatarda has been isolated from the diseased fish showing anaemia, cutaneous lesions and gas filled abscesses in the muscle. Although the mortality was not too high, but the morbidity has been reported to be extreme. Reports are there for *Edwardsiella*causing large scale mortality in spawns of IMC. The young ones showed anatomical *deformities* and greyish white discolouration of the body within 48 hours post-hatching and died within 72 hours. Although treatment with lodophor has been found to be effective, water quality improvement in the hatchery is the most essential component for keeping this disease away.

Important Parasitic diseases

Protozoan diseases

1. **Myxosporidia** are one of the most important groups of pathogen capable of producing diseases in fish causing heavy loss in the juvenile. Myxobolus cysts of varying sizes have been reported on the gills and kidney of *Catla catla.* Larger cysts are located at the distal end of the gill filament where as smaller ones is seen at the proximal end. Infection damages the respiratory surface of the gill and excretory tubules of the kidney. Diagnosis of the disease can be made on the basis of gross appearance of the pin head sized greyish cysts and large number of myxosporidian spores under the microscope.
2. **Ciliophora**: The other groups of protozoa infecting fish belong to ciliophora. The carp fry and fingerlings sometime carry heavy infection of the gills and skin with Trichodinaindica. In heavy infection there is excessive mucus

secretion on the gills which hampers respiration and the fish start surfacing. Some species of trichodina occur inside the urinary bladder. Under microscope they appear as circular transparent organisms with an internal disc-like structure. Fish affected with trichodiniasis eventually turn sluggish, lose weight and become moribund. There may be excessive mucus secretions and epidermal necrosis. A short bath in 2-3% salt solution for 5 minutes can control the infection effectively.

3. **Whirling disease** in Cirrhinusmrigala (Ham) has been attributed to a protozoan parasite Myxosomacerebralis. The gross changes are characterized by deformed spinal column.
4. **Trematodes**: Fishes are infected on their gills and skin by many species of monogenetic and digenetic trematoade and cestodes. The important monogenetic species are Gyrodactylus and Dactylogyrus. They remain attached to the skin or gills by the help of anchors which causes wounds. Their presence causes hypersecretion of mucous.
5. **Ichthyophthiriasis** or "white spot' disease causes large scale mortality in nursery and rearing ponds in different parts of the country. Presence of numerous pinhead shaped white spots on the skin, fins and gills are the common diagnostic features. On microscopic examination these spots look like blisters or nodules. Alternate bath in 1:2000 acetic acid solution followed by bath in 2% sodium chloride solution has been found effective. Some workers have advocated dip treatment in 5% salt solution followed by bath in 1:5000 formalin.

Helminthic Parasites

1. ***Acanthocephala*** have been recorded from major carps and cat fishes in India. The parasite causes lesions in the intestinal wall by its proboscis.
2. **Digenetic trematode** generally infects the alimentary canal but are rarely harmful. However, in the metacercarial stage they are harmful as they get encysted in any part of the body, i.e. skin, muscle, abdominal cavity, liver, eyes and brain. In these cases fish serve as intermediate host while the final host is generally a fish-eating bird. In many.carp fingerlings the 'black spot disease' is due to metacercaria of diplostomatidae. Catla and silver carp are the most susceptible species.
3. **Cestodes** are also internal parasites in alimentary canal. Their larval forms are more harmful than the adults. They are found more in catfishes than in carps. The infected fish show softening of the muscle tissue probably due to accumulation of water. In case of heavy infestation the abdomen bursts resulting in death of the fish.

Crustacean parasites

1. **Parasitic copepoda** are external parasites infecting skin or gills or bucco-pharyngeal cavity. Amongst them species of two genera *Lerneaand Ergasilus* are reported from Indian carps. Lernea is reported to infect fingerlings and adults of major carps especially Catla. The parasite penetrates the muscles and reaches the blood vessels as it feeds on the hosts blood. The fish thus become anaemic and weak. Larger number of these parasites may cause serious problems while a few may not cause any harm, *Ergasilus* sp. Infect the gill lamellae. They attach themselves to the gill by the help of their hook-like second antenna. Due to this there are wounds and haemorrhage in the gill tissue.
2. Heavy mortality due to Argulosis or fish lice infestation has been reported in Indian major carps. Several species of Argulus have been reported to infect cultured fishes. These parasites inhabits the skin, fins and gills of the host and cause extensive pathological lesions in the skin showing circular haemorrhagic patches which become ulcerated. Mucous cells proliferate and copious mucous is produced. These lesions promote secondary infections like fungi and bacteria. Rohu has been observed to be the most susceptible species among the Indian carps. These wound later get infected with pathogenic bacteria and fungi. Weekly applications of malathion @ 0.25 ppm for three consecutive doses. Bath in potassium permanganate @ 500 ppm/minute once a week with the combination of malathion treatment is more helpful in the eradication of parasite.

Fungal Diseases

Saprolegniaparasiticain the fry and fmgerlings of major carps. The disease in fish is characterized by a white to brown cotton like growth consisting of colonies of mycelium and filaments which appear as small to large patches on various parts of the body like fins, gills, mouth, eyes or muscle. The infection starts due to netting injury and over-crowding or lesions caused by other diseases. Saprolegnia often infects the fertilized eggs. They attack the dead eggs first and then spread to the viable living eggs resulting in their spoilage.

Infection of *Cirrhinamrigala* with *Aphanomycespisci* where the scales became black and later fell off. The hyphae of this species were not visible like those *Saprolegnia.* Branchiomyces infection causing gill rot' in fish. These pathogens invade the blood vessels of gills causing necrosis of the surrounding tissues also.

Several treatment measures to counter fungal infection in fish which include dip treatment in 3% common salt solution or 1:2000 copper sulphate solution or 1:1000 potassium permanganate solution for 5-10 minutes. A flush treatment for 1 hour in malachite green (0.1 ppm) is found to be quite effective in curbing fungal infection.

Treatment Chart for commonly encountered disease in Integrated farming system with symptoms

Symptoms	Diagnosis	Treatment
Grayish-white film on skin, damaged fins, ulcers, yellow to gray patches on gills, tissue on head may be eaten away.	Columnaris (Cotton Wool Disease)	Topical application of potassium permanganate or short bath in 500 ppm of KMnO4
Swelling of abdomen, raised scales around swollen area.	Dropsy may be caused by internal bacterial infection	OTC @75mg/kg of feed for 5 days
Ragged or decaying fins.	Finrot	Check pH and correct as needed. If pH is below normal level then treat with lime @50kg/ha followed by bleaching powder @0.5ppm after one week. If level is normal, use OTC antibiotic in feed. Bath treatment in 1:2000 copper sulphate for 1-2 minutes or painting of concentrated copper sulphate solution in the affected fish.
Sluggishness, lack of appetite, fin damage, reddish discoloration, bulging eyes, clamped fins.	Bacterial Septicemia	Antibiotic treatment like OTC in food form is required.
White or gray fungus on eyes.	Fungal disease	Immerse in 3-5% formalin solution or 1-3% salt solution for 20 minutes
White or gray patches resembling cotton, excess mucus.	Mouth or Body Fungus	Immerse in 3-5% formalin solution or 1-3% salt solution for 20 minutes
White cotton-like patches on fins, body, or mouth.	True Fungus (Saprolegnia)	Malchite green @50ppm ie.5-6gm/100l as dip treatment for 30 sec. For bath treatment 100-200mg/100l for 30-60 mints or 10mg/100l for 3-4 days (prolonged treatment). If outbreak is in pond then sprinkle 0.15-0.2ppm malachite green in pond water.
Small string-like worms visible on fish, or burrowed in skin.	Anchor Worm	If the fish is already infected with Lernaea then bath the fingerlings with potassium permanganate solution @10-20ppm for 30mints or change the water quality suddenly like apply 400kg of fermented cow dung or pig manure.
Bluish-white film on body, strained breathing caused by gill damage, peeling skin.	Chilodonella	Salt treatment For bath 1-1.5kg/100l for 1 hour, for dip 2-3kg/100l for 30 sec. for prolonged treatment use 500gm/100l.

(*Contd.*)

Symptoms	Diagnosis	Treatment
White film, reddened areas on body, abnormal swimming, scratching, folded fins.	Costia (Slime Disease)	Must be treated quickly. Use Salt treatment
Sluggishness, flashing, spider web lesions on skin, color loss, reddened fins, drooping fins, fin damage.	Skin Flukes (Gyrodactylus)	Formalin @2.-25ml/100l for 30-60 mints. For prolonged treatment 2ml/100l for 12-24 hrs.
Scratching, white salt-like spots starting on head and spreading over whole body, rapid breathing, cloudiness on eyes or fins	Ich (Ichtyophthirius) very common	Disinfect fingerlings with table salt 1-2% for 2-10 mints or with copper sulphate solution @8ppm for 20-30 mints or with formalin solution @ 30ppm.
Scratching, small worms hanging from body.	Leech	Salt treatment.
Scratching, green to brown lice (up to inch) visible on skin.	Lice	Disinfect fingerlings with table salt 1-2% for 2-10 mints or with copper sulphate solution @8ppm for 20-30 mints or with formalin solution @ 30ppm.
Cloudy appearance on skin, red patches on skin where parasite has bitten.	Trichodina predominately freshwater	Salt treatment.
Red or bloody gills, gasping for air.	Ammonia Poisoning	No treatment. Regular water testing and maintenance will prevent it.
String of feces hanging from fish, swollen abdomen, sluggishness, disinterest in food, off-balance swimming.	Constipation	Stop feeding for 2-3 days and continue with a more varied diet including live and plant-based foods.
Small white spots that get larger over time possibly with black streaks.	Fish Pox	No treatment. Keep up on water maintenance and symptoms should cease after about 10-12 weeks.
Reddening on or under skin, sudden abnormal behavior.	Inflammation	OTC antibiotic treatment.
Unusally bulging of one or both eyes.	Pop-eye (Exophthalmia)	OTC medication for bacterial infections and/or parasites. Check for other symptoms of bacterial or parasitic infections.
Fish struggles to swim, may float with head tipped down, or have difficulty surfacing, no balance, etc. May occur after eating.	Swim Bladder Disease	Stop feeding for 3-4 days. If symptoms persist, feed the affected fish a small amount of fresh spinach or a green pea without the skin (laxatives).
Swelling or distention for internal tumors, external can be seen growing on skin.	Tumors	Usually incurable.
Sluggishness, lack of appetite, open sores with red edges, possible fin rot.	Ulcers	OTC medication for bacterial infections.

Epizootic Ulcerative Syndrome (BUS):One common feature of the disease is that it initially affects the bottom dwelling species like murrels followed by catfishes and weed fishes. Subsequently other fishes including Indian major carps are affected. The lesions start as small grain to pea-sized haemorrhagic spots over the body, which ultimately turn into big ulcers of the size of a coin with greyish slimy central necrotic area surrounded by a zone of hyperemia. The disease affects the fish to such an extent that they start rotting while still alive and eventually die. A number of bacteria viz. *Aeromunashydrophila. A. punctata, Flavohaclerium* sp., *Pseudomonas* sp., *E. tarda, V. parahaemolyticus*and *Streptococcus* sp., have been isolated from the affected specimens. Lime can be applied to prevent the disease. To treat the disease Cifax can be applied @0.1ppm.

Common Symptoms

1. **Bacterial Infections:** Inactivity, loss of color, frayed fins, bloated body, cloudy eyes, open sores, abscesses, red streaks throughout body, reddening or inflammation of the skin, fins or internal organs, bulging eyes, difficulty breathing.

2. **Fungal Infections (often secondary to another type of illness):** Erratic swimming, darting, scratching, visible cotton-like tufts on skin, eyes, or mouth.

3. **Parasitic Infections:** Inactivity, loss of appetite, excess mucus or film on body, visible spots or worms, rapid breathing, scratching.

References

Fish disease management in integrated farming system, Mr. Sudhansu Mishra, CIFRI, ICAR

Fish diseases in India, their causes and control measures, Dr. S.C. Mukherjee, Former Director, Central Institute of Fisheries Education, Mumbai.

www.fishnet.org

❑❑❑

26

Prediction of *Haemonchus contortus* Infections in Goats in a Subtropical Hill Area – A preliminary Study

R. Laha, M. Das and A. Goswami

Abstract

Gastrointestinal parasitic infections particularly Haemonchus contortus infection in goats is very common and found as major nematode infections in farm condition. To chalk out the control programme and to prevent the initiation of anthelmintic resistant parasites, it is essential to know well in advance the prevalence of H. contortus infection. During the year 2009, a study was undertaken on H. contortus infection in the goat farm of ICAR Research Complex for NEH Region, Umiam, Meghalaya, which is a subtropical hill area and bioclimatograph was prepared to predict the prevalence of H. contortus infection in goats of this area. Meteorological data of this area was collected to visualize the effect of temperature, rainfall and humidity. Bioclimatographs were prepared in which total rainfall (TRF) or relative humidity (RH) is plotted against the maximum temperature (T-max) or minimum temperature (T-min) for the months and the resultant points are joined by a closed curve. The lines constituted the windows on these bioclimatographs, indicating the limit of the climatic conditions most suitable for development and survival of exogenous stages of gastrointestinal nematodes. The limits of suitable climatic conditions were taken as total monthly rainfall >50mm, relative humidity more than 50% with average monthly maximum temperature ranging from 18-37°C. The resultant bioclimatographs were compared with the real time incidence of the H.

contortus *infections as observed in faecal egg counts and larval culture. The favourable months predicted by bioclimatographs were March to November; January to December; May to September and June to September when plotted against T-max v/s TRF , RH v/s T-max, RH v/s T-min and TRF v/s T-min respectively. The observed months for higher incidence* H. contortus *were April to September. It can be concluded that like other agroclimatic zones, this subtropical hill area is also favourable for H. contortus infections particularly from the months of April to September and there is a need for further study to confirm this prediction.*

Keywords: *Haemonchus contortus*, Prediction, Subtropical hill area.

Introduction

Gastrointestinal parasitic infections particularly *Haemonchus contortus* infection in goats is very common and found as major nematode infections in farm condition. *H.contortus* is an abomasal dwelling gastrointestinal nematode parasite of goats which has been identified as one of the most pathogenic parasites (Soulsby, 1986). It is commonly known as 'stomach-worm' or 'wire worm' of ruminants with a size of 2-3 cm. inlength. These parasites are responsible for loss of productivity in goats for the blood feeding habit of the adult and the late larval stages of the parasites. An animal which is infected with 5,000 numbers of adult parasite may lose about 250 ml of its blood daily (Urquhart, 1996). This can result in anaemia leading to reduced weight gain, meat production and even cause death of young animals. To chalk out the control programme and to prevent the initiation of anthelmintic resistant parasites, it is essential to know well in advance the prevalence of *H. contortus* infection in a particular. So, to predict the prevalence of *H. contortus* infection in goats in a subtropical hill area, the study was undertaken, as there is a lack of such type of study in this area.

Material and Methods

The study was undertaken by collection of faecal samples of goats maintained in the goat farm of ICAR Research Complex for NEH Region, Umiam, Meghalaya, India during the year 2009. This farm is situated at an altitude of about 900 m above mean sea level where average monthly minimum and maximum temperature were observed as 7.8°C and 28.9°C, respectively, during the study period. The average monthly relative humidity prevailed during the study period were recorded as 67.0% (minimum) to 91.5% (maximum) with average annual total rain fall 2400.4 mm. All meteorological data like average minimum and maximum temperature, relative humidity and rainfall were collected from the Institute Meteorological Station located within the Institute Campus. Meteorological data of this area were used to visualize the effect of temperature, rainfall and humidity.

Rectal faecal samples of goats were collected manually in suitable containers and labeled carefully and these faecal samples were examined to know the presence of nematode eggs. The eggs per gram of faeces (EPG) for nematodes were estimated by the modified McMaster techniques. Gastrointestinal nematode larvae were identified after culture of the pooled faecal samples in B.O.D. at 28°C for 7 days.

For prediction or forecasting of *H. contortus* infection in this goat farm bioclimatographs were prepared as per methods described by Swarnkar and Singh (2011). In brief, Bioclimatographs for this year were prepared in which total rainfall (TRF) or relative humidity (RH) is plotted against the maximum temperature (T-max) or minimum temperature (T-min) for the months and the resultant points are joined by a closed curve. The lines constituted the windows on these bioclimatographs, indicating the limit of the climatic conditions most suitable for development and survival of exogenous stages of gastrointestinal nematodes. The limits of suitable climatic conditions were taken as total monthly rainfall >50mm, relative humidity more than 50% with average monthly maximum temperature ranging from 18-37°C. The resultant bioclimatographs were compared with the real time incidence of the *H. contortus* infections. The real time incidence of the *H. contortus* infections was estimated by collection of faecal samples of goats every month for a period of 12 months and culture of larvae as per methods mentioned above.

Results and Discussion

The results of monthly mean EPG along with meteorological data for the year 2009 are presented in Table 1. It could be observed from the table that gastrointestinal parasitic infections were present through out the year and the EPG count reached its peak in the month of June, dropped in the month of July, then increased up to the month of September and then declined.

The bioclimatograph for prediction of *H. contortus* for the year 2009 have been presented in Fig. 1 and Table 2. The favourable months predicted by bioclimatographs were March to November; January to December; May to September and June to September when plotted against T-max v/s TRF, RH v/s T-max, RH v/s T-min and TRF v/s T-min respectively, for the year 2009. After faecal culture, *H. contortus* has been found predominant species of nematodes. *H. contortus* has been found predominant species in goats of India (Dey *et al.*, 2008; Sharma *et al.*, 2009; Pant *et al.*, 2009). In temperate agroclimatic zone of India *H. contortus* has been recorded as highest prevalent species (Tariq *et al.*, 2010). In the present study, athough EPG was observed throughout the year but more EPG as well as larval culture indicated *H. contortus* observed from April to September. Thus for this agroclimatic zone we can predict *H. contortus* infection from April to September. Thus, observed months for higher incidence *H. contortus* were April to September for the year 2009. The role of bioclimatographs in

Table 1: Monthly prevalence of gastrointestinal parasites in goats and meteorological data during the year 2009

Month	Mean	FEC* (°C)	T_min (°C)	T_max (mm)	TRF RH (%)
January	862	7.92	21.5	0.0	79.2
February	849	9.29	20.0	0.0	76.7
March	460	12.3	26.9	40.2	70.1
April	800	16.7	28.4	56.3	67.0
May	2116	17.4	28.0	381.5	77.5
June	2250	19.5	28.7	352.3	84.9
July	1202	20.7	28.9	419.0	86.5
August	1440	19.6	26.5	452.8	91.5
September	1660	18.6	28.6	410.3	86.7
October	573	15.8	26.9	239.6	86.3
November	237	11.4	24.4	38.3	82.6
December	125	7.8	22.0	11.8	83.3
Total				**2400.4**	

*FEC = Faecal Egg Count

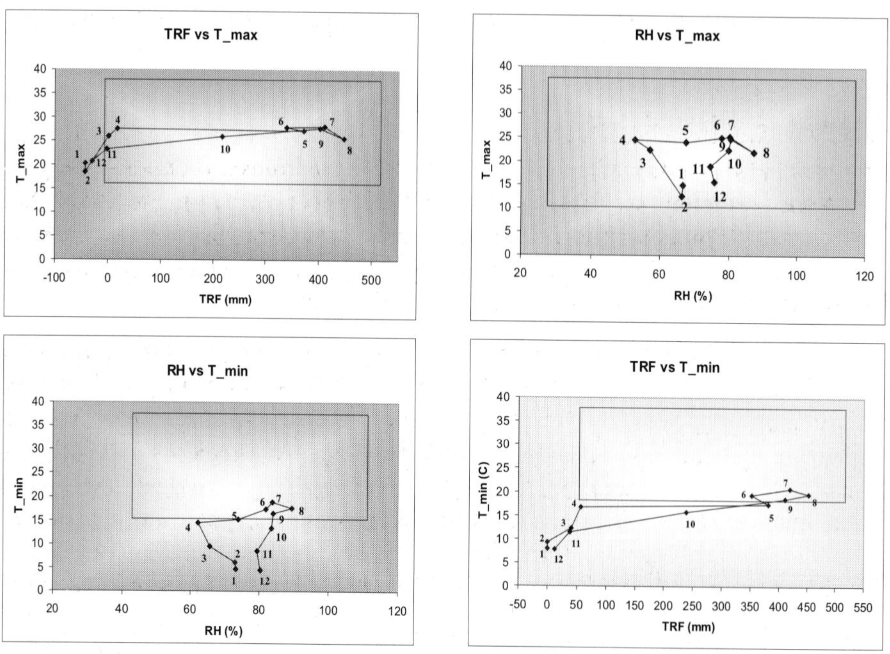

Fig.1: Bioclimatographs for *Haemonchus contortus* for the year 2009
RH = Relative Humidity; T-min = Temperature Minimum; T-max = Temperature Maximum; TRF = Total Rail Fall; 1 to 12 stands for month like 1 = January and 12 = December

Table 2: Comparative prediction of bioclimatographs of *H. contortus* in goats during the year 2009

Parameters of Bioclimatograph	Favourable Months predicted by bioclimatograph	Observed months for higher incidence of *H.contortus*
T-max v/s TRF	March- November	April-September
RH v/s T-max	January - December	April-September
RH v/s T-min	May - September.	April-September
TRF v/s T-min	June - September	April-September

forecasting of strongyle infection particularly H. contortus infection in sheep in India (Rajasthan) was reported by Swarnkar and Singh (2011). Swarnkar and Singh (2011) observed June to mid-September and late June to mid-August in semi-arid and arid zones of India respectively, are suitable for development, survival and dissemination of exogenous stages of *H. contortus*. In the present study, slight variations on monthly prediction of *H. contortus* by bioclimatographs with real time occurrence of *H. contortus* has been observed. In the study of Swarnkar and Singh (2011) the real time incidence was fully in agreement with prediction by bioclimatographs on longterm basis, but minor variation on real time incidence with predicted months by bioclimatographs on annual basis observed by them. So, there is a need for further study on long term basis to confirm this prediction.

Acknowledgements

We are thankful to Indian Council of Agricultural Research, New Delhi to carry out this research work under All India Network Programme on Gastrointestinal Parasitism. The facilities provided by Director, ICAR Research Complex for NEH Region, Umiam, Meghalaya, India, are thankfully acknowledged.

References

Dey, S., Sanyal, P.K., Mukherjee, K., Sarkar, A.K., Patel, N.K., Mandal, S.C., Pal, S. 2008. Caprine parasitic gastroenteritis in semi-organized farming conditions. J. Vet. Parasitol., 22:77-78.

Pant, K., Rajput, M.K.S., Kumar, J., Sahu, S., Rajkumari,V. Gangwar, P. 2009. Prevalence of helminthes in small ruminants in Tarai region of Uttarakhand. Vet. World, 2: 265-266.

Sharma, D.K., Agrawal, N., Mandal, A., Nigam, P. and Bhusan, S. 2009. Coccidia and gastrointestinal nematode infections in semi-intensively managed Jakhrana goats of semi-arid region of India. Trop. Subtrop. Agroecosystems., 11: 135-139.

Soulsby, E.J.L. 1982. Helminths, Arthropods and Protozoa of Domesticated Animals, Bailliere Tindall, London.

Swarnkar, C.P., Singh, D.K. 2011. Role of bioclimatographs in forecasting of strongyle infection in Rajasthan. Indian J. Anim. Sci., 81: 216–223.

Tariq, K.A., Chishti, M.Z., Ahmad, F. 2010. Gastro-intestinal nematode infections in goats relative to season, host sex and age from the Kashmir valley, India. J. Helminthol., 84:93-97.

Urquhart, G.M., Armour, J., Duncan, J.L., Dunn, A.M., Jennings, F.M. 1996. Veterinary Parasitology. 2nd. Edn., Blackwell Science, Oxford.

❑❑❑

Section E: ICT Application in Agriculture

27

Disease Prediction Models — Development and Validation

Suseelendra Desai

Abstract

The disease is an output of the interaction among host, pathogen, and the environment over a period of time. Given that a susceptible host and an aggressive pathogen are available, weather determines the intensity of the disease over time. To capture the dynamics of the interactions among these biophysical variables, efforts have been made in the past to understand the variables per se and their interactions through modeling approaches. The important weather parameters are temperature; rainfall/dew; leaf wetness period, wind, bright sunshine hours all of which are in turn influenced by soil properties. The important parameters of host that can influence disease development are susceptibility; growth stage and form; structure and density of the plant population and overall health. For pathogen, to cause disease, it should be present in the vicinity, pathogenic, adaptable with reproductive fitness and possess efficiency of survival and dispersal.

Availability of a reliable disease prediction model is a prerequisite for integrated plant disease management as these models will help in planning and execution of plant protection strategies through judicious resource management. Unfortunately, monsoon-dependent crop production systems face the challenge of unpredictable rainfall patterns that in turn decide on crop production systems of a given agro-ecological region. In this paper, an attempt has been made to deliberate on various aspects of disease prediction models, opportunities and constraints.

Keywords: Weather based forewarning system, Pathogen host-environment interaction, Crop microclimate and diseases.

Introduction

Despite our efforts since long, we are yet to develop suitable disease forewarning systems especially for monsoon dependent crop production systems. Early references to pest prediction in the Bible (Joseph's prediction for the pharaoh) indicate the importance that our forefathers placed on pest prediction, though as modern scientists we may have difficulty justifying such prediction methods and may question the historical accuracy of this account. The main purpose of plant disease forewarning systems is to help producer's decision-making process such as whether to grow a specific crop in an area or which planting material to be purchased or cost-benefits of fungicide application. The principle behind plant disease forecasting systems is to determine the risk that a disease will occur, or that the intensity of the disease will increase. A successful forewarning system must be reliable, simple, important for a given disease, useful, available, help in multiple-decision making, and cost-effective. Whether a predictive system is a table, a set of printed cards, or elaborate internet-based graphic systems, there is still a basic set of decision rules that should be evaluated in an objective manner. Many of these predictive systems are fallible, and sometimes give incorrect predictions. How often they are incorrect, and how this can affect their usefulness has to be analyzed to arrive at the utility of the predictive system.

Weather-based forewarning systems reduce the cost of production by optimizing the timing and frequency of application of control measures and ensure operator, consumer and environmental safety by reducing chemical usage. Accurate prediction is important to synchronize the use of disease control measures to avoid crop losses. Thus, if a sound forewarning system is developed, the explosive nature of the disease could be prevented by timely application of the control measures.

Requirements for development of forewarning systems

For development of efficient plant disease forewarning systems, information about host, pathogen, environment over a period of time is essential. Each of these components has a significant role to play in development of disease.

a. *The Pathogen:* The presence or absence of a pathogen is the main factor that determines occurrence of disease. For each given agro-climatic region, through co-evolution a set of pathogens get adapted and maintain equilibrium. However, due to commercial agriculture in the recent years, the balance between host and pathogen are disrupted either through introduction of new crop species or new varieties and hence, the disease scenario changes drastically. In such cases, introduction of a pathogen to an area from which it has previously been absent can cause major outbreaks of disease in plant communities. The amount of disease that develops is often determined by pathogenicity of the pathogen. Pathogenicity is nothing but the inherent ability of the pathogen to cause disease and is dependent on the pathogen's reproductive, dispersal and survival fitness.

The adaptability of the pathogen is determines the ability to infect resistant hosts or to survive in the changed environmental conditions. Adaptability is determined by the pathogen's genetic make up and reproductive efficiency. The spread of a disease and the formation of epidemics is reliant on the pathogen's ability to disperse rapidly over long distances. The foliar pathogens such as spores of cereal rusts, for instance, are known to cross continents over a few days time, while soil-borne pathogens have little scope for extensive spread. For a pathogen to cause disease in successive seasons, it must be able to survive during lean phase in absence of host. Some pathogens form spores or sclerotia that can survive in the soil for years, while others colonize alternative plant species until the primary host is cultivated. It is not only the mere presence of the pathogens but also their inoculum density that is crucial in determining the amount of disease. Generally, as the number of propagules increases, the level of disease increases, levelling off when the amount of disease reaches peak levels and when only a few uninfected plants available. The survival of propagules, and therefore the number of propagules available to cause disease, is heavily influenced by environmental factors.

b. *Host:* The development of disease in a plant community relies on the presence of individual hosts that are susceptible to that particular pathogen. If the majority of the population is susceptible to the pathotypes of a pathogen in the vicinity, an epidemic can occur.The best way of controlling disease is by planting species or cultivars that are not susceptible to pathogens of that area. The occurrence of disease can also be influenced by the host plant's growth stage and form. Some diseases are common in seedlings, while others are typical of mature plants. The growth stage of the population can also affect the microclimate around the plants; for example, the humidity and sunlight levels under the canopy. The population structure and density will also affect the development of disease in a plant community. The density of the main host species and the proportion of other plants that are not hosts within the community will determine the rate and extent of epidemic development. Crop plants tend to be densely planted, with no other species in amongst them, making them more susceptible to rapid spread of disease. Extensive, dense plantations can host spectacular epidemics, particularly if a new pathogen is introduced to the area. In addition, the general health of the host plant before infection is important in determining the success of a disease. Necrotrophs do well on poorly growing plants, while biotrophs thrive on a healthy host plant.

c. *The Environment:* The presence of a pathogen against a particular plant will generally not cause serious disease unless the environmental conditions are favourable. This includes the aerial environment and the soil (*edaphic*) environment. Human attempts at controlling disease usually involve manipulating the environment in some way. For example, breeding wheat cultivars to tolerate dry conditions allows Australian farmers to plant the crop in areas that are not favourable for pathogens such as powdery mildew and leaf rust. Properties of the aerial environment that influence disease development include moisture levels, temperature and pollution.

Moisture is particularly important to pathogenic bacteria and fungi. Rain splash plays an important role in the dispersal of some fungi and nearly all bacteria, and a period of leaf wetness is necessary for the germination of most airborne spores. By using water for dispersal, propagules are dispersed at a time when they are likely to be able to germinate as well. Because the process of germination and infection takes time, the duration of leaf wetness also influences the success of the infection. The duration necessary for infection varies with temperature. Usually, a longer period of leaf wetness is needed to establish an infection in cooler temperatures, as germination and infection are generally accelerated in warmer conditions.

Temperature also affects the incubation, or latent, period (the time between infection and the appearance of disease symptoms), the generation time (the time between infection and sporulation), and the infectious period (the time during which the pathogen keeps producing propagules). The disease cycle speeds up at higher temperatures, resulting in faster development of epidemics. The period of leaf wetness, combined with temperature information can be used to predict outbreaks of some diseases (infection periods) and be used to time preventative treatments, such as spraying. A recently recognised aspect of the aerial environment that can influence disease in plants is air pollution. A high concentration of pollutants can affect disease development and, in extreme cases, damage the plants directly by causing acid rain.

The edaphic (soil) environment affects soil-borne diseases, largely by determining the amount of moisture available to pathogens for germination, survival and motility. Germination and infection success also rely on the temperature of the soil. The fertility and organic matter content of the soil can affect the development of disease. Plant defences are weakened by nutrient deficiency, although some pathogens, such as rusts and powdery mildews, thrive on well-nourished plants. Other diseases thrive in soils that are specifically low in organic matter.

d. *Interaction between host, pathogen and weather:* The pathogen, the host and the environment interact, usually in ways that are difficult to quantify and predict. Control measures can include sowing of a crop species early, to avoid exposing seedlings to a disease during the time of year that provides the best environmental conditions for the pathogen.

Prediction of disease outbreaks enables the effective use of control measures, such as chemical or biological treatments, the prediction of crop yields and of the market potential for that crop. Disease forecasting involves the use of weather data and biological data to predict disease incidence. Usually, disease forecasting is only performed on economically important diseases, and as a method of cost reduction. If controlling a particular disease involves an expensive or time-consuming treatment, being able to predict outbreaks of the disease allows the treatment to be timed correctly, increasing its effectiveness, and reducing the cost compared to repeated treatments. Because environmental conditions vary from season to season, disease forecasting is necessary to predict the chance of disease in a certain set of conditions.

Disease can be forecast using computer modelling and empirical correlations relating to weather conditions, levels of inoculum, test plots and site factors and the predictions can then be communicated to growers. Computer modelling of plant diseases uses systems analysis to accumulate all the factors that affect the development of a certain disease into a computer-based model, and make predictions of disease under different environmental conditions. A disease needs to be well understood in order to formulate an accurate model, and models based on diseases that we know little about are generally not very accurate. The more straightforward approach of developing empirical correlations between particular weather factors and disease has had considerable success. This does not attempt a complete modelling of all factors involved in a disease, but only those most important in affecting the disease. The accuracy of the model can then be measured statistically by comparing its predictions to what actually happens.

Monitoring the weather is the most important consideration in disease forecasting, because of the overriding effect that weather has on disease development. While broad scale weather data has been used for disease forecasting, it is well known that the microclimate within the crop has a more direct impact on disease. Devices have been developed to monitor microclimate factors such as duration of leaf wetness and temperature, and with time, they will be affordable and accurate enough for widespread use on individual farms. Synoptic weather forecasting charts can be used to predict 'critical periods' - the occurrence of conditions favourable for disease development - so that farmers can spray their crop before it happens. There are now several self-calculating disease forecasting monitors available commercially that use environmental data and past season data to predict outbreaks of particular diseases.

e. Crop Microclimate and disease: The microclimate of the crop is the weather in the immediate vicinity of the crop, which is also sometimes referred to as weather at canopy level. Usually, the weather at the observatory and crop canopy is not same as at canopy level the weather is influenced by soil parameters. Since, the pathogens survive on plant parts, the weather in this region is more important than weather at the observatory level. For instance many foliar pathogens need a thin film of moisture on the leaf surface for development of infection structures which is usually termed as leaf wetness. The duration of leaf wetness determines the successful establishment of the pathogen on the host.

Accounting for the microclimate can help in developing reliable weather based prediction systems for pathogens. However, it is not always possible to record microclimate in the crop canopy due to lack of infrastructure and trained manpower. An alternative could be to develop reliable indices that can explain the relationship between observatory and canopy weather data. An effort was made in groundnut to establish such relationship. The preliminary relationships are given below.

$$\text{Tmax} = y = \frac{30.5922}{1 + 4631.89e^{-0.3160x}}$$

$$\text{Tmin} = y = \frac{20.9361}{1 + 14.9403e^{-0.2367x}}$$

$$\text{RH (morning)} = y = 93.641e^{-0.0511x}$$

$$\text{RH (evening)} = y = 67.272e^{-0.0143x}$$

As seen from the above, the temperatures followed logistic relationship whereas humidity followed an exponential relationship. Such relationships should be standardized for different crops, spacing and soil types.

Microclimate also influences rate of disease development. It was observed that even when susceptible cultivar and aggressive inoculum were provided, the variable weather conditions resulted in different apparent infection rates (‘r’) and area under disease progress curve (AUDPC) over two years. Fro instance, in groundnut-late leafspot system, though the ‘r’ in kharif 2004 (0.0664) was similar to ‘r’ in kharif 2005 (0.0759), the AUDPC was relatively less in 2005 and compared to 2004, which indirectly signifies the role of microclimate which in turn is influenced by other crop husbandry practices including date of sowing. Similarly, analysis of disease progression data from three locations of Tamil Nadu showed that the ‘r’ values differed over seasons and across years.

Many foliar pathogens require free water to germinate; therefore, disease pressure should favor plants that are able to repel water. For a suite of 18 sympatric clover species, leaf traits were evaluated affecting leaf wetness and susceptibility to infection by the fungal pathogen *Stemphylium* sp., causal agent of *Stemphylium* leaf spot. Spore germination increased with time in free water, and the relative susceptibility of host plants to infection was proportional to the duration of water retention on leaves. Larger leaves captured more water and retained it longer. Unexpectedly, trichomes and leaf wettability did not affect water capture. For clovers planted within natural clover populations at two sites, infection was threefold greater at the wetter site. At the drier site, water retention on the leaf surface was an important predictor of infection rates across host species, but persistent fog and dew at the wetter site reduced the importance of rapid leaf drying. Hence it can be suggested that plant adaptations that reduce water retention on leaves may also reduce disease incidence, but the selective advantage of these traits will vary among habitats.

Because moisture and temperature influence the growth of fungi, characterizing weather conditions favorable for fungi may be used to predict the abundance and richness of fungi in habitats with different climate conditions. To estimate habitat favorability to fungi, the relationship of fungal abundance and species richness to various weather and environmental parameters was studied in the Intermountain West. It was observed that fungal richness was positively correlated with fungal abundance ($r = 0.75$). Measures of moisture availability, such as relative humidity and vapor pressure deficit, explained more of the variance in fungal abundance and richness than did temperature. Weather variables that took into account the proportion of time habitats experienced favorable or unfavorable relative humidity and temperatures were the best predictors, explaining up to 56% of the variation in fungal abundance and 72% for fungal richness.

The persistence of leaf surface moisture, a condition critical for the development of most foliar fungal pathogens (Jones, 1986), plays a key role in the epidemiology of fungal diseases. Numerous studies have shown an increase in the incidence and severity of disease with increasing duration of leaf wetness (Cowling and Gilchrist, 1982; Evans *et al.*, 1992; Filajdic and Sutton, 1992; Montesinos et al., 1995; Basallote-Ureba *et al.*, 1999; Suheri and Price, 2000). Many foliar pathogens require extended periods in free water for spore germination, germ tube growth, and host penetration (Everts and Lacy, 1990; Wadia and Butler, 1994; Vloutoglou *et al.*, 1996; Gilles *et al.*, 2000). Therefore, plant traits that reduce the time that leaves remain wet after rain or dew could reduce susceptibility to foliar pathogens and potentially slow disease spread. In natural ecosystems, foliar pathogens can exert important selection pressures on plants (Alexander and Burdon, 1984; Esquivel and Carranza, 1996; Gilbert, 2002), suggesting that the ability to repel water from a leaf surface may be an adaptation for the prevention of fungal infection. The mechanistic links between leaf water retention and fungal infection can provide a unifying framework for understanding variation in infection among plant species and across different habitats.

High relative humidity and several hours of free surface water are critical for both spore germination and successful infection (Huber and Gillespie, 1993; Cook and Whipps, 1993; Harrison *et al.*, 1994). In addition, infection i.e., invasion of plant tissue by the fungus and disease i.e., the expression of symptoms such as lesions or necrosis (Agrios, 1998) on plants due to air-borne fungi are favored by temperatures of 15–40?C (Cook and Whipps, 1993; Griffin, 1994). Field studies on plant pathogens have demonstrated that the growth of fungi is favored by high moisture and moderate temperatures (Frolich and Snow, 1986; Griffin, 1994; Colhoun, 1973; Taylor, 1979; Rowan *et al.*, 1999) and that low relative humidity and extreme temperatures inhibit growth and spore germination (Harrison *et al.*, 1994; Juniper, 1991). Other studies on soil fungi also show that prevalence differs among habitats and seasons and correlates positively with moisture and negatively with temperature (Harrison *et al.*, 1994; Johansen and Rushforth, 1985). This apparent positive relationship between moisture and fungal growth and abundance may result from the high surface-to-volume ratio of fungi, making them vulnerable to water loss.

Types of disease forewarning systems: Starting from simple thumb rules, the disease prediction models can go to very complex models involving several analytical tools. The purpose of a model is to help grower to arrive at a decision about crop health management. Some of these models also could go as sub-routines in crop growth models which are useful o ultimately workout the predicted yields. However, due to complexity of the processes involved in disease development, very few crop growth models have considered including the disease sub-routines. Even such models like DSSAT do not take all parameters of disease development and thereby leaving unaccounted variability. Various techniques of computer modeling and simulation like artificial neural networks, fuzzy logic or Montecarlo analysis and the conventional multiple regression approaches are being used to help synthesize and develop scientists' understanding of this complex plant-pathogen-environment relationship. The resultant models enable exploration of the factors that govern disease epidemics. The same models have potential to guide breeding programs and work to develop strategies that will prolong the usefulness of disease-resistance genes.

Thumb rules: Thumb rules are qualitative knowledge-based models that could be used for disease prediction. These thumb rules are derived based on past experience of definitive relationships among host, pathogen and environment. 'Dutch rules' for prediction of late blight of potato, Mill's rules for prediction of apple scab are some examples. These rules are not quantifiable and also not fit for numerical analysis.

a. *Empirical or mechanistic models:* These models use developed based on elaborate experiments and deriving simple relationships between various parameters associated with disease development. These are mainly descriptive models covering various weather parameters and their influence on disease development. However, these models cannot be used for quantification or analysis of disease purposes. They also include geophytopathological maps showing disease spread.

b. *Mathematical models:* Mathematical models are developed by applying the mathematical relationships between host, pathogen and environment interactions. The analytical models use the epidemic analysis on a theoretical basis without taking into consideration the external factors. These include parameters such as apparent rate of growth of infections, disease growth under non-restrictive environment, equations developed for fitness and survival of the pathogens etc. The statistical models include use of statistical tools such as correlation, regression methods. These models could be linear or non-linear relationships between various interacting parameters of a disease.

The simulation models include the art and science of constructing composite models and mimicking one system with the other. Simulation models, when run on the computer, generate growth parameters by considering algorithms based on a detailed flow diagram of the events of disease development. The first computer-based plant disease simulator is EPIDEM for early blight of potato and tomato. Later several simulation models were developed either as a stand-alone model or as a sub-routine of the crop growth model. With the advent of internet, web-based decision support systems are also in place for various diseases. The web-based models are easy to develop as they do not need expertise of computer programs and the web pages could be written with a relative ease as compared to complicated programs such as Java, J-Builder etc. Magarey *et al.* (2005) developed a generic infection model based on temperature and wetness duration. The model is generic in the sense that it was developed to describe any pathosystem when appropriate parameters are supplied.

The production and availability of decision support systems (DSS) as aids in applying or using disease control methods have become common in contemporary agriculture, perhaps because it has become easier to obtain and process the data that are needed to provide this kind of advice. Another factor affecting adoption of DSS is the ability to reach large audiences with relatively little effort. While increased availability of this kind of information is undoubtedly a benefit for most users, the quality of such information should be sufficient to allow maximization of the economic and environmental benefits that can result from the use of such systems. Thus, other ways of development of such systems could be useful. In addition, clear and objective methods are needed to evaluate and compare different DSS, although the ultimate criterion is adoption by the end-users.

With the development of more rapid computing technologies, it is possible to compute a large number of simulations to support new modeling methods for plant disease forecasting. The development of disease forecasting models is based on determining the value of unknown parameters for an appropriate model. One method for estimation of parameters is maximum likelihood estimation (MLE), first developed by R.A. Fisher. The basic idea behind MLE is that one can calculate the probability of observing a particular set of data given the model of interest and this is called the likelihood of a sample. The likelihood of observing the actual data for each of the potential parameter values is calculated to determine which set of parameter values maximizes the likelihood.

c. *Stochastic models:* These models are based on probability distributions and generally adopted when a reliable estimate of the expected value is desired and when there is a curvilinear relationship between stochastic characteristic and one of the state variables. From these models generally variance of the model is expected output. These models include heavy calculations needing electronic calculation aids. These models are more common in human and animal diseases where the target organisms are mobile without a continuum and the probability is involved in such cases.

Validation - Where are we and What Next

In our early days, we started with correlations and regressions and developed relations between host and disease. Unfortunately, most of the disease forewarning systems developed in India during early era of plant disease epidemics, adapted the successful procedures used in developed nations. However, for monsoon-influenced crop production systems, these procedures do not hold good as they need a different logic to understand disease development. Hence, most of the mathematical models developed by those methods failed to validate across locations. With our enhanced understanding about disease development, we accept that simple mathematical models do not help to understand the epidemics. Simialrly, we tried to develop simple thumb rules and here also regionality parameter was neglected. Also, in many of such models, the pathogen parameters and their biocontrol agents were neglected and there by leaving a large amount of variability un-accounted for. Today, our understanding of the epidemic has greatly improved due to availability of modern scientific tools such as GIS, remote sensing, a better knowledge about the pathogenicity factors, host-pathogen interactions, and more importantly our ability to measure weather parameters in a more precise way. Most disease forecasts are based on favorability of weather for disease development. Input data for the forecasts are usually derived from automatic weather stations (AWS) that may be irregularly and incompletely dispersed across a region. Therefore, interpolation or extrapolation from nearby AWS sites is often necessary. Simple interpolation from a network of AWS sites may produce disease forecasts with enhanced on-farm resolution. However, complex terrain can make such interpolation impractical and inaccurate. Large AWS networks also require maintenance and substantial financial support. We do have access to satellite datasets which could be used for better understanding of the disease environment. Mesoscale weather models are capable of simulating weather conditions such as air temperature and humidity at a resolution of as high as 100 meter, and then clustering as a supercomputer, overcoming some of these previous limitations.

The successful development of a plant disease forecasting system also requires the proper validation of a developed model. There is increased interest among plant disease modelers and researchers to improve producer profitability through validation based on quantifying the cost of a model making false predictions (positive and/or negative). An economic validation of a plant disease forecasting system requires the examination of false positive and false negative predictions.

Conclusions

Reformulating the question from How much disease will occur?, with a quantitative answer, to a question like Should I spray my field?, with a qualitative answer entails use of different techniques. Disease prediction models are useful tools for the benefit of the farmers to make farm decisions. However, to develop an effective and efficient disease prediction model, a careful analysis of the disease environment is required. A multi-disciplinary approach to the problem solving would help in better understanding of the disease in a production system mode by taking into consideration a holistic overview of the disease. In India, due to limited manpower trained to analyze the disease epidemic is a major limitation and hence, human resource development in this faculty is badly needed to address this important area of research.

References

Agrios, G.N. 1998. Plant Pathology New York: Academic Press.

Ahlbom, A., 1993. Biostatistics for Epidemiologists. Lewis Publishers.

Alexander H. M. and J. J. Burdon 1984. The effect of disease induced by Albugo candida (white rust) and Peronospora parasitica (downy mildew) on the survival and reproduction of Capsella bursa-pastoris (shepherds purse). Oecologia 64: 314-318.

Basallote-Ureba M. J. A. M. Prados-Ligero J. M. Melero-Vera 1999. Aetiology of leaf spot of garlic and onion caused by Stemphylium vesicarium in Spain. Plant Pathology 48: 139-145.

Campbell, C. L. and Madden, L., 1990. Introduction to Plant Disease Epidemiology. John Wiley and Sons.

Colhoun, J. 1973. Effects of environmental factors on plant disease. Annu Rev Phytopath 1973, 11:343-363

Cooke RC, Whipps JM 1993. Ecophysiology of Fungi. Oxford: Blackwell Scientific Publications.

Cowling W. A. D. G. Gilchrist 1982 Effect of light and moisture on Stemphylium leaf spot of alfalfa. Plant Disease 66: 291-294.

Esquivel R. E. J. Carranza 1996. Pathogenicity of Phylloporia chrysita (Aphyllophorales: Hymenochaetaceae) on Erythrochiton gymnanthus (Rutaceae). Revista de Biología Tropical 44: 137-145.

Evans K. J. W. E. Nyquist R. X. Latin 1992. A model based on temperature and leaf wetness duration for establishment of Alternaria leaf blight of muskmelon. Phytopathology, 82: 890-895.

Everts K.L.M.L. Lacy 1990. The influence of dew duration, relative humidity, and leaf senescence on conidial formation and infection of onion by Alternaria porri. Phytopathology 80:1203-1207.

Filajdic N.T.B. Sutton 1992. Influence of temperature and leaf wetness duration on infection of apple leaves and virulence of different isolates of Alternaria mali. Phytopathology, 82: 1279-1283.

Froelich, R.C., Snow, G.A. 1986. Predicting site hazard to fusiform rust. For Sci 32:21-35.

Gilbert, G.S. 2002. Evolutionary ecology of plant diseases in natural systems. Annual Review of Phytopathology, 40:13-43.

Gilles, T.B.D.L. Fitt R. Kennedy S. J. Welham M. J. Jeger 2000. Effects of temperature and wetness duration on conidial infection, latent period and asexual sporulation of Pyrenopeziza brassicae on leaves of oilseed rape. Plant Pathology, 49: 498-508.

Griffin, D.H. 1994. ed: Fungal Physiology New York: Wiley-Liss.

Harrison, J.G., Lowe R, Williams NA 1994. Humidity and fungal diseases of plants – problems. In: Ecology of plant pathogens (Edited by: JP Blakeman, B Williamson) Wallingford: CAB International pp. 79-97.

Huber, L., Gillespie, T.J. 1993. Modeling leaf wetness in relation to plant disease epidemiology. Annual Review of Phytopathology, 30:553-577.

Hughes, G., McRoberts, N., and Burnett, F., 1999. Decision-making and diagnosis in disease management. Plant Pathology 48:147–153.

Johansen, J.R. and Rushforth, S.R. 1985. Cryptogamic soil crusts: seasonal variation in algal populations in the Tintic Mountains, Juab county, Utah. Great Basin Nat, 45:14-21.

Jones A. L. 1986. Role of wet periods in predicting foliar diseases. In K. J. Leonard and W. E. Fry [eds.], Plant disease epidemiology, 87–100. McGraw-Hill, New York, New York, USA.

Juniper BE 1991. The leaf from the inside and the outside: a microbe's perspective. In: Microbial Ecology of Leaves (Edited by: JH Andrews, SS Hirano) New York: Springer-Verlag pp. 21-42.

Magarey, R. D., Sutton , T. B., and Thayer, C. L. 2005. A simple generic infection model for foliar fungal plant pathogens. Phytopathology 95: 92-100.

Mills, W., 1944. Efficient use of sulfur dusts and sprays during rain to control apple scab. Bulletin 630, New York Agricultural Experiment Station Ithaca Extension.

Montesinos E. C. Moragrega I. Llorente P. Vilardell A. Bonaterra I. Ponti R. Bugiani P. Cavanni A. Brunelli 1995. Development and evaluation of an infection model Stemphylium vesicarium on pear based on temperature and wetness. Phytopathology 85: 586-592.

Rowan, N.J., Johnstone, C.M., McLean, R.C., Anderson, J.G. and Clarke, J.A. 1999. Prediction of toxigenic fungal growth in buildings by using a novel modelling system. Appl Environ Microbiol, 65:4814-4821.

Suheri H.T.V. Price 2000. Infection of onion leaves by Alternaria porri and Stemphylium vesicarium and disease development in controlled environments. Plant Pathology 49: 375-382.

Taylor, E.C. 1979. Seasonal distribution and abundance of fungi in two desert grassland communities. J Arid Environ 1979, 2:295-312.

Vanderplank, J., 1963. Plant Diseases: Epidemics and Control. Academic Press.

Vloutoglou I. B. D. L. Fitt J. A. Lucas 1996. Germination of Alternaria linicola conidia on linseed: effects of temperature, incubation time, leaf wetness and light regime. Plant Pathology 45: 529-539.

Wadia K. D. R. and D. R. Butler 1994. Infection efficiency of Phaeoisariopsis personata and the influence of different wetness patterns on germ-tube growth of the pathogen. Plant Pathology 43: 802-812.

Zadoks, J. C. and Schein, R.D., 1979. Epidemiology and Plant Disease Management. Oxford University Press.

❑❑❑

28

Comparative Study of Hit-ratio (Accuracy Percentage) of Rice Blast Forecasting Using Various Forecasting Models Based on Different Statistical Techniques: A Case Study

R. Bhagawati, Kaushik Bhagawati, G. Bhagawati, U.D. Singh, S. Baruah and S.P. Singh

Abstract

Rice blast disease (Pyricularia grisea cavara) which is very much weather sensitive, continues to be an enigmatic problem of West Siang District of Arunachal Pradesh, owing to its geographical location and climate. There is an urgent need of a forewarning system which can provide advance information for outbreak of the diseases attack so that protection measures can be implemented before the actual economic damage of the crop occurs. Two most important statistical techniques for disease forecasting viz. regression analysis and artificial neural network were chosen and forecasting models has been developed for the area and analyzed with local weather data to find best fit for the area and also compared with the Support Vector Machine (SVM) based model RB-Pred developed by Institute of Microbial Technology, Chandigarh. The model based on artificial neural network was found to be best with accuracy of 75-80%, with the present data and algorithm. The accuracy can be improved with long archive records and proper implementation and algorithm selection of the hidden layer, of both ANN and SVM based (being special case of ANN) models, for site-specific forecasting.

Keywords: Rice blast forcasting, Artificial neural network, MATLAB software, Support vector machine.

Introduction

Climate change and agriculture

It is now doubtless that recent human induced changes in the atmospheric composition have changed the changing trend of climate change. Climate change present challenges to all sectors, particularly agriculture because this sector is heavily reliant on natural resources and ecosystem services, which are influenced by climatic conditions. Although the magnitude of these changes is uncertain, particularly at regional level, the threat is very real. The current climate change trend is likely to lead to an increase in variability and in the occurrence of extreme events such as exposure to prolonged high temperature, severe storms and dry conditions. Climate change impact will vary between regions—some impacts may pose threats to the variability of agriculture in some regions, while in other regions changes in climate may improve viability or create new opportunities. In response to climate change, farmers and researchers will be required to develop new approaches to manage climate related risks in a very location specific manner.

The distribution, abundance and management of pathogens have been affected by climate change. The impact of pathogen and pests are influenced by changes to the host and the pests and their interactions with each other, as well as the environment so change in any of these factors will affect the incidence and severity of a pest. Temperature, rainfall, humidity, radiation and dew all contribute to the occurrence of plant diseases. As dispersal of some pathogens is controlled by rain and winds, changes to these factors could also affect the spread of pathogens.

For example, crops growing in a marginal climate and experiencing an environmental stress could be exposed to new or increased levels disease outbreaks (Coakley *et al*, 1999). Similarly, a plant host stressed by the environment could have an altered physiology and chemical composition, which may render it more susceptible to the diseases due to altered concentrations of defensive compounds. Host-pathogen relationships may also change if the host crops are being grown in areas altered by climate change.

These very uncertainties and unpredictable situations can be dealt and studied under the umbrella of climate resilient agriculture. Resilience is used to describe the magnitude of disturbance that a system can withstand without crossing a threshold into a new structure or dynamics. For natural systems, it is a measure of how much disturbance an ecosystem can handle without shifting into a qualitatively different state. Therefore, the ability of institutions and individuals to avoid potential damage and to take advantage of opportunities will be critical factor in building resilience to climate change. The lens of resilience is useful in analyzing climate change because it is founded on the recognition that human existence within ecological system is complex, unpredictable, and dynamic and that institutional measures and responses should be based on these principles.

Current Research

The present research work is concentrated to West Siang District of Arunachal Pradesh. The district is located between 93°57'- 95°23' E latitude and 27°69'- 29°27' E longitude with an altitude range between 150 to 6000 metres above mean sea level. The district of West Siang comes under the Eastern Himalayas, warm per humid eco-region under zone two. It comes under temperate sub alpine climate zone. The normal annual rainfall is around 2550 mm with mean rainy days of 144.

Rice is the single most important food crops for almost cent per cent population of West Siang District of Arunachal Pradesh. Out of the various diseases limiting rice productivity, blast disease (casual organism, *Pyricularia grisea* cavara) continues to be an enigmatic problem in this area owing to its geographical location and climate, and its intensity that is magnifying with variation/change in climate. It is a serious constraint in realizing the full yield potential of rice cultivars. Weather plays important role in the appearance, multiplication and spread of blast fungus. The spores of this fungus are present all the year round, and strike the crop when environmental conditions favor. Hence, the disease is regarded as one of the greatest pathological threats to the rice crop and it is weather and host nutrition sensitive. Its progress is driven by low night temperature (22-28°C), high relative humidity (>95%), dew deposit, extended leaf wetness period (>10 hrs), cloudy and drizzling weather, soil fertility (high nitrogen), degree of host susceptibility, presence of alternate hosts and straw of the previously infected crop heaped nearby. Blast occurs under divergent agro-meteorological conditions in India. In high rainfall zones (rainfall 1500 mm, temp: 20-24°C) north and north eastern India, rice crop suffers due to this disease during June to September. Early appearance of the disease was observed during April-July in Arunachal Pradesh.

The climate of West Siang District display a great variation in the recent past as depicted in the Fig.1 (a&b) which clearly show the erratic nature of the rainfall [Fig.1(a)] and constant rise in the minimum temperature [Fig.1(b)]. These erratic natures of climate adversely affect the rice production as it continuously favor increases in the incidences and severity of blast disease. Such losses can be reduced to a great extent if their occurrence is known in advance so that timely remedial measures can be taken. Thus there is an urgent need to develop forewarning systems which can provide advance information for outbreak of disease attack so that protection measures can be implemented before the actual onset of the damage. However reliable and well-timed forecast are of vital importance for appropriate, foresighted and up-to-date planning especially for blast which is very environment oriented. The present work aims at comparative study of the two most important statistical models used and developed for rice blast forecasting in the ICAR Research Complex for NEH Region, Arunachal Pradesh Centre, Basar viz. Regression Analysis and Artificial Neural Network (ANN) and also compared with Support Vector Machine (SVM) based model RB-Pred (Developed by Institute of Microbial Technology, Chandigarh) to determine the best fit for the area and to explore the necessary improvement required for site-specific forecasting. The paper also discusses some necessary steps required to improve the rice variety for durable resistance to blast.

Methods

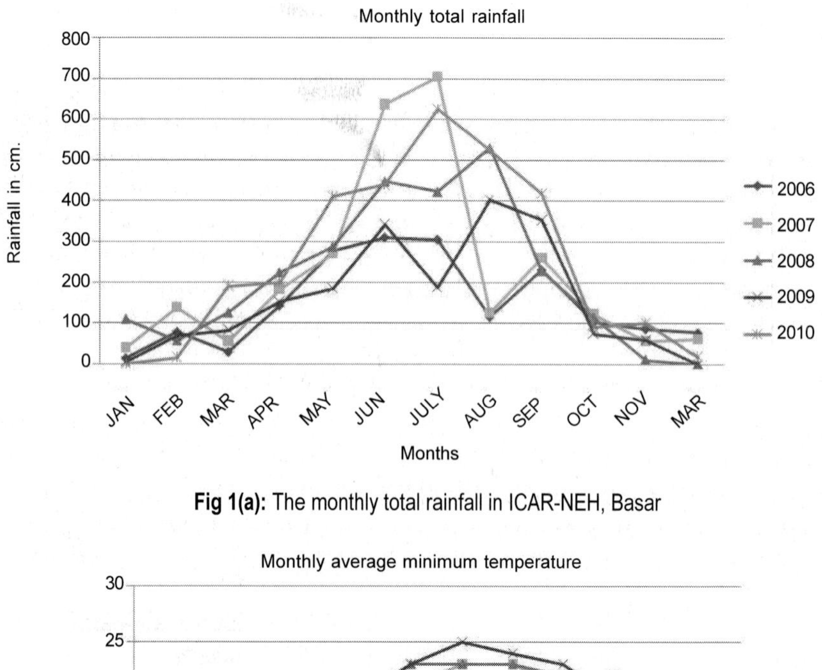

Fig 1(a): The monthly total rainfall in ICAR-NEH, Basar

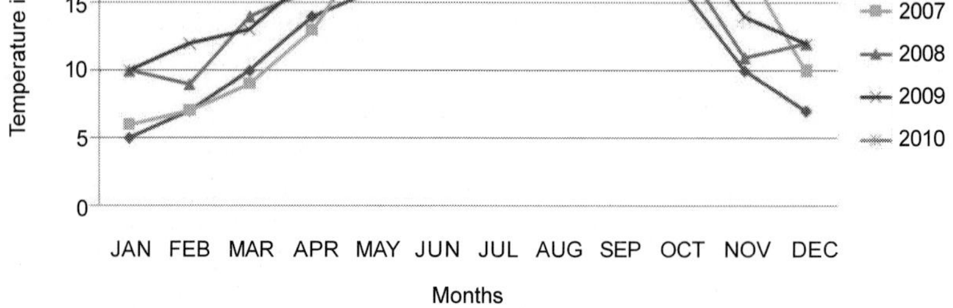

Fig 1(b): The monthly average minimum temperature in ICAR-NEH, Basar

In the current study we are concerned with the epidemic status such as occurrence/non-occurrence, low/high incidence etc of the disease rather than its severity. This information could be obtained through modeling qualitative data.

1. Regression Analysis: As we are concerned with qualitative information, we use the binary logistic regression model. It is used to describe the relationship of several independent variables to the binary (say, named 0 & 1) dependent variable.

The model is of the form:

$P(E=1) = 1/[1+\exp(-Z)]$, where Z is a function of weather variables.

If $P(E=1) \geq 0.5$, then there is more chances of occurrence of disease and if $P(E=1) < 0.5$, then probability of occurrence of the disease is minimum.

The model is given by

$$P(Y=1) = 1/[1+\exp\{-\beta_1 + \beta_2 X2 + \beta_3 X_3\}]$$

The models were developed using the maximum likelihood estimation procedure in SPSS. Consider 2006-2009 model based on second week of august average data using which forewarning probability is obtained for the year 2010. Feed the 2007-2009 data upon Y, X and X, in SPSS data editor, then choose Analyse-> Regression->Binary Logistic. Under the output see for "Variables in the Equation" table and note the parameter estimates corresponding to intercepts, X_1 and X_2 as β_1=-62.55; β_2=1.90; and β_3=0.25. Then the model becomes

$$P(Y=1) = 1/[1+\exp\{-(-62.55) + (1.90\ X_1) + (0.25X_2)\}]$$

Putting the value of X_1 (minimum temperature) = 21.8°C and X_2 (rainfall) = 89mm of the year 2010 it can be seen that P(Y=1) = 0.606, which is e"0.5, so positive prediction. The prediction with the model is found to be 39% accurate (on an average) for occurrence of disease. As shown in the Fig.2, the very low accuracy is shown in the year of maximum variability in the weather from the normal.

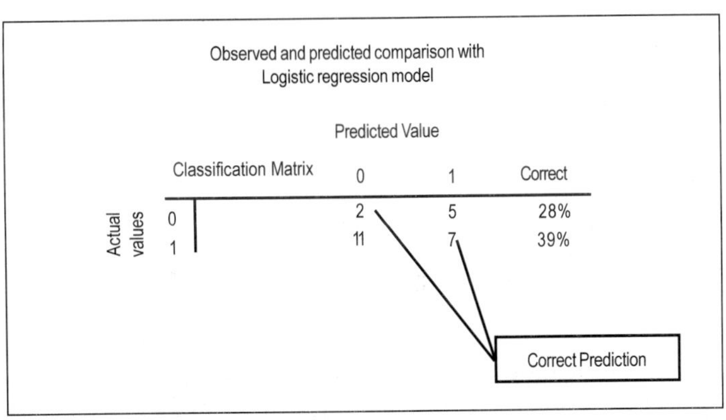

Fig. 2: Comparison of the observed and predicted values using logistic regression model

2. Artificial Neural Network (ANN): ANN has been developed as generalizations of mathematical models of biological nervous systems. The basic processing elements of neural networks are called artificial neurons, or simply neurons or nodes. A typical artificial neuron and the modeling of a multilayered neural network are illustrated in Fig.3. Referring to the Fig.3, the signal flow from inputs x_1, x_2,........., x_n is considered to be unidirectional, which are indicated by arrows, as is a neuron's output signal O is given by the following relationship:

$$O = f(\sum_{j=1}^{n} wjxj) \quad \text{(i)}$$

Where w_j is the weight vector, and the function *f(net)* is referred to as an activation (transfer) function. The variable *net* is defined as a scalar of the weight and input vectors,

$$net = w^T x = w_1 x_1 + w_2 x_2 + \dots\dots\dots\dots\dots + w_n x_n \quad (2)$$

Where T is the transpose matrix and in the simplest case, the output value O is computed as

$$O = f(net) = \begin{cases} 1, & if\ wT \geq \theta \\ 0, & otherwise \end{cases} \quad (3)$$

Where 5Ø±ß is called the threshold; and this type of node is called a linear threshold unit.

Here we use the MATLAB software to generate the network and do the required analysis with the qualitative analysis. The advantage of ANN lies on the fact that it can handle both linear and non-linear model approaches together and so it can get the correlation between variables on a more valid basis. ANN is accepted as a strong method that learns the structures of the current data, establishes a new relations network in the real world and conducts many statistical processes such as making parameter estimation, classification, optimization and time series analysis in a more determined way.

Here using MATLAB software to generate the model and forecasting. By trial and error we choose the hidden layer to be of 10 nodes. The MATLAB network is shown below in Fig.4. The model display accuracy of 78% on an average of occurrence.

3. Support Vector Machine (SVM): A support vector machine (SVM) performs classification by constructing an N-dimensional hyperplane that optimally separates the data into two categories. SVM models are closely related to neural network. In fact, a SVM model using a sigmoid kernel function is equivalent to a two-layer, perceptron neural network. Using a kernel function, SVMs' are an alternative training method for polynomial, radial basis function and multi-layer perceptron classifiers in which the weights of the network are found by solving a quadratic

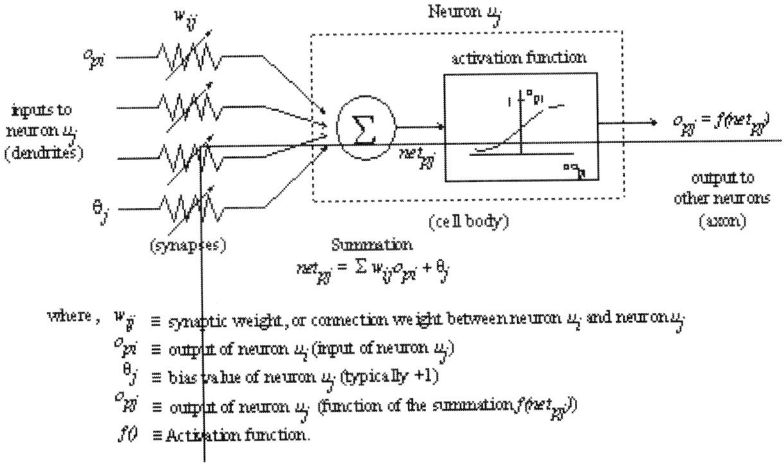

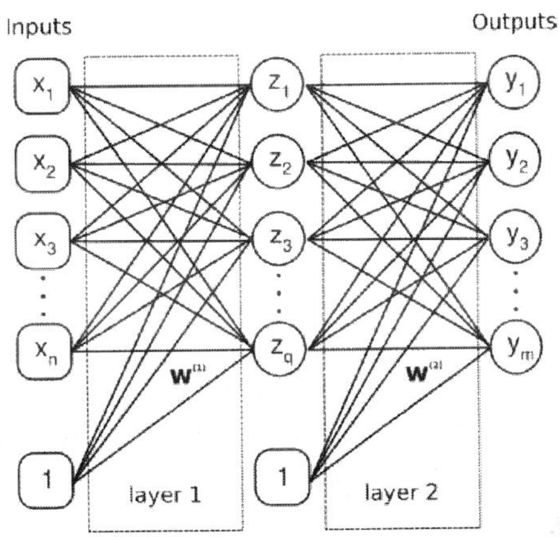

Fig 3: Multilayered artificial neural network

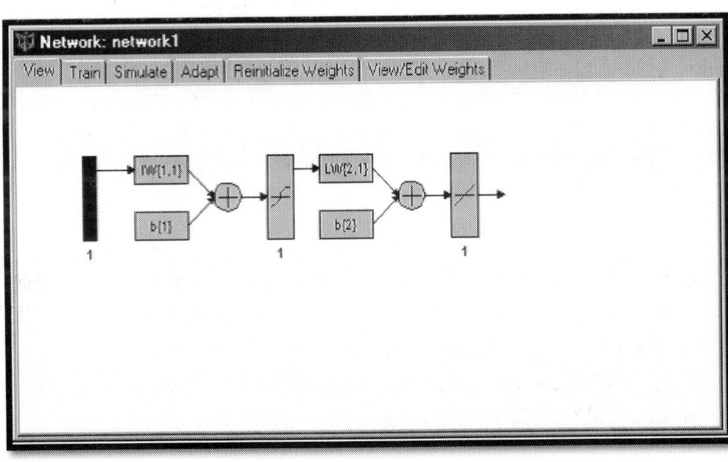

Fig 4: ANN developed in MATLAB

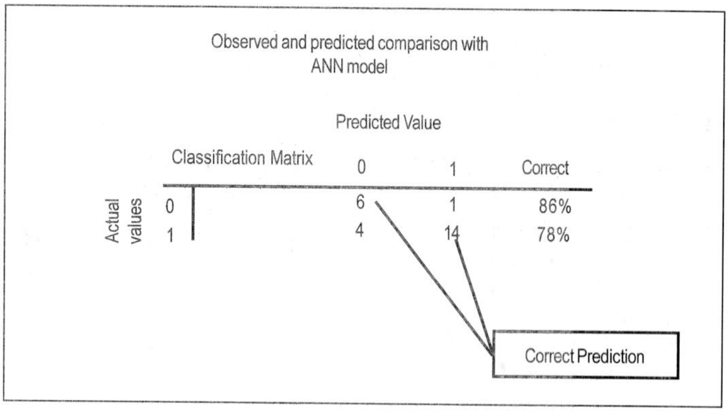

Fig 5: Comparison of the observed and predicted values using ANN model

programming problem with linear constraints, rather than by solving a non-convex, unconstrained minimization problem as in standard neural network training. In SVM, a predictor variable is called an attribute, and a transformed attribute that is used to define the hyperplane is called a feature. The task of choosing the most suitable representation is known as feature selection. A set of features that describes one case (i.e. the roe of predictor values) is called a vector. So the goal of SVM modeling is to find the optimal hyperplane that separates clusters of vector in such a way that cases with one category of the target variable are one side of the plane and cases with the other category are on the other size of the plane. The vector near the hyperplane is the support vectors. The flow chart and architecture of general SVM is depicted in the Fig.6 (a).

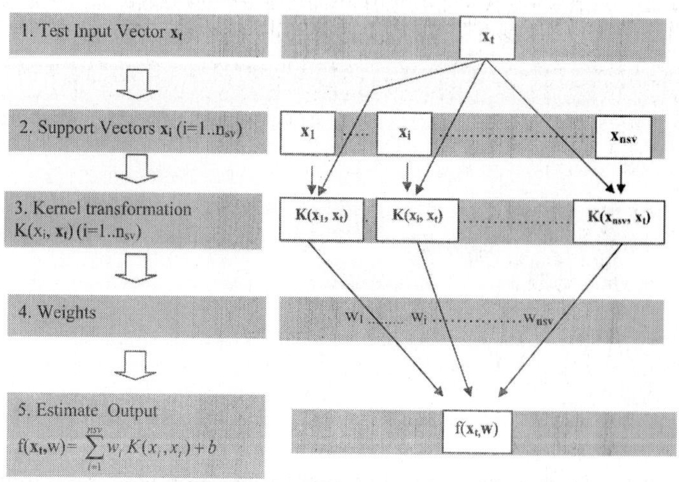

Fig 6 (a): (SVM flowchart and architecture)

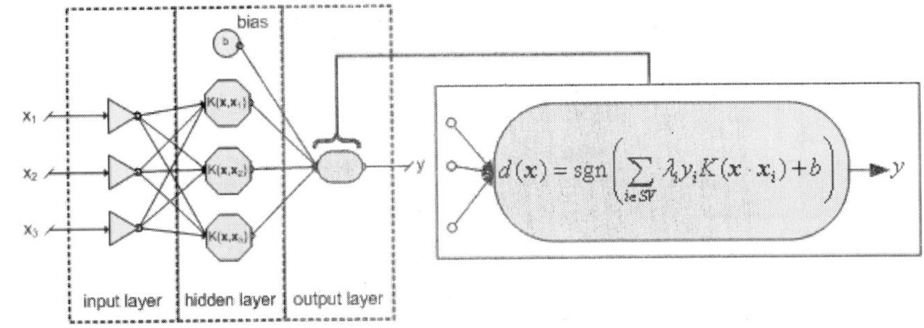

Fig.6 (b): The layers and function of SVM

The SVM analysis is done by feeding the data directly into the RB-Pred web-based model developed by Institute of Microbial Technology, Chandigarh. The web-page is shown in the Fig.7, which on feeding the location specific data gives the predicted severity. Comparing with the real data and considering the severity d”3 of the BL(SES 0-9)-scale of IRRI as below Economic Threshold Level (non-occurrence), the model display the accuracy percentage of 58% on an average of occurrence in four years for the area as shown in the Fig.8.

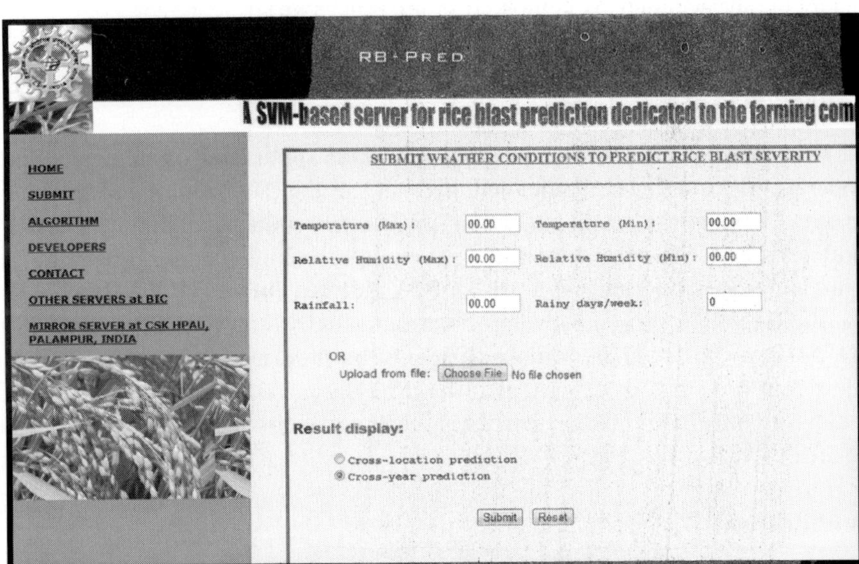

Fig 7: The RB-Pred web page

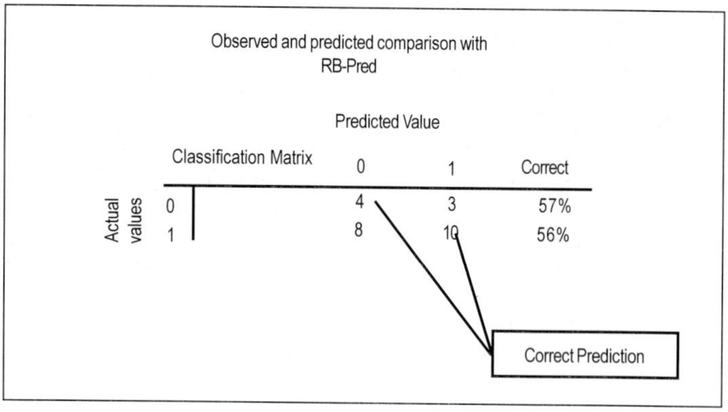

Fig 8: Comparison of the observed and predicted values using SVM based RB-Pred model

The RB-Pred model is based on SVM, being special case of ANN, the low performance of the model may be because of the model has been developed for very different location and with very different training data.

Results and Comparision

Yearly data are taken for two rice crops, one sown on mid-March and other first week of July. The accuracy is nearly same for both the crops as depicted in the bar graph shown in the Fig.6 (a & b). The ANN based model, in both the cases display higher accuracy as compared to the other two.

Success Story with current forecasting with the developed ANN model .

The timely and specific forecasting help reduce the application of fungicides with proper timing and doges. Pathogenic diversity in the pathogen populations and the easy loss of resistance in major and single genes could account for the lack of durable resistance based on host-pathogen-environment interaction. However gene pyramiding lines on Vandana-V_3(P_i2 + P_i9), Kalinga (iii)-K, (P_i2 + P_i9), K_9,K_6, RIL 10, RIL 29 background developed through ARBN lines like 129, 45, 46, 139, 169, 141 showed promising resistance against blast during 2007-2010 with L.S.I below 3 in 0-9 scale in experimental field of Basar, Arunachal Pradesh. These findings coupled with precise forecasting techniques proved to be promising for the area.

Conclusion

The study clearly revealed that the current weather pattern (i.e. no definite pattern) pose a great challenge in predicting plant diseases that are very much weather sensitive. Artificial Neural Network, owing to its ability to learn from the training data, may become a good statistical technique to predict the variable and erratic conditions with proper algorithm selection and training data for that particular location. The SVM based RB-Pred display low accuracy may be because of the fact that it was not developed for the current area of research. The experiment made it very clear that for site-specific forecasting the location specific precise data and proper implementation of the algorithm is very critical. The present weather data are taken with available ordinary rain gauge and normal thermometers. Improved data collection techniques and instruments such as Automatic Weather Station may increase the accuracy of the model and its training data. The present study is a prominent step towards the development of a proper forecasting model.

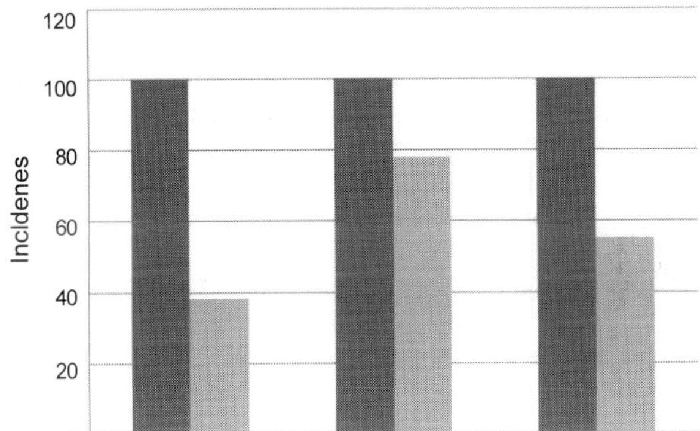

	Regression	ANN	SVM
■ Observed	100	100	100
■ Forecasted	38	78	55

Fig 9 (a): The hit-ratio for mid-March sown crop.

	Regression	ANN	SVM
■ Observed	100	100	100
■ Forecasted	34	79	50

Fig 9 (b): Hit-ratio for first week of July sown crop.

References

Tylor, M.C., Hardwick, N.V., Bradshaw, N.J. and Hall, A.M. 2003. Relative performance of five forecasting schemes for potato late blight (Phytophthora infestans) I. Accuracy of infection warning and reduction of unnecessary, theoretical, fungicide applications. Crop Protection, 22:275-83.

Ishiguro K, Hashimoto A: Computer based forecasting of rice blast epidemics in Japan. Rice blsat modeling and forecasting. Selected papers from the International Rice Research Conference, 27-31 August 1990, Seoul, Korea Republic: 39-51.

Kaundal, R. Kapoor, A.S. and Raghava, G.S: Machine learning techniques in disease forecasting: a case study on rice blast prediction, BMC Bioinformatics 2006.

Agrawal, Ranjana, and Mehta, S.C.: Weather Based Forecasting of Crop Yields, Pests and Diseases – IASRI Models, Journal of Indian Society of Agricultural Statistics 2007.

Baker, K.M. , Kirk, W.W. and Wharton, P. 1999. Inclusion of Synoptic Weather Forecast Data in Decision Support Systems for Agriculture, International Journal of Forecasting 12.

Wilson, R., Sharda, R., 1992. Neural Networks. OR/MS Today, August, 36-42.

Cox, D.R. 1958. The regression analysis of binary sequences (with discussion). Journal of the Royal Statistical Society B, 20, 215-242.

Smola, A., Scholkopf, B., A tutorial on support vector regression. NeuroCOLY Technical Report 1998.

Rocal *et al.* Developing Durable Resistance to Rice Blast: Productivity and Environment Considerations. Seminar Paper, ISNAR Biotechnology Service 1996.

Coakley, S. M., Scherm, H. and Chakraborty, S. 1999. 'Climate change and plant disease Management', Annual Review of Phytopathology, 37.

BRS, 2007, Australian Agriculture, Fisheries and Forestry At A Glance, Australian Government Bureau of Rural Sciences, Canberra, Australia.

Patterson, D.T., Westbrook, J. K., Joyce, R. J. V., Lingren, P. D. and Rogasik, J. 1999. 'Weeds, insects and diseases', Climate Change, 43.

Fuhrer, J. 2003. 'Agroecosystem responses to combinations of elevated CO_2, ozone, and global climate change', Agriculture, Ecosystems & Environment, 97: 1-20.

❑❑❑

29

Useful Softwares for Climate Change Research

Roop Singh Maitry, Hemavati R, M. Datta and D. Daschaudhuri

Abstract

Climate change is today's major concern throughout the world. All sectors have been influenced with the impact of climate change. Similarly, agriculture sector has also been affected with the worse climatic changes. All components of agricultural production system like water, soil, vegetation, organism and microorganism etc. affected greatly due to climatic abruptions. And thus, the productivity is hampering day by day. Today, we requires to predict the trend of changes in climatic parameters like maximum and minimum temperature, humidity, wind speed, sunshine hours etc, to project future change in these climatic parameters, to assess the effect of theses climatic parameters on agricultural components like water resources, soil, vegetation, organism and microorganism and, also to find the adaptation and mitigation strategies. All these works have beenmainly done by manual methods earlier. Developed country like USA, Japan, UK etc adopted the computer modelling to climate change research. Computer modelling and software assistance have been adopted slowly in developed countries. Since, climatic changes requires processing of very large number of data, they needs computer and software assistance. Without computer assistance climate change research seems to be impossible. Several softwares have been developed and continue to be developed for making climate change research easier and accurate.Here, a review of useful software related to climate change research in agriculture has been done to enlist the major useful and authentic software. In this review,

32 important climate change software has been presented with their applicability, utility, reliability and reliable source. These includes AgrometShell (AMS)-software, Aquacrop,Clim WAT, Crop Monitoring (CM) Box software, CROPWAT, EPIC model, ET calculator,Ex-ACT,Local Climate Estimate Tool, MOSAICC, FAO-Rain Fall Estimate Routine, Climpag,Global Climate Map,InfoCrop, Lingo software,MASICC,MOSAICC, PAR_CSDLL, SFTEWR_RegCM3, SIM Clime, Soil Para-software, WinDisp51 software, MAGICC/SCENGEN-CENTURY, MAACU, WTGROWS, ORYZA, CERES V4.0 MODEL FOR WHEAT, APSIM MODEL FOR SOYABEAN, SWAT, Rainfall simulator for Africa and ARIM/SERIM- Time Series Analysis etc. The papers conclude with enabling the scientist and researchers for the climate change researches.

Keywords: Climate change researches, Software, Crop Monitoring, Aquacrop, CropWat.

Introduction

Climate change is likely to be the defining issue of the 21st Century. Recent studies indicate that climate change is accelerating, confirming the more pessimistic of scenarios identified by climate scientists. Our past use of fossil fuels commit the world to around 2°C average temperature rise since the pre-industrial era, and, unless urgent and drastic cuts are made, further heating is likely to trigger any of a number of climate change tipping points. The results will be a dramatic reduction of food production and water supplies, more extreme weather events, the spread of disease, sea level rise, ocean acidification, and mass extinctions. We are faced with the twin challenges of mitigation (avoiding the worst climate change effects by rapidly transitioning the world to a low-carbon economy) and adaptation (re-engineering the infrastructure of modern society so that we can survive and flourish on a hotter planet). These challenges are global in nature, and pervade all aspects of society. To address them, we will need researchers, engineers, policymakers, and educators from many different disciplines to come the table and ask what they can contribute. There are both short term challenges (such as how to deploy, as rapidly as possible, existing technology to produce renewable energy; how to design government policies and international treaties to bring greenhouse gas emissions under control) and long term challenges (such as how to complete the transition to a global carbon-neutral society by the latter half of this century). In nearly all these challenges, software has a major role to play as a critical enabling technology. Climate change research is full of high computation, complex inter-related systems and thus huge complex simulation and these can only be handled precisely with the application of software.

The software can make possible to meet the challenges of tackling climate change. Software is a critical enabling technology in nearly all aspects of climate change, from the computational models used by climate scientists to improve our understanding of the impact of human activities on earth systems, through to the information and control systems needed to build an effective carbon-neutral society (SRCC, 2010).

At present many software has been developed and is continue by different national and international reputed research centers like FAO, Italy; Mac Hardey, UK; NOAA, USA. These softwares are basically developed for climate change researches on modeling, projecting future climate change, to tackle the problems of mitigation and adaptation of climate change, to evaluate the climate change research and many more aspects of climate change research.

Descriptions

There are many softwares has been developed to project future climate change (modelingsoftwares), to study the effect of climate change on natural resources (water, soil, flora and fauna) and crops, to adopt and mitigate the climate change, to evaluate the climate change research and many more aspects. Even, for climate change study, the required climatic data can also be generated with the help of softwares applications. Many softwares can handle more than one task also.

These softwares are listed here and briefly discussed about the works or tasks performed by the related softwares, sources of their availability and developing parent organization for its authenticity.

List of Softwares for climate change studies

1. AgrometShell (AMS)-software
2. Aquacrop
3. Clim WAT
4. Crop Monitoring (CM) Box software
5. CROPWAT
6. EPIC model
7. ET calculator
8. Ex-ACT
9. Local Climate Estimate Tool
10. MOSAICC
11. FAO-Rain Fall Estimate Routine
12. Climpag

13. Global Climate Map
14. InfoCrop (Kalra, 2012)
15. Lingo software
16. MASICC
17. MOSAICC
18. PAR_CSDLL
19. SFTEWR_RegCM3
20. SIM Clime
21. SoilPara-software
22. WinDisp51 software
23. MAGICC/SCENGEN-
24. CENTURY
25. MAACU
26. WTGROWS (Kalra, 2012)
27. ORYZA (Kalra, 2012)
28. CERES V4.0 MODEL FOR WHEAT (Singh *et a.l,* 2012)
29. APSIM MODEL FOR SOYABEAN (Mohanty*et al.,* 2012)
30. SWAT
31. Rainfall simulator for Africa
32. ARIM/SERIM- Time Series Analysis

Important softwares

1. **AgroMetShell:** AgroMetShell is an integrated toolbox of several software packages which helps professionals in National Agrometeorological Services, agronomic research and food security units etc. to assess the impact of climatic conditions on crops, to analyze climate risks and to perform regional crop forecasting using statistical and crop modelling approaches. It allows an integrated analysis of ground data and low resolution satellite information under a common interface. It can be best use for the assessment of current and future climatic impacts on crops, derivation of management options for the growing season possible. *Sources*: *http://www.fao.org/nr/climpag/aw_6_en.asp*

2. **AquaCrop:** AquaCrop is a crop water productivity model developed by the Land and Water Division of FAO. It simulates yield response to water of herbaceous crops, and is particularly suited to address conditions where water is a key limiting factor in crop production. AquaCrop attempts to balance accuracy, simplicity, and robustness. It uses a relatively small number of explicit and mostly-intuitive parameters and input variables requiring simple methods for their determination. It can be used for

    - assessing water-limited, attainable crop yields at a given geographical location
    - as a benchmarking tool, comparing the attainable yields against actual yields of a field, farm, or region, to identify the yield gap and the constraints limiting crop production
    - assessing rainfed crop production on the long term
    - developing irrigation schedules for maximum production (seasonal strategies and operational decision-making), and for different climate scenarios
    - scheduling deficit and supplemental irrigation
    - evaluating the impact of fixed delivery irrigation schedules on attainable yields
    - simulating crop sequences
    - carrying out future climate scenario analyses
    - optimizing a limited amount of water available (economic, equitability, and sustainability criteria)
    - evaluating the impact of low fertility and of water-fertility interactions on yields
    - assessing actual water productivity (biological and/or economic) at the field and higher scales, up to regions
    - supporting decision making on water allocation and other water policy actions
    - appraising the role of various water-related crop responses in yield determination for ideotype design
      Sources: *www.fao.org/nr/water/aquacrop.html*

3. **ClimWat:** ClimWat is a climatic database to be used in combination with the computer program CROPWAT. and allows the calculation of crop water requirements, irrigation supply and irrigation scheduling for various crops

for a range of climatological stations worldwide. It provides long-term monthly mean values of seven climatic parameters, namely: Mean daily maximum and minimum temperature in °C, Mean relative humidity in %, Mean wind speed in km/day, Mean sunshine hours per day, Mean solar radiation in MJ/m2/day, Monthly rainfall in mm/month, Monthly effective rainfall in mm/month, Reference evapotranspiration calculated with the Penman-Monteith method in mm/day. The data can be extracted for a single or multiple stations in the format suitable for their use in CROPWAT. *Source: www.fao.org/nr/water/ infores_databases_climwat.html.*

4. **Crop monitoring and forecasting:** The "CM Box" (Crop Monitoring Box) is a toolbox for agrometeorological crop monitoring and forecasting. It is an automated software suite with a "visual menu" that offers easy access to database that holds all the data needed to analyse the impact of weather on crops. The tool is useful for risk analysis, monitoring and forecasting crop production, which is an essential input to food security planning. The tool can compare maps of current yield expectations with historical average conditions.The CM Box is meant to offer an easy solution to rapidly setting up an operational crop monitoring and forecasting system. In the initial phase, reference data as well as real-time satellite and weather data can be provided by FAO based on international sources, but over the period, more and more national data can be used. Interested countries receive a combination of training, hardware, software customized for local use, as well as the real-time data required to operate the system in-country. The package can be tailored to suit the countries' specific requirements, based on national preferences as well as available expertise, methods and data. *Sources:http://www.fao.org/nr/climpag/ aw_6_en.asp*

5. **CROPWAT**: CROPWAT is a decision support tool developed by the Land and Water Development Division of FAO. It can be used for the calculation of crop water requirements and irrigation requirements based on soil, climate and crop data. In addition, it allows the development of irrigation schedules for different management conditions and the calculation of scheme water supply for varying crop patterns. CROPWAT 8.0 can also be used to evaluate farmers' irrigation practices and to estimate crop performance under both rainfed and irrigated conditions. *Sources: http://www.fao.org/nr/water/ infores_databases_cropwat.html*

6. **Erosion Productivity Impact Calculator (EPIC):** EPIC is an IBM, Macintosh, or Sun based generalized crop model that simulates daily crop growth on a hectare scale. Like most process plant growth models, it predicts plant biomass by simulating carbon fixation by photosynthesis, maintenance respiration, and growth respiration. Several different crops may be grown in rotation within one model execution. It uses the concept of light-use

efficiency as a function of photosynthetically available radiation (PAR) to predict biomass. EPIC has been modified to simulate the direct effects of atmospheric carbon dioxide on plant growth and water use. Crop management is explicitly incorporated into the model. This approach is useful for evaluating a limited number of agronomic adaptations to climate change, such as changes in planting dates, modifying rotations (i.e., switching cultivars and crop species), changing irrigation practices, and changing tillage operations. The parameter files are extremely sensitive to local conditions and EPIC can give grossly misleading results when relying on default settings as it is being tailored to different locations and cropping systems. *Sources: http://unfccc.int/adaptation/ nairobi_work_programme/knowledge_resources_and_publications/items/5454.php*

7. **ETo Calculator:** ETo calculator is a software developed by the Land and Water Division of FAO. Its main function is to calculate Reference evapotranspiration (ETo) according to FAO standards. *Sources: http://www.fao.org/nr/water/eto.html*

8. **EX-Ante Carbon balance Tool:** EX-ACT (EX-Ante Carbon Balance Tool) is a tool jointly developed from three FAO divisions (TCS, TCI and ESA and it is aimed at providing ex-ante estimations of the impact of agriculture and forestry development projects on GHG emissions and Carbon (C) sequestration, indicating its effects on the C balance. It is an easy tool to be used in the context of ex -ante project/program formulation, it is cost effective, it requires a minimum amount of data, and it has resources (tables, maps) which can help finding the information required to run the model. Also, EX-ACT works at project level but it can easily be up-scaled at programme/sector level. The Version 3.3 (September 2011) of the EX-ACT software is available on the Web site: http://www.fao.org/tc/exact/ex-act-home/en/.

9. **Local Climate Estimate Tool:** The Local Climate Estimate Tool (New_locClim), a software program and database, provides estimates of average climatic conditions at any location on earth based on the FAOCLIM database. The programme can create climatic maps, extract data in various formats from the database for further processing and can display graphs showing the annual cycle of monthly climate and the crop calendar. The tool provides growing season characteristics based on a comparison of rainfall and potential evapotranspiration and estimates of monthly, 10-day and daily values of common climate variables. The programme includes the current updated version of the FAOCLIM database of almost 30 000 stations worldwide, but users can also process their own data and prepare maps at any spatial resolution. Computer application programs (in Microsoft Excel) are included in the CD-ROM to help simplify complex calculations. Also

called FAOCLIM.There is also a java-based web version of LocClim named as "Web_LocClim". It also called Local Monthly Climate Estimator. *Sources: http://www.fao.org/nr/climpag/data_5_en.asp*

10. **MOSAICC** (FAO-MOSAICC): Climate change impact assessment on agriculture Information on the impacts of climate change is fundamental for designing policies and making informed decisions in all areas. With this in mind, FAO in partnership with other specialized European Institutions has developed an integrated package of models to carry out climate change impact assessment studies at national level. The "FAOMOSAICC" (Modelling System for Agricultural Impacts of Climate Change) package combines different models in an integrated modeling system. Although it can be used for a wide range of analyses - climate change impacts on water resources, crop yields, etc. - the ultimate objective of MOSAICC is to assess the effect of changing crop yields on national economies and which adaptation strategies would be most adequate to reduce the potentially adverse effects of climate change on national food security.

11. **FAO-Rain Fall Estimate Routine:** FAO Rain Fall Estimate (FAO-RFE) for Africa is a new independent method to estimate the rainfall amount, particularly, for certain regions where the coverage of the weather stations is scarce. FAO RFE is based on the Meteosat Second Generation IR channel combined with data coming from ECMWF global forecast model and EUMETSAT MPE. A local calibration is performed using the ground gauges, directly received as SYNOP messages and after a data validation. FAO-RFE offers 10-day and monthly rainfall totals for whole of Africa and for four regions. The methodology is currently transferred to the Sudanese Meteorological Authority. *Sources: http://geonetwork3.fao.org/climpag/FAO-RFE.php*

12. **Climpag:**Climpag (climate impact on agriculture) is a web portal bringing together the various aspects and interactions between weather, climate and agriculture in the context of food security. Climpag contains data, maps, methodologies and tools for better understanding and analysis of the effect of the variability of weather and climate on agriculture. The web portal covers six major thematic areas: advice and warnings, climate change, climate indicators, data and maps, hotspots and natural disasters. User friendly drop-down menu provides access to all publications, tools and methods relevant to all the thematic areas. *Sources:http://www.fao.org/nr/climpag/*

References

SRCC, 2010. Second International Workshop on Software Research and Climate Change. Cape Town, South Africa, 3 May, 2010.

Kalra, N. 2012. International Conference on Climate Change, Sustainable Agriculture & Public Leadership, New Delhi. Pg-130.

Singh, P.K., Singh, K.K., Baxla,A.K. and Rathore, L.S. 2012. International Conference on Climate Change, Sustainable Agriculture & Public Leadership, New Delhi. Pg-119.

Mohanty, S., Sultan, T., Panda, D.K. and Kumar, A. 2012. International Conference on Climate Change, Sustainable Agriculture & Public Leadership, New Delhi. Pg-131

2012. International Conference on Climate Change, Sustainable Agriculture & Public Leadership, New Delhi.

❑❑❑

30

Disease Forecasting Models: Their Role in Agriculture and Climate Change

Hemavati Ranebennur, Roop Singh Maitry, B.L. Meena, M. Datta and D. Daschaudhuri

Abstract

Climate change is a long-term change in the statistical distribution of weather patterns over periods of time that range from decades to millions of years. It may be a change in the average weather conditions or a change in the distribution of weather events with respect to an average. Climate change may be limited to a specific region, or may occur across the whole Earth.Since, time immemorial human beings have attempted to predict the disease occurrence informally by mere observing the changes in pattern of weather parameters. But the present scenario has been very much altered with the incessant and irrational use of natural resources exposing the atmosphere to the enormous encumbrance. Excessive CO_2 emissions, deforestation and injudicious use of chemicals in agricultural fields have ultimately led to the climate change, which needs an absolute engrossment. Climate change has also invited the emergence and resurgence of many of the pest and diseases posing a grave challenge to manage them. So there is a need for immediate action through Government policies and scientific community to combat the situation. Disease forecasting models may be one of the rescuing options to mitigate the problems of agriculture. Forecasting models can come to rescue to save the crop by giving the forewarning of the occurrence of the disease or pest and recommending the suitable management practice well before ahead.

Keywords: Climate change, Disease forecasting.

Introduction

Climate change means a long term change in the earth's climate, i.e global weather parameters especially the increase in average temperature and the storm activity, melting glacial ice and rise in sea levels.Intergovernmental Panel on Climate Change (IPCC) indicated that during the 21st century the global surface temperature is likely to rise a further 1.1 to 2.9°C (2 to 5.2°F) for their lowest emissions scenario and 2.4 to 6.4°C (4.3 to 11.5°F) for their highest. The ranges of these estimates arise from the use of models with differing sensitivity to greenhouse gas concentrations. This type of change in the global weather parameters will also show their vagaries mainly on agriculture and allied sectors.

The predicted medium-run impact is negative and statistically significant: It is found that projected climate change over the period 2010-2039 reduces major crop yields by 4.5 to nine percent. The long-run (2070-2099) impact is dramatic, reducing yields by 25 percent or more in the absence of long-run adaptation. These results suggest that climate change is likely to impose significant costs on the Indian economy unless farmers can quickly recognize and adapt to increasing temperatures. Such rapid adaptation may be less plausible in a developing country, where access to information and capital is limited. (Raymond Guiteras, 2007).

Floods and heavy rains

Increased moisture benefits epidemics and prevalence of leaf fungal pathogens.

1. Rice leaf blight caused great famine in Bengal (1942), 2 million people died.
2. Wheat stripe rust outbreak in major production regions of China contributed to the 1960s famine.
3. Fungal epidemics in corn, soybean, alfalfa, and wheat (U.S. Midwest, 1993).
4. Mycotoxin (produced by *Fusarium* spp.) reached a record high (U.S. Great Plains, 1993);
5. Mycotoxin increases are related to high humidity during harvest (East Africa and South America, 1990s).
6. Humid summers drive epidemics of gray leaf spot of maize (Iowa and Illinois, 1996).
7. Water induced soil transport increases dissemination of soilborne pathogens to non-infected areas.
8. Outbreaks of soybean sudden death syndrome in the north central U.S. (1993).
9. Continuous soil saturation causes long-term problems related to rot development and increased damage by pathogens.
10. In maize, crazy top and common smut outbreak

Drought

1. Water stress diminishes plant vigour and alters C/N lowering plant resistance to nematodes, and insects.
2. Attack by fungal pathogens of stems and roots are favoured by weakened plant conditions.
3. Dry and warm conditions promote growth of insect vector populations, increasing viral epidemics.
4. Outbreak of soybean cyst nematode correlated to drought conditions in north central U.S. (1990).
5. Summer locust outbreak correlated to drought in Mexico (1999).
6. Increased incidence of *Aspergillusflavus*(producer of aflatoxin) in southern U.S. (1977 and 1983).

Air currents

Air currents provide large-scale transportation for disease agents (e.g., spores of fungi) or insects from overwintering areas to attacking areas.

1. The spread of the stem rust that overwinters in Mexico and Texas is favoured by moist southern air currents.
2. The southern leaf blight of corn spread from Mississippi to the Midwest by air currents of a tropical storm in the Gulf of Mexico during 1970.

Warm winters Increase overwintering populations of all pests and insect vectors.

1. Data reported for the European corn borer; wheat scab; wheat rust; and potato leafhopper.
2. Increase population of aphids that carry the soybean mosaic virus.
3. Increase population and number of generations of Mexican bean beetle and bean leaf beetle in the U.S. (Rosenzweig *et al.*, 2000.)

In this context it becomes necessary to give emphasis on the insect pests and pathogens which are more adapted to the mechanisms to overcome the adverse effect of the environment unlike the plants which takes much longer time to adapt to the vagaries of the climate. In such an unsuitable environmental situation it becomes harder to the host plants to perform, which can make them much more vulnerable to the pathogens. In addition to that this environment may become much more favourable to the virulent pathogen. On the other hand the vectors which spread the disease also may appreciate the increased temperature for their breeding and overcome the winter with cold temperatures with much ease.

The situation may even become much more conducive for the pathogen to attack the plants; the pathogen may gain the entry to the current restricted area, which is prohibited may be due to the lower temperature not favouring the pathogen. So when the pathogen is accepted into this new area with the favourable climatic conditions, this will lead to larger epidemics and pandemics if neglected at this initial stage.

To combat the situation to address the burning situation, there has to be a disease forecasting models. The principle behind the forecasting models is to know the probability of the diseases occurrence, the intensity of the disease and warn the farming systems well before hand to mitigate the losses and damage it would cause if left otherwise. Forecasting models mainly aim at reducing the initial inoculums or reducing the rate of infection or sometimes both.

Forecasting systems are based on the assumptions that the three components if the triangle mainly the host, environment and the pathogen are present in such a condition which is capable of causing the maximum damage to the crop and incur severe losses beyond the economic threshold level. In such circumstances forecasting models can come to rescue to save the crop by giving the forewarning of the occurrence of the disease or pest and recommending the suitable management practice well before ahead.

Such forecasting models can become a boon to farming community if it is made reliable, simple, and cost effective and can be applied to the multiple diseases. At present there are already developed models, such models have to be validated and standardised in locality wise predictions

Forecasting of pests and diseases-pest modelling pest modelling developed by Srinivasa Rao (2004), CRIDA has mentioned the different approaches for monitoring the occurrence of the pest:

In Geographic method the appearance of the pest in aregion is taken as aindicator for the subsequent occurrence of the pest in the geographically contiguous areas provided the weather conditions are likely to show a definite and discernible progression in space with time. For example aphids on mustard crop in northern parts of India

In phenological technique, phenology deals with the time and occurrence or completion in nature of the various but distinct growth and development phases of biological entities.

Biological indicators pathogen population in the inoculated fields is directly correlated with the disease severity and hence can be used for forecasting the occurrence of the diseases.

Therefore any bioclimatic analysis requires good forecasting models may gain the significant role with the trend of change in the climate. Predictions of the epidemics in the areas not known for such disease is a challenge for the scientific community and disease forecasting models may be the answer for this.

The research has to be directed in such a way that leads to the development of the databases on different parameters of weather the occurrence of the disease and pests and the correlation between them for effective management of the disease.

References

Raymond Guiteras., The Impact of Climate Change on Indian Agriculture, Department of Economics, MIT December, 2007. Job Market Paper Draft.

Rosenzweig, C., A. Iglesias, X.B. Yang, P.R. Epstein, and E. Chivian. 2000. Implications of Climate Change for U.S. Agriculture: Extreme Weather Events,Plant Diseases, and Pests. Cambridge, Massachusetts: Centre for Health and the Global Environment, Harvard Medical School. Cambridge, MA. 56 pp.

❑❑❑